1,001 Geometry Practice Problems

FOR DUMMIES®

A Wiley Brand

by Allen Ma and Amber Kuang

1,001 Geometry Practice Problems For Dummies®

Published by: **John Wiley & Sons, Inc.,** 111 River Street, Hoboken, NJ 07030-5774, www.wiley.com

Copyright © 2015 by John Wiley & Sons, Inc., Hoboken, New Jersey

Media and software compilation copyright © 2015 by John Wiley & Sons, Inc. All rights reserved.

Published simultaneously in Canada

For general information on our other products and services, please contact our Customer Care Department within the U.S. at 877-762-2974, outside the U.S. at 317-572-3993, or fax 317-572-4002. For technical support, please visit www.wiley.com/techsupport.

Wiley publishes in a variety of print and electronic formats and by print-on-demand. Some material included with standard print versions of this book may not be included in e-books or in print-on-demand. If this book refers to media such as a CD or DVD that is not included in the version you purchased, you may download this material at http://booksupport.wiley.com. For more information about Wiley products, visit www.wiley.com.

Library of Congress Control Number: 20149456253

ISBN 978-1-118-85326-9 (pbk); ISBN 978-1-118-85305-4 (ebk); ISBN 978-1-118-85302-3 (ebk)

Manufactured in the United States of America

10 9 8 7 6 5 4 3

Table of Contents

Introduction 1

Part I: The Questions 3

Chapter 1: Diving into Geometry 5
The Problems You'll Work On 5
What to Watch Out For 5
Understanding Basic Geometric Definitions 6
Applying Algebra to Basic Geometric Definitions 7
Recognizing Geometric Terms 8
Properties and Postulates 8
Adjacent Angles, Vertical Angles, and Angles That
  Form Linear Pairs 10
Complementary and Supplementary Angles 10
Angles in a Triangle 11

Chapter 2: Constructions 13
The Problems You'll Work On 13
What to Watch Out For 13
Creating Congruent Constructions 14
Constructions Involving Angles and Segments 14
Parallel and Perpendicular Lines 15
Creative Constructions 16

Chapter 3: Geometric Proofs with Triangles 17
The Problems You'll Work On 17
What to Watch Out For 17
Triangle Congruence Theorems 18
Completing Geometric Proofs Using Triangle
  Congruence Theorems 21
Overlapping Triangle Proofs 24
Indirect Proofs 28

Chapter 4: Classifying Triangles 31
The Problems You'll Work On 31
What to Watch Out For 31
Classifying Triangles by Their Sides 32
Properties of Isosceles, Equilateral, and Right Triangles 33
Classifying Triangles by Their Angles 34
Understanding the Classification of Triangles 35
Geometric Proofs Involving Isosceles Triangles 36

Chapter 5: Investigating the Centers of a Triangle.................41
The Problems You'll Work On 41
What to Watch Out For 41
The Incenter of a Triangle 42
Understanding the Orthocenter 42
Understanding Centroids 43
Finding the Centroid of a Triangle...44
The Circumcenter of a Triangle 45
Recognizing Triangle Centers 45
Constructing the Centers of a Triangle...................................47
The Euler Line...48

Chapter 6: Similar Triangles.......................................49
The Problems You'll Work On 49
What to Watch Out For 49
Understanding Similar Triangles 50
Midsegments 50
Creating Similar Triangles 51
Similar-Triangle Word Problems 52
Proving That Two Triangles Are Similar to Each Other 53
Proving That Corresponding Sides Are in Proportion....................54
Proving with the Means and Extremes 55

Chapter 7: The Right Triangle.....................................59
The Problems You'll Work On 59
What to Watch Out For 59
Pythagorean Theorem 60
Right Triangle Proportions..60
Word Problems Involving Right-Triangle Proportions 62
Working with Special Right Triangles......................................63
Application of Special Right Triangles.....................................63
Trigonometric Ratios 65
Applying the Trigonometric Ratios to Word Problems....................66

Chapter 8: Triangle Inequalities 67
The Problems You'll Work On 67
What to Watch Out For 67
Relationships between the Sides and Angles of a Triangle...............68
Triangle Inequality Theorem..69
Finding the Missing Side Length..69
Isosceles Triangles 70
Using the Exterior Angle Theorem for a Triangle.........................70
Geometric Proofs Involving Triangle Inequality Theorems...............72

Chapter 9: Polygons 75
The Problems You'll Work On 75
What to Watch Out For 75
Naming Polygons 76
Understanding Angles of a Polygon 76
The Sum of the Interior and Exterior Angles of a Polygon...............77
Finding the Area of Regular Polygons 79

Chapter 10: Properties of Parallel Lines 81

The Problems You'll Work On 81
What to Watch Out For 81
Alternate Interior and Alternate Exterior Angles 82
Classifying Triangles by Their Angle Measurements 83
Finding Angle Measures Involving Parallel Lines 83
Reviewing Corresponding, Adjacent, and Vertical Angles 84
More Practice with Angles Involving Parallel Lines 85
Geometric Proof Incorporating Parallel Lines 86
Geometric Proof Incorporating Parallel Lines 87

Chapter 11: Properties of Quadrilaterals 89

The Problems You'll Work On 89
What to Watch Out For 89
Properties of Parallelograms 90
Word Problems with Parallelograms 91
Properties of Rectangles 91
Finding the Diagonal of a Rectangle 92
Reviewing the Properties of a Rhombus 93
Diagonal Properties of a Rhombus 93
Properties of a Square 94
Properties of a Trapezoid 96

Chapter 12: Coordinate Geometry 97

The Problems You'll Work On 97
What to Watch Out For 97
Determining Distance 98
Using the Midpoint Formula 98
Using the Slope Formula 99
Parallel and Perpendicular Lines 100
Writing the Equation of a Line in Slope-Intercept Form 101
Coordinate Geometry Proofs 102

Chapter 13: Transformational Geometry 105

The Problems You'll Work On 105
What to Watch Out For 105
Rigid Motion 106
Reflecting Points over the x- and y-axes 109
Writing Equations for Lines of Reflection 110
Understanding Point Symmetry 110
Triangle Translations 111
Translating Points 112
Finding Translation Rules 113
Doing Dilations 113
Practicing with Rotations 114
Understanding the Rules for Rotations 115
Rigid Motion of Triangles 115
Compositions of Transformations 116
Glide Reflections and Direct and Indirect Isometries 117
Transformations of a Segment 117
Trying Rigid Motion Constructions 118

Chapter 14: Exploring Circles............................121

The Problems You'll Work On 121
What to Watch Out For 121
Working with the Circumference of a Circle 122
Understanding the Area of a Circle 122
Working with Sectors 123
Arc Length 124
The Equation of a Circle in Standard Form 124

Chapter 15: Circle Theorems 127

The Problems You'll Work On 127
What to Watch Out For 127
Central Angles and Arcs 128
Inscribed Angles and Arcs 128
Angles Formed by Intersecting Chords of a Circle 129
Angles Formed by Secants and Tangents 130
The Intersecting Chord Theorem 132
Lengths of Tangents and Secants 134
Tangent and Radius 136
"BIG" Circle Problems 136
Circle Proofs 138

Chapter 16: Three-Dimensional Geometry.......................141

The Problems You'll Work On 141
What to Watch Out For 141
Understanding Points, Lines, and Planes 142
Surface Area of Solid Figures 143
Calculating the Volume of Solid Figures 145
Rotations of Two-Dimensional Figures 146

Chapter 17: Locus Problems 149

The Problems You'll Work On 149
What to Watch Out For 149
Basic Locus Theorems 150
Loci Using Coordinate Geometry 150
The Locus of Points Equidistant from One or Two Lines 150
The Locus of Points Equidistant from Two Points 151
Writing the Equation of a Circle 152
Compound Locus in Coordinate Geometry 152
Compound and Challenging Locus Problems 153

Part II: The Answers 155

Chapter 18: Answers and Explanations 157

Index 445

Introduction

This book is intended for anyone who needs to brush up on geometry. You may use this book as a supplement to material you're learning in an undergraduate geometry course. The book provides a basic level of geometric knowledge. As soon as you understand these concepts, you can move on to more complex geometry problems.

What You'll Find

The 1,001 geometry problems are grouped into 17 chapters. You'll find calculation questions, construction questions, and geometric proofs, all with detailed answer explanations. If you miss a question, take a close look at the answer explanation. Understanding where you went wrong will help you learn the concepts.

Beyond the Book

This book provides a lot of geometry practice. If you'd also like to track your progress online, you're in luck! Your book purchase comes with a free one-year subscription to all 1,001 practice questions online. You can access the content with your computer, tablet, or smartphone whenever you want. Create your own question sets and view personalized reports that show what you need to study most.

What you'll find online

The online practice that comes free with the book contains the same 1,001 questions and answers that are available in the text. You can customize your online practice to focus on specific areas, or you can select a broad variety of topics to work on — it's up to you. The online program keeps track of the questions you get right and wrong so you can easily monitor your progress.

This product also comes with an online Cheat Sheet that helps you increase your geometry knowledge. Check out the free Cheat Sheet at (www.dummies.com/cheatsheet/1001geometry) (No PIN required. You can access this info before you even register.)

How to register

To gain access to additional tests and practice online, all you have to do is register. Just follow these simple steps:

1. **Find your PIN access code:**

   • **Print-book users:** If you purchased a print copy of this book, turn to the inside front cover of the book to find your access code.

   • **E-book users:** If you purchased this book as an e-book, you can get your access code by registering your e-book at www.dummies.com/go/getaccess. Go to this website, find your book and click it, and answer the security questions to verify your purchase. You'll receive an email with your access code.

2. **Go to Dummies.com and click** Activate Now.

3. **Find your product (*1,001 Geometry Practice Problems For Dummies (+ Free Online Practice*)) and then follow the on-screen prompts to activate your PIN.**

Now you're ready to go! You can come back to the program as often as you want — simply log on with the username and password you created during your initial login. No need to enter the access code a second time.

For Technical Support, please visit http://wiley.custhelp.com or call Wiley at 1-800-762-2974 (U.S.), +1-317-572-3994 (international).

Where to Go for Additional Help

This book covers a great deal of geometry material. Because there are so many topics, you may struggle in some areas. If you get stuck, consider getting some additional help.

In addition to getting help from your friends, teachers, or coworkers, you can find a variety of great materials online. If you have Internet access, a simple search often turns up a treasure trove of information. You can also head to www.dummies.com to see the many articles and books that can help you in your studies.

1,001 Geometry Questions For Dummies gives you just that — 1,001 practice questions and answers to improve your understanding and application of geometry concepts. If you need more in-depth study and direction for your geometry courses, you may want to try out the following *For Dummies* products:

✔ *Geometry For Dummies,* **by Mark Ryan:** This book provides an introduction into the most important geometry concepts. You'll learn all the principles and formulas you need to analyze two- and three-dimensional shapes. You'll also learn the skills and strategies needed to write a geometric proof.

✔ *Geometry Workbook For Dummies,* **by Mark Ryan:** This workbook guides you through geometric proofs using a step-by-step process. It also provides tips, shortcuts, and mnemonic devices to help you commit some important geometry concepts to memory.

Part I
The Questions

In this part . . .

The best way to become proficient in geometry is through a lot of practice. Fortunately, you now have 1,001 practice opportunities right in front of you. These questions cover a variety of geometric concepts and range in difficulty from easy to hard. Master these problems, and you'll be well on your way to a solid foundation in geometry.

Here are the types of problems that you can expect to see:

- Geometric definitions (Chapter 1)
- Constructions (Chapter 2)
- Geometric proofs with triangles (Chapter 3)
- Classifying triangles (Chapter 4)
- Centers of a triangle (Chapter 5)
- Similar triangles (Chapter 6)
- The Pythagorean theorem and trigonometric ratios (Chapter 7)
- Triangle inequality theorems (Chapter 8)
- Polygons (Chapter 9)
- Parallel lines cut by a transversal (Chapter 10)
- Quadrilaterals (Chapter 11)
- Coordinate geometry (Chapter 12)
- Transformations (Chapter 13)
- Circles (Chapters 14 and 15)
- Surface area and volume of solid figures (Chapter 16)
- Loci (Chapter 17)

Chapter 1

Diving into Geometry

. . . .

Geometry requires you to know and understand many definitions, properties, and postulates. If you don't understand these important concepts, geometry will seem extremely difficult. This chapter provides practice with the most important geometric properties, postulates, and definitions you need in order to get started.

The Problems You'll Work On

In this chapter, you see a variety of geometry problems. Here's what they cover:

✔ Understanding midpoint, segment bisectors, angle bisectors, median, and altitude

✔ Working with the properties of perpendicular lines, right angles, vertical angles, adjacent angles, and angles that form linear pairs

✔ Noting the differences between complementary and supplementary angles

✔ Using the addition and subtraction postulates

✔ Understanding the reflexive, transitive, and substitution properties

What to Watch Out For

The following tips may help you avoid common mistakes:

✔ Be on the lookout for when something is being done to a segment or an angle. Bisecting a segment creates two congruent segments, whereas bisecting an angle creates two congruent angles.

✔ The transitive property and the substitution property look extremely similar in proofs, making them very confusing. Check whether you're just switching the congruent segments/angles or whether you're getting a third set of congruent segments/angles after already being given two pairs of congruent segments/angles.

✔ Make sure you understand what the question is asking you to solve for. Sometimes a question asks only for a particular variable, so as soon as you find the variable, you're done. However, sometimes a question asks for the measure of the segment or angle; after you find the value of the variable, you have to plug it in to find the measure of the segment or angle.

Understanding Basic Geometric Definitions

1–3 Fill in the blank to create an appropriate conclusion to the given statement.

1. If *M* is the midpoint of $\overline{AB}$, then $\overline{AM} \cong$ ___ .

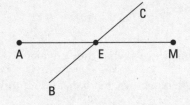

2. If $\overline{BC}$ bisects $\overline{AM}$ at *E*, then $\overline{AE} \cong$ ___ .

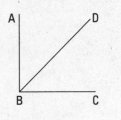

3. If $\overline{AB} \perp \overline{BC}$, then ____ is a right angle.

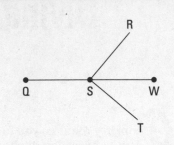

4–9 In the following figure, $\overrightarrow{QW}$ bisects $\angle RST$ and $\overline{RS} \perp \overline{TS}$. Determine whether each statement is true or false.

4. $\angle RST$ is a right angle.

5. $\angle RSQ \cong \angle RSW$.

6. $\angle QSR$ and $\angle WSR$ form a linear pair.

7. $\angle WST \cong \angle WSR$.

8. $\angle RSQ$ is an obtuse angle.

9. If Point *S* is the midpoint of $\overline{QW}$, then it's always true that $\overline{RS} \cong \overline{TS}$.

10–14 *Use the following figure and the given information to draw a valid conclusion.*

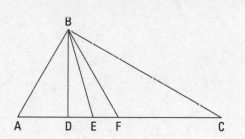

10. $\overline{BF}$ is the median of $\triangle ABC$.

11. $\overline{BD}$ is the altitude of $\triangle ABC$.

12. $\overline{BE}$ bisects $\angle ABC$.

13. F is the midpoint of $\overline{AC}$.

14. F is the midpoint of $\overline{AC}$. What type of angle does $\angle BFC$ have to be in order for $\overline{BF}$ to be called a perpendicular bisector?

Applying Algebra to Basic Geometric Definitions

15–18 *Use the figure and the given information to answer each question.*

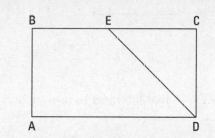

15. E is the midpoint of $\overline{BC}$. If $BE = 50$ and $CE = 2x + 25$, find the value of x.

16. $\overline{ED}$ bisects $\angle ADC$. If $m\angle ADE$ is represented by $3x - 5$ and $m\angle CDE$ is represented by $x + 25$, find $m\angle ADC$.

17. If $\overline{BA} \perp \overline{DA}$ and $m\angle BAD$ is represented by $5x - 20$, find the value of x.

18. $\overline{DE}$ bisects $\overline{BC}$. If $BC = 5x - 3$ and $CE = x + 12$, find the length of $\overline{BE}$.

Recognizing Geometric Terms

19–26 Write the geometric term that fits the definition.

19. Two adjacent angles whose sum is a straight angle: _____

Supp.

20. Two lines that intersect to form right angles: _____

21. An angle whose measure is between 0° and 90°: _____

22. A type of triangle that has two sides congruent and the angles opposite them also congruent: _____

23. Divides a line segment or an angle into two congruent parts: _____

24. An angle greater than 90° but less than 180°: _____

25. A line segment connecting the vertex of a triangle to the midpoint of the opposite side: _____

26. The height of a triangle: _____

Properties and Postulates

27–34 Refer to segment $\overline{DREW}$ to fill in the blank.

D   R   E   W

27. $\overline{DR} + \overline{RE} = $ ___

28. $\overline{DR} + \overline{RW} = $ ___

29. $\overline{DW} - \overline{EW} = $ ___

30. $\overline{DE} - \overline{RE} = $ ___

31. The _____ would be the reason used to prove that $\overline{RE} \cong \overline{RE}$.

32. If $\overline{DR} \cong \overline{WE}$, then $\overline{DR} + \overline{RE} \cong \overline{WE} + \underline{\quad}$.

33. If $\overline{DE} \cong \overline{RW}$, then $\overline{RW} - \overline{RE} \cong \underline{\quad} - \overline{RE}$.

34. Assuming the figure is not drawn to scale, if $\overline{DR} \cong \overline{RE}$ and $\overline{EW} \cong \overline{RE}$, then you can prove that $\overline{DR} \cong \overline{EW}$. The _____ postulate can be used to draw this conclusion.

35–40 *In the given diagram, $\angle JOM \cong \angle NOA$. Use the basic geometric postulates to answer each question.*

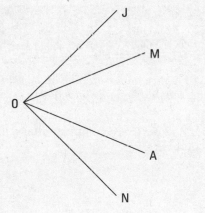

35. Which property or postulate is used to show that $\angle MOA \cong \angle MOA$?

36. $\angle JOM + \angle MOA \cong \angle MOA + \underline{\quad}$

37. $m\angle JON - m\angle JOA = \underline{\quad}$

38. What information must be given in order for the following to be true?

$$\angle JOA - \angle MOA \cong \angle NOM - \angle MOA$$

39. If $\overline{OM}$ bisects $\angle JON$, you can conclude that $2(\angle JOM) \cong \underline{\quad}$.

40. If $\overline{OM}$ bisects $\angle JON$, you can conclude that $\frac{1}{2}(\angle JON) \cong \underline{\quad}$.

Adjacent Angles, Vertical Angles, and Angles That Form Linear Pairs

41–47 In the following figure, $\overline{MET}$ intersects $\overline{AEH}$ at E. Fill in the blank to make the statement true.

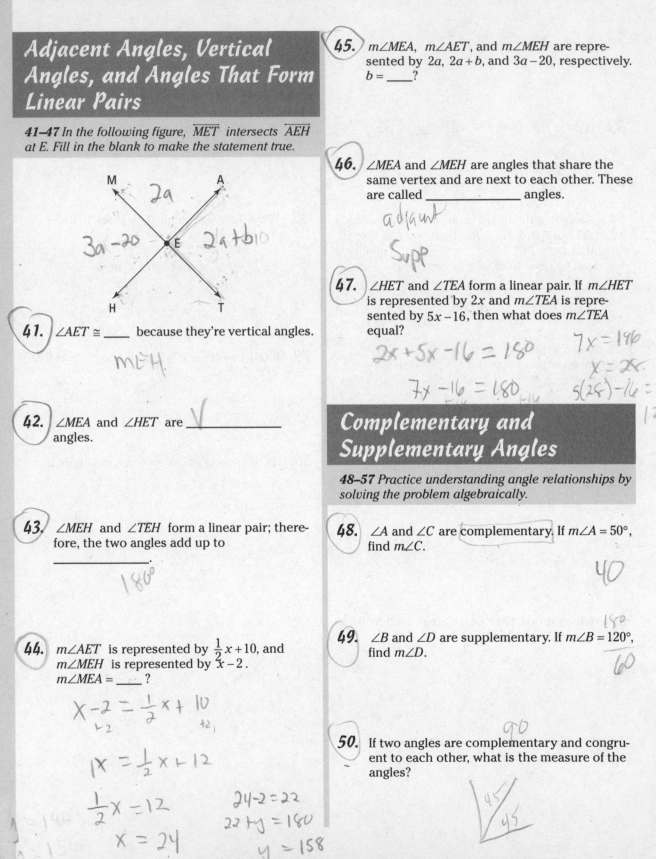

Handwritten on figure: M, A, 2a, 3a−20, E, 2a+b 10, H, T

41. $\angle AET \cong$ ____ because they're vertical angles.

MEH

42. $\angle MEA$ and $\angle HET$ are ____ angles.

43. $\angle MEH$ and $\angle TEH$ form a linear pair; therefore, the two angles add up to ____.

180°

44. $m\angle AET$ is represented by $\frac{1}{2}x+10$, and $m\angle MEH$ is represented by $x-2$. $m\angle MEA =$ ____ ?

$$x-2 = \tfrac{1}{2}x+10$$
$$+2 \qquad\qquad +2$$
$$1x = \tfrac{1}{2}x+12$$
$$\tfrac{1}{2}x = 12$$
$$x = 24$$

24−2 = 22
22 + y = 180
y = 158

45. $m\angle MEA$, $m\angle AET$, and $m\angle MEH$ are represented by $2a$, $2a+b$, and $3a-20$, respectively. $b =$ ____?

46. $\angle MEA$ and $\angle MEH$ are angles that share the same vertex and are next to each other. These are called ____ angles.

adjacent
Supp

47. $\angle HET$ and $\angle TEA$ form a linear pair. If $m\angle HET$ is represented by $2x$ and $m\angle TEA$ is represented by $5x-16$, then what does $m\angle TEA$ equal?

$$2x+5x-16 = 180 \qquad 7x = 196$$
$$7x-16 = 180 \qquad x = 28$$
$$5(28)-16 = 124$$

Complementary and Supplementary Angles

48–57 Practice understanding angle relationships by solving the problem algebraically.

48. $\angle A$ and $\angle C$ are complementary. If $m\angle A = 50°$, find $m\angle C$.

40

49. $\angle B$ and $\angle D$ are supplementary. If $m\angle B = 120°$, find $m\angle D$.

180
60

50. If two angles are complementary and congruent to each other, what is the measure of the angles?

90
45 / 45

51. Two angles are supplementary and congruent. What type of angles must they be?

52. The ratio of two angles that are supplements is 2:3. Find the larger angle.

53. If two angles are supplementary and one angle is 40° more than the other angle, find the smaller angle.

54. If two angles are complementary and one angle is twice the measure of the other, find the measure of the smaller angle.

$$1x + 2x = 90 \qquad x = 30$$
$$3x = 90$$

90.

55. If two angles are complementary and one angle is 6 less than twice the measure of the other angle, find the larger angle.

90

$$x + 2x - 6 = 90$$
$$\quad +6 \quad +6$$
$$3x = 96 \qquad x = 32$$
$$2(32) - 6 = 58$$

56. If two angles form a linear pair, what is their sum?

180

57. The ratio of two angles that are complements of each other is 5:4. Find the measure of the smaller angle.

Angles in a Triangle

58–60 Use the following figure and the given information to solve each problem. $\overline{TV}$ and $\overline{RI}$ intersect at E.

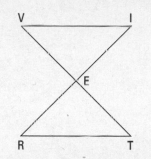

58. If $m\angle IVE = 50°$, $m\angle VIE = 70°$, and $m\angle RET = 2x - 10$, find the value of x.

59. If $m\angle IVE = 50°$ and $m\angle VIE = 70°$, find the degree measure of $\angle TEI$.

60. If $m\angle VEI$ is represented by $2a + b$, $m\angle RET$ is represented by $a - b$, and $m\angle REV$ is represented by $4a + 2b$, find the value of a.

Chapter 2

Constructions

. . . .

*O*ne of the most visual topics in geometry is constructions. In this chapter, you get to demonstrate some of the most important geometric properties and definitions using a pencil, straight edge, and compass.

The Problems You'll Work On

In this chapter, you see a variety of construction problems:

- Constructing congruent segments and angles
- Drawing segment, angle, and perpendicular bisectors
- Creating constructions involving parallel and perpendicular lines
- Constructing $30°$-$60°$-$90°$ and $45°$-$45°$-$90°$ triangles

What to Watch Out For

The following tips may help you avoid common mistakes:

- If you're drawing two arcs for a construction, make sure you keep the width of the compass (or radii of the circles) consistent.
- Make your arcs large enough so that they intersect.
- Sometimes you need to do more than one construction to create what the problem is asking for. This idea is extremely helpful when you need to construct special triangles.

Creating Congruent Constructions

61–65 Use your knowledge of constructions (as well as a compass and straight edge) to create congruent segments, angles, or triangles.

61. Construct $\overline{CD}$, a line segment congruent to $\overline{AB}$.

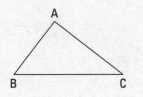

62. Construct $\angle D$, an angle congruent to $\angle A$.

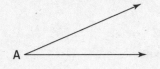

63. Construct $\triangle DEF$, a triangle congruent to $\triangle BCA$.

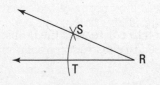

64. Is the following construction an angle bisector or a copy of an angle?

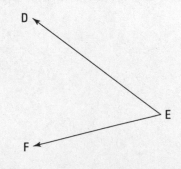

65. Construct $\triangle ABC$, a triangle congruent to $\triangle STR$.

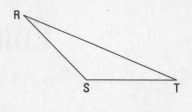

Constructions Involving Angles and Segments

66–70 Apply your knowledge of constructions to angles and segments.

66. Construct segment $\overline{CD}$, whose measure is twice the measure of $\overline{AB}$.

67. Given $\angle B$, construct $\overline{BG}$, the bisector of $\angle B$.

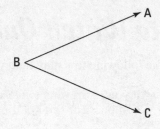

68. Construct the angle bisector of $\angle DEF$.

69. What type of construction is represented by the following figure?

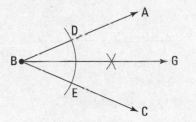

70. True or False? The construction in the following diagram proves that $\angle ABG \cong \angle GBC$.

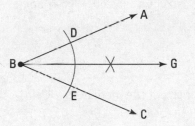

Parallel and Perpendicular Lines

71–77 Apply your knowledge of constructions to problems involving parallel and perpendicular lines.

71. Place Point E anywhere on $\overline{AB}$. Construct $\overline{DE}$ perpendicular to $\overline{AB}$ through Point E.

72. Use the following diagram to construct a line perpendicular to $\overline{AB}$ through Point C.

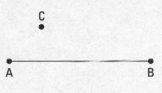

73. Construct the perpendicular bisector of $\overline{AB}$.

74. Which construction is represented in the following figure?

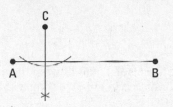

75. Construct a line parallel to $\overline{AB}$ that passes through Point C.

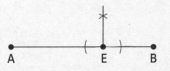

76. True or False? The construction in the following diagram proves that $\overline{AE} \cong \overline{EB}$.

77. True or False? The following diagram is the correct illustration of the construction of a line parallel to $\overline{AB}$.

Creative Constructions

78–85 Apply your knowledge of constructions to some more creative problems.

78. Construct a 30° angle.

79. True or False? The following diagram shows the first step in constructing a 45° angle.

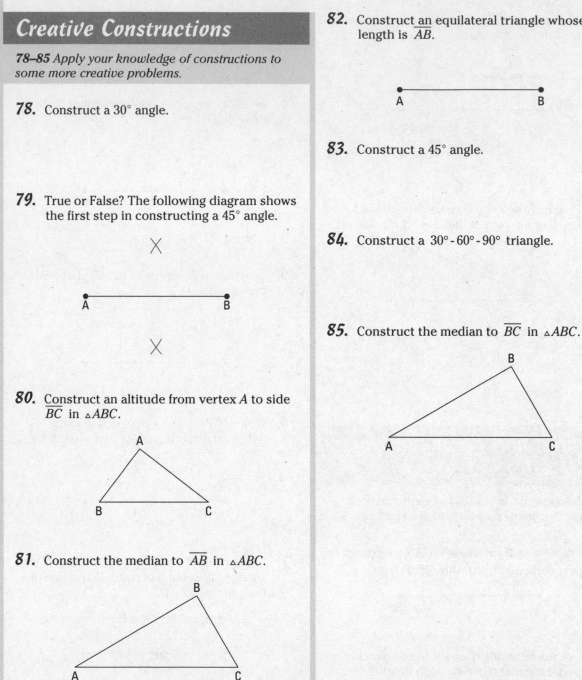

80. Construct an altitude from vertex *A* to side $\overline{BC}$ in △*ABC*.

81. Construct the median to $\overline{AB}$ in △*ABC*.

82. Construct an equilateral triangle whose side length is $\overline{AB}$.

83. Construct a 45° angle.

84. Construct a 30°-60°-90° triangle.

85. Construct the median to $\overline{BC}$ in △*ABC*.

Chapter 3

Geometric Proofs with Triangles

· ·

In geometry, you're frequently asked to prove something. In this chapter, you're given specific information and asked to prove specific information about triangles. You do this by using various geometric properties, postulates, and definitions to generate new statements that will lead you toward the information you're looking to prove true.

The Problems You'll Work On

In this chapter, you see a variety of problems involving geometric proofs:

✔ Using SAS, SSS, ASA, and AAS to prove triangles congruent

✔ Showing that corresponding parts of congruent triangles are congruent

✔ Formulating a geometric proof with overlapping triangles

✔ Using your knowledge of quadrilaterals to complete a geometric proof

✔ Completing indirect proofs

What to Watch Out For

Remember the following tips as you work through this chapter:

✔ The statement that needs to be proven has to be the last statement of the proof. It can't be used as a given statement.

✔ You must use all given information to formulate the proof. Each given should be used separately to draw its own conclusion.

✔ If you've used all your given information and still require more to prove the triangles congruent, look for the reflexive property or a pair of vertical angles.

✔ After you find angles or segments congruent, mark them in your diagram. The markings make it easier for you to see what other information you need to complete the proof.

✔ To prove parts of a triangle congruent, you'll first need to prove that the triangles are congruent to each other using the proper triangle congruence theorems.

Triangle Congruence Theorems

86–102 Use your knowledge of SAS, ASA, SSS, and AAS to solve the problem.

86. What method can you use to prove these two triangles congruent?

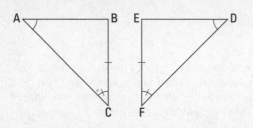

87. What method can you use to prove these two triangles congruent?

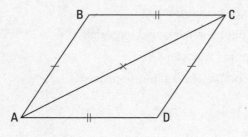

88. What method can you use to prove these two triangles congruent?

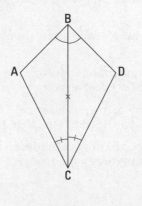

89. What method can you use to prove △ABC ≅ △EDC ?

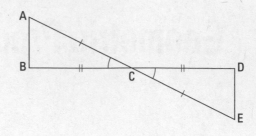

90. Which pair of segments or angles would need to be proved congruent in order to prove these triangles congruent using the SSS method?

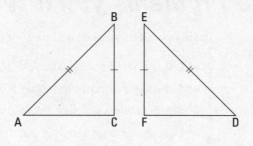

91. Which pair of segments or angles would need to be proved congruent in order to prove these triangles congruent using the SAS method?

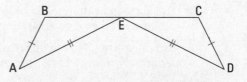

92. Which pair of segments or angles would need to be proved congruent in order to prove these triangles congruent using the AAS method?

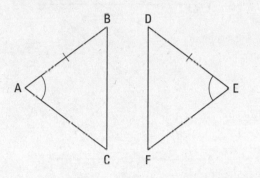

93. Which pair of segments or angles would need to be proved congruent in order to prove these triangles congruent using the ASA method?

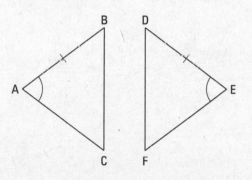

94. Which pair of segments or angles would need to be proved congruent in order to prove these triangles congruent using the SSS method?

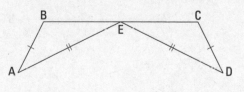

95. Which pair of segments or angles would need to be proved congruent in order to prove the triangles congruent using the SAS method?

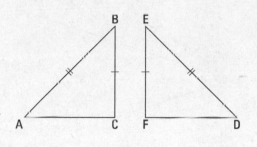

96. Given: $\overline{SE}$ bisects $\angle RST$ and E is the midpoint of $\overline{RT}$. Is it possible to prove $\triangle RSE \cong \triangle TSE$ using only the given information and the reflexive property?

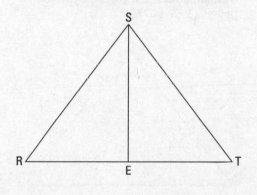

97. Given: $\overline{AC}$ bisects $\angle BAD$ and $\overline{AC}$ bisects $\angle BCD$. Which method of triangle congruence would you use to prove $\triangle ABC \cong \triangle ADC$?

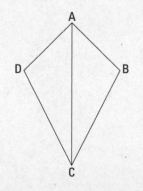

98. Given: $\overline{QR} \cong \overline{QT}$ and $\overline{ST} \cong \overline{SR}$. Which method of triangle congruence would you use to prove △QRS ≅ △QTS?

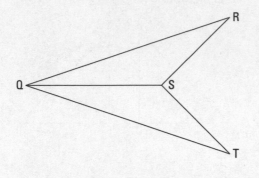

99. Given: $\overline{SE}$ is the altitude drawn to $\overline{RT}$, and $\overline{SE}$ bisects ∠RST. Which method of triangle congruence would you use to prove △RES ≅ △TES?

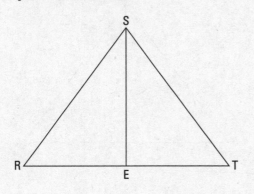

100. Given: $\overline{AB} \parallel \overline{DE}$ and $\overline{AB} \cong \overline{DE}$. Which method of triangle congruence would you use to prove △ABC ≅ △EDC?

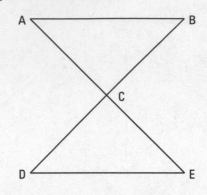

101. Given: $\overline{AD} \cong \overline{FC}$, $\overline{BC} \cong \overline{ED}$, and ∠ADC ≅ ∠FCD. Which method of triangle congruence would you use to prove △ABD ≅ △FEC?

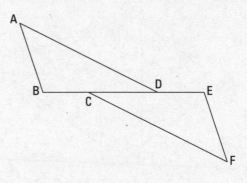

102. Given: Quadrilateral ABCD, $\overline{AD} \parallel \overline{BC}$, and ∠BAD ≅ ∠BCD. Which method of triangle congruence would you use to prove △ABD ≅ △CBD?

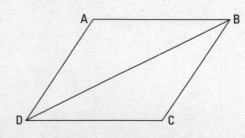

Completing Geometric Proofs Using Triangle Congruence Theorems

103–107 *Use the following figure to answer each question.*

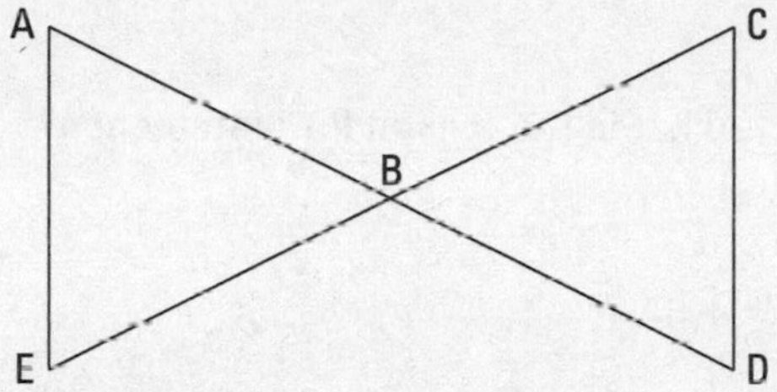

Given: $\overline{AD}$ and $\overline{CE}$ bisect each other at *B*.

Prove: $\overline{AE} \cong \overline{DC}$

Statements	*Reasons*
1. $\overline{AD}$ and $\overline{CE}$ bisect each other at *B*.	1. Given
2. $\overline{AB} \cong \overline{BD}$ and $\overline{EB} \cong \overline{BC}$	2.
3.	3. Intersecting lines form vertical angles.
4. $\angle ABE \cong \angle DBC$	4.
5. $\triangle ABE \cong \triangle DBC$	5.
6. $\overline{AE} \cong \overline{DC}$	6.

103. What is the reason for Statement 2?

104. What is the statement for Reason 3?

105. What is the reason for Statement 4?

106. What is the reason for Statement 5?

107. What is the reason for Statement 6?

108–111 *Use the following figure to answer each question.*

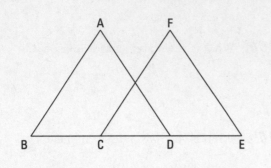

Given: $\overline{AB} \cong \overline{FC}$, $\angle ABC \cong \angle FCE$, and $\overline{BC} \cong \overline{DE}$

Prove: $\angle A \cong \angle F$

108. What is the reason for Statement 2?

109. What is the reason for Statement 3?

110. What is the reason for Statement 4?

111. What is the reason for Statement 5?

Statements	Reasons
1. $\overline{AB} \cong \overline{FC}$, $\angle ABC \cong \angle FCE$, $\overline{BC} \cong \overline{DE}$	1. Given
2. $\overline{CD} \cong \overline{CD}$	2.
3. $\overline{BC} + \overline{CD} \cong \overline{CD} + \overline{DE}$ or $\overline{BD} \cong \overline{CE}$	3.
4. $\triangle ABD \cong \triangle FCE$	4.
5. $\angle A \cong \angle F$	5.

112–116 *Use the following figure to answer each question.*

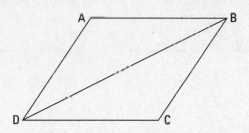

Given: $\overline{AB} \parallel \overline{CD}$ and $\angle CBD \cong \angle ADB$

Prove: $\overline{AB} \cong \overline{CD}$

112. What is the statement for Reason 2?

113. What is the reason for Statement 3?

114. What is the statement for Reason 4?

115. What is the reason for Statement 5?

116. What is the reason for Statement 6?

Statements	Reasons
1. $\overline{AB} \parallel \overline{CD}$ and $\angle CBD \cong \angle ADB$	1. Given
2.	2. When two parallel lines are cut by a transversal, alternate interior angles are formed.
3. $\angle ABD \cong \angle CDB$	3.
4.	4. Reflexive property
5. $\triangle ABD \cong \triangle CDB$	5.
6. $\overline{AB} \cong \overline{CD}$	6.

117 Complete the following proof.

117.

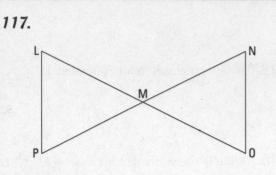

Given: $\overline{LP} \parallel \overline{ON}$, and M is the midpoint of $\overline{LO}$.

Prove: $\overline{PM} \cong \overline{NM}$

Statements	Reasons
1. $\angle TRL \cong \angle TMS$ and $\overline{RT} \cong \overline{MT}$	1. Given
2. $\angle RTL \cong \angle MTS$	2.
3. $\triangle RTL \cong \triangle MTS$	3.
4. $\overline{TL} \cong \overline{TS}$	4.

118. What is the reason for Statement 2?

119. What is the reason for Statement 3?

120. What is the reason for Statement 4?

Overlapping Triangle Proofs

118–120 Use the following figure to answer the question regarding overlapping triangles.

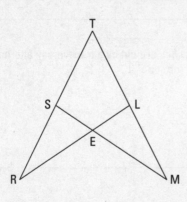

Given: $\angle TRL \cong \angle TMS$ and $\overline{RT} \cong \overline{MT}$

Prove: $\overline{TL} \cong \overline{TS}$

121–125 *Use the following figure to answer each question regarding overlapping triangles.*

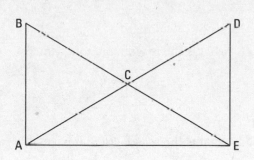

Given: $\overline{AB} \perp \overline{AE}$, $\overline{DE} \perp \overline{AE}$, and $\angle DAE \cong \angle BEA$

Prove: $\angle ABE \cong \angle EDA$

121. What is the statement for Reason 2?

122. What is the reason for Statement 3?

123. What is the reason for Statement 4?

124. What is the reason for Statement 5?

125. What is the reason for Statement 6?

Statements	Reasons
1. $\overline{AB} \perp \overline{AE}$, $\overline{DE} \perp \overline{AE}$, and $\angle DAE \cong \angle BEA$	1. Given.
2.	2. Perpendicular lines form right angles.
3. $\angle BAE \cong \angle DEA$	3.
4. $\overline{AE} \cong \overline{AE}$	4.
5. $\triangle ADE \cong \triangle EBA$	5.
6. $\angle ABE \cong \angle EDA$	6.

126–130 Use the following figure to answer the questions regarding overlapping triangles.

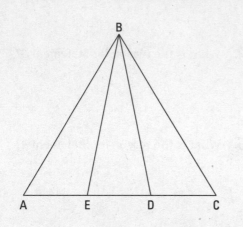

Given: $\overline{AE} \cong \overline{CD}$ and $\overline{BE} \cong \overline{BD}$

Prove: $\overline{AB} \cong \overline{CB}$

Statements	Reasons
1. $\overline{AE} \cong \overline{CD}$ and $\overline{BE} \cong \overline{BD}$	1. Given
2. $\overline{ED} \cong \overline{ED}$	2.
3. $\overline{AE} + \overline{ED} \cong \overline{CD} + \overline{ED}$ or $\overline{AD} \cong \overline{CE}$	3.
4. $\angle BDE \cong \angle BED$	4.
5. $\triangle BDA \cong \triangle BEC$	5.
6. $\overline{AB} \cong \overline{CB}$	6.

126. What is the reason for Statement 2?

127. What is the reason for Statement 3?

128. What is the reason for Statement 4?

129. What is the reason for Statement 5?

130. What is the reason for Statement 6?

131–136 Use the following figure to answer each question regarding overlapping triangles.

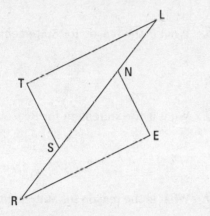

Given: $\overline{RS} \cong \overline{LN}$, $\angle LRE \cong \angle RLT$, and $\angle TSR \cong \angle ENL$

Prove: $\overline{TL} \cong \overline{ER}$

131. What is the reason for Statement 2?

132. What is the reason for Statement 3?

133. What is the reason for Statement 4?

134. What is the reason for Statement 5?

135. What is the reason for Statement 6?

136. What is the reason for Statement 7?

Statements	Reasons
1. $\overline{RS} \cong \overline{LN}$, $\angle LRE \cong \angle RLT$, and $\angle TSR \cong \angle ENL$	1. Given
2. $\overline{SN} \cong \overline{SN}$	2.
3. $\overline{LN} + \overline{NS} \cong \overline{RS} + \overline{SN}$ or $\overline{LS} \cong \overline{RN}$	3.
4. $\angle LNE$ and $\angle ENR$ are supplementary angles. $\angle TSR$ and $\angle TSL$ are supplementary angles.	4.
5. $\angle TSL \cong \angle ENR$	5.
6. $\triangle TSL \cong \triangle ENR$	6.
7. $\overline{TL} \cong \overline{ER}$	7.

Indirect Proofs

137–143 Use the following figure to answer each question regarding this indirect proof.

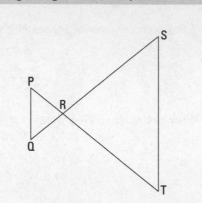

Given: $\overline{ST} \not\cong \overline{PQ}$, $\overline{QS}$ and $\overline{PT}$ are straight lines.

Prove: $\overline{QS}$ and $\overline{PT}$ do not bisect each other.

137. What is the statement for Reason 2?

138. What is the reason for Statement 3?

139. What is the statement for Reason 4?

140. What is the reason for Statement 5?

141. What is the reason for Statement 6?

142. What is the reason for Statement 7?

143. What is the reason for Statement 8?

Statements	Reasons
1. $\overline{ST} \not\cong \overline{PQ}$, $\overline{QS}$ and $\overline{PT}$ are straight lines.	1. Given
2.	2. Assumption
3. $\overline{PR} \cong \overline{RT}$ and $\overline{QR} \cong \overline{RS}$	3.
4.	4. Intersecting lines form vertical angles.
5. $\angle PRQ \cong \angle SRT$	5.
6. $\triangle PRQ \cong \triangle SRT$	6.
7. $\overline{ST} \cong \overline{PQ}$	7.
8. $\overline{QS}$ and $\overline{PT}$ do not bisect each other.	8.

144–149 *Use the following figure to answer each question regarding this indirect proof.*

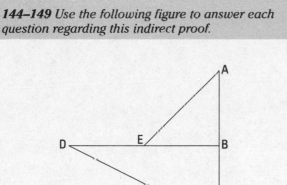

Given: $\overline{BD} \perp \overline{AC}$, $\overline{BC} \cong \overline{EB}$, and $\overline{CD} \not\cong \overline{AE}$

Prove: $\angle DCB \not\cong \angle AEB$

144. What is the statement for Reason 2?

145. What is the reason for Statement 3?

146. What is the reason for Statement 4?

147. What is the reason for Statement 5?

148. What is the reason for Statement 6?

149. What is the reason for Statement 7?

Statements	Reasons
1. $\overline{BD} \perp \overline{AC}$, $\overline{BC} \cong \overline{EB}$, and $\overline{CD} \not\cong \overline{AE}$	1. Given
2.	2. Assumption
3. $\angle DBA$ and $\angle DBC$ are right angles.	3.
4. $\angle DBA \cong \angle DBC$	4.
5. $\triangle EAB \cong \triangle CDB$	5.
6. $\overline{CD} \cong \overline{AE}$	6.
7. $\angle DCB \not\cong \angle AEB$	7.

150 *Complete the following indirect proof.*

150.

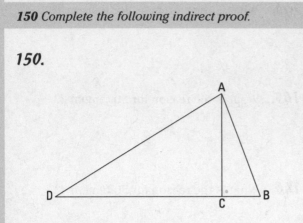

Given: $\overline{AC} \perp \overline{DB}$, $\overline{AB} \not\cong \overline{AD}$

Prove: $\overline{BC} \not\cong \overline{DC}$

Chapter 4

Classifying Triangles

· ·

Understanding triangles is a very important part of geometry. Triangles can be classified by the measures of their sides or angles. When given a specific type of triangle, you can use its special properties to solve various math problems. Having knowledge of these properties helps you set up a problem algebraically or complete a geometric proof.

The Problems You'll Work On

In this chapter, you see a variety of geometry problems:

- ✔ Identifying equilateral, isosceles, scalene, and right triangles by their side measurements
- ✔ Identifying equiangular, isosceles, scalene, right, acute, and obtuse triangles by their angle measurements
- ✔ Doing geometric proofs with isosceles triangles
- ✔ Using the hypotenuse-leg (HL) theorem to prove triangles congruent

What to Watch Out For

Don't let common mistakes trip you up. Some of the following suggestions may be helpful:

- ✔ A triangle can be classified as more than one type of triangle. For example, a triangle can be both isosceles and right.
- ✔ If you want to use HL as a method of proving triangles congruent, you must first be able to show that you have a right angle in each of the triangles.
- ✔ A lot of these questions require algebraic solutions. Be sure to read the questions carefully. Sometimes a question asks you to solve for the variable, and other times the question asks for the actual measurement of the side or angle, which means you need to plug in the variable to get your solution.
- ✔ Remember that it's the degree measure of the three angles of a triangle, not the three sides of a triangle, that must add up to 180.
- ✔ Remember that the congruent angles of an isosceles triangle are found opposite the congruent sides of the triangle.

Classifying Triangles by Their Sides

151–157 The given numbers represent the three sides of a triangle. Classify the triangle as isosceles, equilateral, scalene, and/or right.

151. $\{4, 4, 7\}$

152. $\{7, 9.2, 11.5\}$

153. $\{10.4, 10.4, 10.4\}$

154. $\{6, 6, 6\sqrt{2}\}$

155. $\{7, 7\sqrt{3}, 14\}$

156. $\{a, a, a\}$

157. $\{\sqrt[3]{8}, 2, \sqrt{2}\}$

158–164 Refer to △PER. Use the following information to calculate the length of each side of the triangle and classify the triangle as isosceles, equilateral, scalene, and/or right.

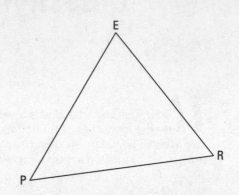

158. The perimeter of △*PER* is 108 units. The three sides of the triangle are represented by

$$PE = 5x + 11$$
$$ER = 7x + 1$$
$$RP = 8x - 4$$

Classify this triangle.

159. The perimeter of △*PER* is 210 units. The three sides of the triangle are represented by

$$PE = x + 80$$
$$ER = 2x + 60$$
$$RP = x - 10$$

Classify this triangle.

160. The perimeter of $\triangle PER$ is 60 units. The three sides of the triangle are represented by

$PE = 2a + 8$

$ER = 2a - 6$

$RP = 3a + 2$

Classify this triangle.

161. The perimeter of $\triangle PER$ is 24 units. The three sides of the triangle are represented by

$PE = x$

$ER = x + 2$

$RP = 2x - 2$

Classify this triangle.

162. The perimeter of $\triangle PER$ is 34 units. The three sides of the triangle are represented by

$PE = x$

$ER = 4x$

$RP = 3x + 2$

Classify this triangle.

163. The perimeter of $\triangle PER$ is $19\sqrt{2}$ units. The three sides of the triangle are represented by

$PE = \sqrt{50}$

$ER = 9\sqrt{2}$

$RP = x$

Classify this triangle.

164. The perimeter of $\triangle PER$ is 42 units. The three sides of the triangle are represented by

$PE = x^2$

$ER = 4x$

$RP = 3x - 2$

Classify this triangle.

Properties of Isosceles, Equilateral, and Right Triangles

165–169 Use the properties of isosceles, equilateral, and right triangles to solve the problems.

165. $\triangle DEF$ is isosceles with vertex E. If $DE = 2x + 10$ and $FE = 3x - 4$, find x.

166. $\triangle LIN$ is an equilateral triangle. If $IN = 2.5x - 14$ and $LN = x + 22$, find the length of $\overline{LI}$.

167. $\triangle HYP$ is a right triangle with $m\angle P = 90°$. If PY is 3 more than HP and if HY is 3 less than two times HP, find the length of $\overline{HP}$.

168. $\triangle SHE$ is isosceles with vertex H. If SH is represented by x^2 and HE is represented by $x + 12$, find the positive value of x.

169. △*ALM* is an equilateral triangle. If *AL* is represented by x^3 and *AM* = 64, find the positive value of *x*.

Classifying Triangles by Their Angles

170–176 The given numbers represent the three angles of a triangle. Classify the triangle as acute, obtuse, equiangular, or right.

170. $\{40°, 50°, x\}$

171. $\{60°, 60°, x\}$

172. $\{50°, 70°, x\}$

173. $\{20°, 60°, x\}$

174. $\{10°, 80°, x\}$

175. $\{2x, x+30, 3x-30\}$

176. $\{a, b, a+b\}$

177–182 △WXY is drawn with $\overline{WY}$ extended to $\overline{WYZ}$. Use the diagram and the given information to classify each triangle as acute, obtuse, equiangular, or right.

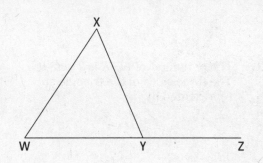

177. $m\angle XWY = 50°$ and $m\angle WXY$ is 20 more than $m\angle WYX$.

178. $m\angle WXY = 60°$, $m\angle XWY$ is represented by $2x$, and $m\angle XYW$ is represented by $x+30$.

179. $m\angle XYZ$ is twice $m\angle XYW$. Is it possible to classify this triangle?

180. $m\angle X = x+40$
$m\angle W = x+10$
$m\angle WYX = 2x-10$

181. $m\angle XYZ$ is 30 more than twice $m\angle WYX$, and $m\angle W = 40°$.

182. $m\angle X = x$
$m\angle W = 2x$
$m\angle WYX = 3x$

Understanding the Classification of Triangles

183–190 Use the properties of triangles to solve the following problems.

183. $\triangle DEF$ is isosceles with vertex E. If $m\angle D = 4x - 40$ and $m\angle F = x + 20$, find the degree measure of $\angle F$.

184. $\triangle RST$ is equiangular. If $m\angle R$ is represented by $x + 30$ and $m\angle S = 2x$, find the value of x.

185. In right $\triangle ABC$, $\angle A$ is a right angle. If $m\angle B$ is represented by $x + 10$ and $m\angle C$ is represented by $2x - 10$, find the degree measure of $\angle C$.

186. In right $\triangle ABC$, $\angle A$ is a right angle. If $m\angle B$ is represented by x^2 and $m\angle C$ is represented by $13x$, find the degree measure of $\angle C$.

187. True or False? All equilateral triangles are similar to each other.

188. True or False? From the vertex of an isosceles triangle, the altitude, angle bisector, and median are all the same segment.

189. True or False? In an equilateral triangle, the angle bisector is always perpendicular to the opposite side.

190. True or False? There can be two obtuse angles in a triangle.

Geometric Proofs Involving Isosceles Triangles

191–194 Complete the following geometric proof by filling in the statement or reason.

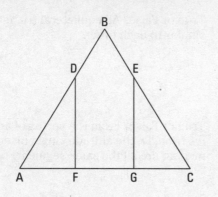

Given: △*ABC* is an isosceles triangle with vertex *B*, $\overline{BD} \cong \overline{BE}$, and ∠*ADF* ≅ ∠*CEG*

Prove: △*ADF* ≅ △*CEG*

191. What is the statement for Reason 2?

192. What is the statement for Reason 3?

193. What is the reason for Statement 4?

194. What is the reason for Statement 5?

Statements	Reasons
1. △*ABC* is an isosceles triangle with vertex *B*, $\overline{BD} \cong \overline{BE}$, and ∠*ADF* ≅ ∠*CEG*	1. Given
2.	2. If a triangle is isosceles, then its base angles are congruent.
3.	3. If a triangle is isosceles, then its legs are congruent.
4. $\overline{BA} - \overline{BD} \cong \overline{BC} - \overline{BE}$ or $\overline{DA} \cong \overline{EC}$	4.
5. △*ADF* ≅ △*CEG*	5.

195–199 *Complete the following geometric proof by filling in the statement or reason.*

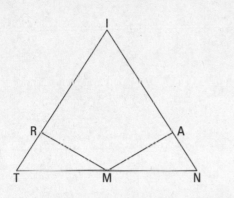

Given: *M* is the midpoint of $\overline{TN}$, $\overline{TR} \perp \overline{MR}$, $\overline{NA} \perp \overline{MA}$, and $\angle TMA \cong \angle NMR$

Prove: △*TIN* is isosceles

195. What is the reason for Statement 2?

196. What is the statement for Reason 4?

197. What is the reason for Statement 7?

198. What is the statement for Reason 8?

199. What is the reason for Statement 9?

Statements	Reasons
1. *M* is the midpoint of $\overline{TN}$, $\overline{TR} \perp \overline{MR}$, $\overline{NA} \perp \overline{MA}$, and $\angle TMA \cong \angle NMR$	1. Given
2. $\overline{TM} \cong \overline{NM}$	2.
3. $\angle TRM$ and $\angle NAM$ are right angles	3. Perpendicular lines form right angles.
4.	4. All right angles are congruent.
5. $\angle RMA \cong \angle RMA$	5. Reflexive property
6. $\angle TMA - \angle RMA \cong \angle NMR - \angle RMA$ or $\angle RMT \cong \angle AMN$	6. Angle subtraction postulate and substitution
7. △*TRM* $\cong$ △*NAM*	7.
8.	8. CPCTC
9. △*TIN* is isosceles	9.

200 Complete the following geometric proof.

200.

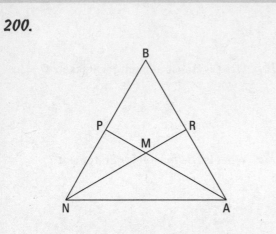

Given: △*NBA* is isosceles with vertex *B*, and △*NMA* is isosceles with vertex *M*

Prove: △*PNA* ≅ △*RAN*

201–205 *Complete the geometric proof by filling in the statement or reason.*

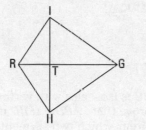

Given: $\overline{RI} \perp \overline{GI}$, $\overline{RH} \perp \overline{GH}$, and $\triangle RHI$ is an isosceles triangle with vertex R

Prove: $\triangle RIG \cong \triangle RHG$

201. What is the reason for Statement 2?

202. What is the statement for Reason 4?

203. What is the statement for Reason 5?

204. What is the reason for Statement 6?

205. What is the reason for Statement 7?

Statements	Reasons
1. $\overline{RI} \perp \overline{GI}$, $\overline{RH} \perp \overline{GH}$; $\triangle RHI$ is isosceles with vertex R	1. Given
2. $\angle RIG$ and $\angle RHG$ are right angles	2.
3. $\angle RIG \cong \angle RHG$	3. All right angles are congruent.
4.	4. If a triangle is isosceles, then its legs are congruent.
5.	5. Reflexive property
6. $\angle RIG$ and $\angle RHG$ are right triangles	6.
7. $\triangle RIG \cong \triangle RHG$	7.

206 *Complete the following geometric proof.*

206.

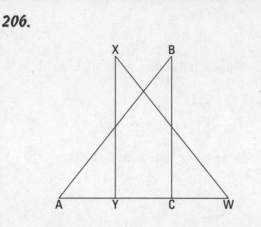

Given: $\overline{AY} \cong \overline{WC}$, $\overline{XY} \perp \overline{AW}$, $\overline{BC} \perp \overline{AW}$, and $\overline{AB} \cong \overline{WX}$

Prove: $\overline{BC} \cong \overline{XY}$

207–210 *Use the diagram and the given information to answer the question.*

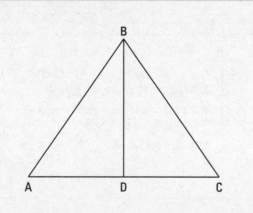

207. $\overline{AD} \cong \overline{CD}$ and $\overline{BD}$ is an altitude. If you want to prove that $\triangle ABD \cong \triangle CBD$ using the hypotenuse-leg theorem, which two angles or segments must be shown to be congruent?

208. $\triangle ABC$ is isosceles with vertex B. If you want to prove that $\triangle ABD \cong \triangle CBD$ using the SAS theorem, which two segments must be shown to be congruent?

209. $\triangle ABC$ is isosceles. If you want to prove that $\triangle ABD \cong \triangle CBD$ using the AAS theorem, which two angles or segments must be shown to be congruent?

210. A student marked in the diagram that $\angle ADB$ and $\angle CDB$ are right angles, $\overline{AB} \cong \overline{CB}$, and $\overline{BD} \cong \overline{BD}$. With these marks, which triangle congruence theorem can you use to prove that $\triangle ABD \cong \triangle CBD$?

Chapter 5

Investigating the Centers of a Triangle

●●

Triangles have four types of centers that are common in geometry: The *incenter* is the point where the angle bisectors of a triangle intersect, the *orthocenter* is the point where the altitudes intersect, the *centroid* is the point where the medians intersect, and the *circumcenter* is the point where the perpendicular bisectors intersect. These centers can appear in algebraic problems and in problems dealing with constructions.

The Problems You'll Work On

In this chapter, you see a variety of geometry problems:

- ✔ Applying the definitions of incenter, orthocenter, centroid, and circumcenter
- ✔ Locating the coordinates of a triangle's centers
- ✔ Constructing the incenter, orthocenter, centroid, and circumcenter of a triangle
- ✔ Determining the equation of the Euler line

What to Watch Out For

To avoid common mistakes, remember the following points:

- ✔ Make sure you understand the definitions of the angle bisector, perpendicular bisector, median, and altitude of a triangle.
- ✔ Knowing the midpoint and slope formulas is necessary for solving many of the problems involving the centers of a triangle.
- ✔ Writing the equation of the Euler line requires you to determine the equations of two centers of a triangle and see where they intersect.
- ✔ Keep in mind that it's possible for the orthocenter and circumcenter to lie outside the triangle.

The Incenter of a Triangle

211–217 Point I is the incenter of △CEN. Use the following figure and the given information to solve the problem.

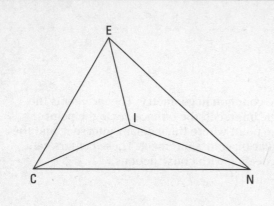

211. If $m\angle ICN = 38°$, find $m\angle ICE$.

212. If $m\angle CEN = 60°$ and $m\angle ECN = 68°$, find $m\angle CNI$.

213. If $m\angle CEI = 28°$, $m\angle ENI = 40°$, $m\angle CNI$ is represented by $2c$, and $m\angle NEI$ is represented by $2b + c$, find the value of b.

214. If $m\angle ENC = 70°$ and $m\angle ECN = 60°$, find $m\angle CIN$.

215. If $m\angle CIE = 100°$ and $m\angle CEN = 76°$, find $m\angle ECI$.

216. If $m\angle EIN = 120°$ and $m\angle CNE = 50°$, find $m\angle CEN$.

217. If $m\angle CIE = 110°$, $m\angle CIN = 120°$, and $m\angle IEN = 30°$, find $m\angle ENC$.

218 Solve the problem using your knowledge of the incenter.

218. In isosceles △INC, $\angle NIC \cong \angle NCI$. If T represents the triangle's incenter and $m\angle NIT = 25°$, find the measure of $\angle CNT$.

Understanding the Orthocenter

219–223 Use your knowledge of the orthocenter of a triangle to solve the following problems.

219. The coordinates of △ABC are $A\ (0, 2)$, $B\ (-2, 6)$, and $C\ (4, 0)$. Find the coordinates of the orthocenter of this triangle.

220. The coordinates of △*ORT* are *O* (0, 0), *R* (6, 0), and *T* (–2, 4). Find the coordinates of the orthocenter of this triangle.

221. The coordinates of △*AND* are *A* (0, 0), *N* (6, 0), and *D* (–2, 8). Find the coordinates of the orthocenter of this triangle.

222. In which type of triangle does the orthocenter lie on the vertex of the triangle?

223. In which type of triangle does the orthocenter lie outside of the triangle?

Understanding Centroids

224–231 Use the given information to solve for the missing side.

224. Point *C* represents the centroid of △*RST*. If *SC* = 4, find *DC*.

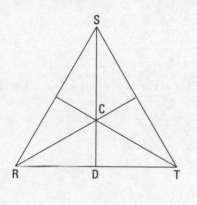

225. Point *C* represents the centroid of △*RST*. If *DC* = 2, find *SC*.

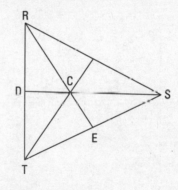

226. Point *C* represents the centroid of △*RST*. If *SD* = 21, find *SC*.

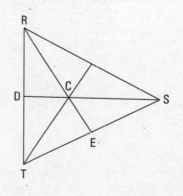

227. Point *C* represents the centroid in △*RST*. If *RE* = 24, find *CE*.

228. Point *C* represents the centroid of △*RST*. If *SD* = 54, find *DC*.

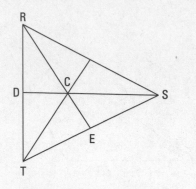

229. Point *C* represents the centroid in △*RST*. If *RE* = 27, find *RC*.

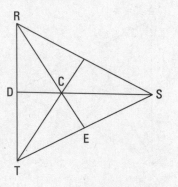

230. Point *C* represents the centroid of △*RST*. If *DC* = 15, find *DS*.

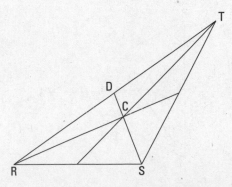

231. Point *C* represents the centroid of △*RST*. If *DC* = 5.5, find *DS*.

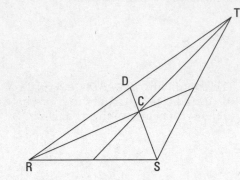

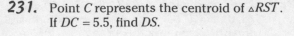

Finding the Centroid of a Triangle

232–235 Use your knowledge of the centroid of a triangle to solve the following problems.

232. The vertices of a triangle are (0, –2), (4, 0), and (2, 8). Find the coordinates of the centroid of the triangle.

233. The vertices of a triangle are (0, 0), (0, 9), and (9, 0). Find the coordinates of the centroid of the triangle.

234. The vertices of a triangle are (–6, 2), (–2, 8), and (4, 4). Find the coordinates of the centroid of the triangle.

235. The vertices of a triangle are $(1, 1)$, $(6, 0)$, and $(4, 3)$. Find the sum of the x and y coordinates of the centroid.

240. Solve for the angle that is represented by b.

241. Solve for the angle that is represented by d.

The Circumcenter of a Triangle

236–241 In the following figure, Point C represents the circumcenter of the triangle.

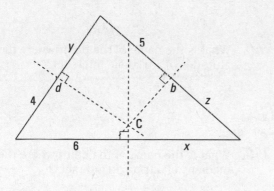

Recognizing Triangle Centers

242–251 Identify the type of center.

242. State whether Point O represents the incenter, orthocenter, centroid, or circumcenter.

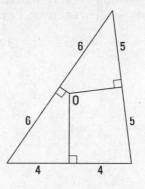

236. Solve for x.

237. Solve for y.

243. State whether Point O represents the incenter, orthocenter, centroid, or circumcenter.

238. Solve for z.

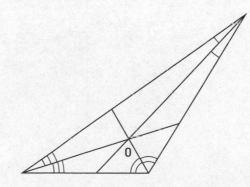

239. Solve for the perimeter of the triangle.

244. State whether Point *O* represents the incenter, orthocenter, centroid, or circumcenter.

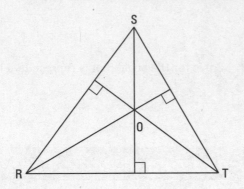

245. State whether Point *O* represents the incenter, orthocenter, centroid, or circumcenter.

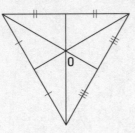

246. State whether Point *O* represents the incenter, orthocenter, centroid, or circumcenter.

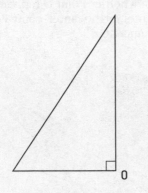

247. Point *O* represents the intersection of the three angle bisectors of a triangle. State whether Point *O* represents the incenter, orthocenter, centroid, or circumcenter.

248. Point *O* represents the intersection of the three perpendicular bisectors of a triangle. State whether Point *O* represents the incenter, orthocenter, centroid, or circumcenter.

249. What is the name of the point where the altitudes of a triangle intersect?

250. What is the name of the point where the medians of a triangle intersect?

251. The center of a circle inscribed in a triangle coincides with which of the triangle's centers?

Constructing the Centers of a Triangle

252–256 Use your knowledge of constructions to answer the following questions.

252. What is the following figure a construction of?

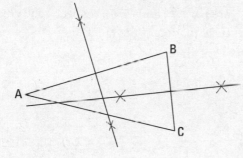

253. What is the following figure a construction of?

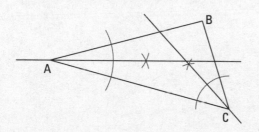

254. What is the following figure a construction of?

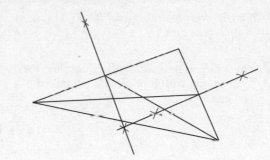

255. What is the following figure a construction of?

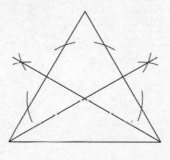

256. Construct the incenter for the following triangle.

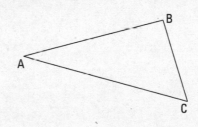

The Euler Line

257–260 Use your knowledge of the Euler line to answer each question.

257. In order to determine the Euler line for a triangle, which points do you need to know?

258. The equation for the Euler line of a triangle is $y = \frac{3}{4}x - 1$. If the x coordinate of the circumcenter is 5, what is the y coordinate for the circumcenter?

259. The centroid of a triangle is (6, 2), and the orthocenter of the triangle is $\left(\frac{32}{3}, \frac{8}{3}\right)$. What is the equation of the Euler line for this triangle?

260. The centroid and orthocenter of a triangle are $\left(2, \frac{-2}{3}\right)$ and (4, 0), respectively. What is the equation of the Euler line?

Chapter 6

Similar Triangles

. . . .

An important concept in geometry is the difference between congruence and similarity. This chapter focuses on the similarity of triangles. If two triangles are *similar,* their sides are in proportion. Here, you practice setting up proportions and solving them algebraically. You also work on geometric proofs dealing with similar triangles.

The Problems You'll Work On

In this chapter, you see a variety of geometry problems:

- ✔ Determining side lengths of similar triangles
- ✔ Connecting the midpoints of two sides of a triangle to create a segment parallel to the third side
- ✔ Finding the perimeter and area of similar triangles
- ✔ Writing geometric proofs with similar triangles
- ✔ Understanding that the product of the means equals the product of the extremes

What to Watch Out For

Don't let common mistakes trip you up. Some of the following suggestions may be helpful:

- ✔ Remember that if two triangles are similar, the corresponding sides are in proportion.
- ✔ If you're looking to prove that the sides of a triangle are in proportion, you must first prove the triangles similar by AA (angle-angle).
- ✔ Similar triangles are different only in size. The corresponding angles still have the same measure.
- ✔ If two triangles are similar, the ratio of the perimeters of the triangles is the same as the ratio of the sides; however, the ratio of the area of the triangles is the square of the ratio of the sides of the triangles.
- ✔ Make sure you know what the question is asking. Sometimes a question asks you to solve for a variable. In other cases, you have to find the value of a variable and then plug it in to find the measure of a segment or angle.

Understanding Similar Triangles

261–267 Use the given information to answer the question regarding similar triangles △ABC and △DEF.

261. △ABC ~ △DEF. If $AB = 4$, $DF = 9$, and $DE = 12$, what is the length of $\overline{AC}$?

262. △ABC ~ △DEF. If $AB = 10$, $AC = 8$, and $DF = 16$, what is the length of $\overline{DE}$?

263. △ABC ~ △DEF. If $AB = 8$, $BC = 12$, $AC = 6$, and $DE = 16$, find the perimeter of △DEF.

264. △ABC ~ △DEF. If $AB = 8$, $BC = 12$, $AC = 6$, and the perimeter of △DEF $= 13$, find the length of $\overline{EF}$.

265. △ABC ~ △DEF. If $m\angle A = 78°$, find $m\angle D$.

266. △ABC ~ △DEF. If $m\angle D = 31°$ and $m\angle E = 88°$, find $m\angle C$.

267. △ABC ~ △DEF. If $DE = 22$, $DF = 16$, and $AC = 24$, find the length of $\overline{AB}$.

Midsegments

268–274 In △SHE, R is the midpoint of $\overline{SH}$ and Y is the midpoint of $\overline{EH}$. Use the given information to solve the problem.

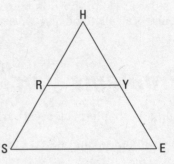

268. If $RY = b + 4$ and $SE = 20$, find the value of b.

269. If $SE = 2b - 8$ and $RY = 50$, find the value of b.

270. If $HR = 7x - 28$ and $RS = 2x - 3$, find the length of $\overline{RS}$.

271. If $RY = 8b + 4$ and $SE = 17b + 1$, solve for b.

272. If $RY = 8$ and $SE = b^2$, find the positive value of b.

273. If $RY = b + 24$ and $SE = b^2$, find the positive value of b.

274. If $RY = 128$ and $SE = b^4$, find the positive value of b.

Creating Similar Triangles

275–278 *In △ABC, D is a point on $\overline{AB}$ and E is a point on $\overline{AC}$. If $DE \parallel BC$, use the given information to solve the problem.*

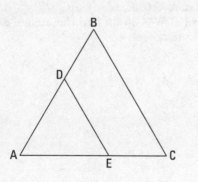

275. $BD = 3$, $AD = 6$, and $DE = 8$. Find the length of $\overline{BC}$.

276. $BD = 7$, $AD = 7$, and $DE = 8$. Find the length of $\overline{BC}$.

277. $AD = 3$, $AE = 5$, and $AB = 9$. Find the length of $\overline{AC}$.

278. $AD = 25$, $AE = 24$, and $AB = 30$. Find the length of $\overline{EC}$.

279–284 *In △MAT, B is a point on $\overline{MA}$ and C is a point on $\overline{AT}$. If $\overline{BC} \parallel \overline{MT}$, use the given information to solve the problem.*

279. $AB = 30$, $BC = 6$, and $MT = 8$. Find the length of $\overline{BM}$.

280. $BC = 2.5$, $BM = 2$, and $MT = 3$. Find the length of $\overline{AB}$.

281. $AB = 2x + 12$, $BC = 8$, $BM = 27$, and $MT = 17$. Find the value of x.

282. $AB = 3x$, $BC = 3$, $BM = 12$, and $MT = 6$. Find the value of x.

283. $AB = x$, $BC = 6$, $BM = x + 6$, and $MT = x + 3$. Find the positive value of x.

284. $AB = x$, $BC = 4$, $BM = x + 4$, and $MT = x + 2$. Find the value of x.

Similar-Triangle Word Problems

285–290 Apply your knowledge of similar triangles to solve each problem.

285. The lengths of the sides of a triangle are 3, 4, and 5. If the longest side of a similar triangle is 50, find the length of that triangle's shortest side.

286. The lengths of the sides of a triangle are 16, 23, and 31. If the perimeter of a similar triangle is 280, find the length of that triangle's longest side.

287. The length of the altitude of a triangle is 21, and the triangle's perimeter is 28. If the length of the corresponding altitude of a similar triangle is 12, find the perimeter of this triangle.

288. In $\triangle RST$, W is a point on $\overline{RS}$ and Y is a point on $\overline{ST}$ such that $\overline{RT} \parallel \overline{WY}$. If $WS = 5$, $RW = x + 7$, $YS = x$, and $TY = x + 3$, solve for x.

289. $\triangle LIN \sim \triangle DER$, $LI = 4$, and $DE = 3$. If the area of $\triangle LIN$ is 32 ft^2, find the area of $\triangle DER$ in square feet.

290. $\triangle LIN \sim \triangle DER$, $LI = 3$, and $DE = 5$. If the area of $\triangle LIN$ is 36 ft^2, find the area of $\triangle DER$ in square feet.

Proving That Two Triangles Are Similar to Each Other

291–294 Complete the proof by giving the statement or reason.

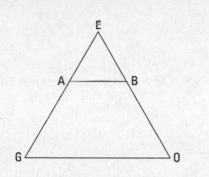

Given: $\overline{AB} \parallel \overline{GO}$

Prove: $\triangle GEO \sim \triangle AEB$

291. What is the missing angle in Statement 2?

292. What is the reason for Statement 3?

293. What is the statement for Reason 4?

294. What is the reason for Statement 5?

Statements	Reasons
1. $\overline{AB} \parallel \overline{GO}$	1. Given
2. $\angle EAB$ and $\angle$ _____ are corresponding angles.	2. If two parallel lines are cut by a transversal, then they form corresponding angles.
3. $\angle EAB \cong \angle EGO$	3.
4.	4. Reflexive property
5. $\triangle GEO \sim \triangle AEB$	5.

Proving That Corresponding Sides Are in Proportion

295–299 Complete the proof by giving the statement or reason.

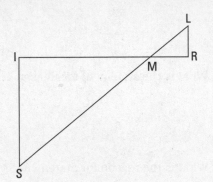

Given: $\overline{SL}$ intersects $\overline{IR}$ at M; $\overline{SI} \perp \overline{RI}$ and $\overline{LR} \perp \overline{RI}$

Prove: $\dfrac{IS}{IM} = \dfrac{RL}{RM}$

295. What is the missing angle in Statement 2?

296. What is the statement for Reason 3?

297. What is the reason for Statement 4?

298. What is the reason for Statement 6?

299. What is the reason for Statement 7?

Statements	Reasons
1. $\overline{SL}$ intersects $\overline{IR}$ at M; $\overline{SI} \perp \overline{RI}$ and $\overline{LR} \perp \overline{RI}$	1. Given
2. $\angle SMI$ and $\angle$ ____ are vertical angles.	2. Intersecting lines form vertical angles.
3.	3. Vertical angles are congruent.
4. $\angle SIM$ and $\angle LRM$ are right angles.	4.
5. $\angle SIM \cong \angle LRM$	5. All right angles are congruent to each other.
6. $\triangle SIM \sim \triangle LRM$	6.
7. $\dfrac{IS}{IM} = \dfrac{RL}{RM}$	7.

Proving with the Means and Extremes

300–305 Complete the proof by giving the statement or reason.

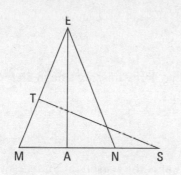

Given: $\overline{MNS}$, $\overline{ME} \cong \overline{NE}$, $\overline{EA} \perp \overline{MN}$, and $\overline{ST} \perp \overline{EM}$

Prove: $AN \times SM = EN \times TM$

300. What is the statement for Reason 2?

301. What is the reason for Statement 3?

302. What is the reason for Statement 4?

303. What is the statement for Reason 5?

304. What is the reason for Statement 6?

305. What is the statement for Reason 7?

Statements	Reasons
1. $\overline{MNS}$, $\overline{ME} \cong \overline{NE}$, $\overline{EA} \perp \overline{MN}$, and $\overline{ST} \perp \overline{EM}$	1. Given
2.	2. If two sides of a triangle are congruent, then the angles opposite those sides are congruent.
3. $\angle EAN$ and $\angle STM$ are right angles	3.
4. $\angle EAN \cong \angle STM$	4.
5.	5. AA
6. $\dfrac{AN}{TM} = \dfrac{EN}{SM}$	6.
7.	7. In a proportion, the product of the means is equal to the product of the extremes.

306–311 Complete the proof by giving the statement or reason.

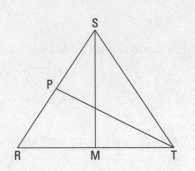

Given: △*RST* is isosceles with vertex *S*; ∠*RPT* ≅ ∠*SMT*

Prove: $RP \times SM = TP \times MT$

306. What is the statement for Reason 2?

307. What is the statement for Reason 3?

308. What is the missing side in Statement 4?

309. What is the reason for Statement 4?

310. What is the missing side in Statement 5?

311. What is the reason for Statement 5?

Statements	Reasons
1. △*RST* is isosceles with vertex *S*; ∠*RPT* ≅ ∠*SMT*	1. Given
2.	2. If a triangle is isosceles, then its base angles are congruent.
3.	3. AA
4. $\dfrac{RP}{TP} = \dfrac{TM}{SM}$	4.
5. $RP \times SM = TP \times$ ___	5.

312–314 Identify the errors in the following geometric proof.

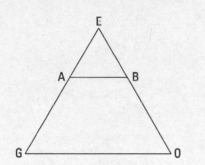

Given: $\angle EAB \cong \angle EGO$

Prove: $EA \times GO = EG \times AB$

312. What is the correct statement for Reason 2?

313. What is the correct reason for Statement 3?

314. What is the correct statement for Reason 4?

Statements	Reasons
1. $\angle EAB \cong \angle EGO$	1. Given
2. $\angle G \cong \angle G$	2. Reflexive property
3. $\triangle EAB \sim \triangle EGO$	3. AAS
4. $\dfrac{EA}{GO} = \dfrac{EG}{AB}$	4. Similar triangles have corresponding sides that are in proportion.
5. $EA \times GO = EG \times AB$	5. In a proportion, the product of the means is equal to the product of the extremes.

315 Complete the following proof.

315.

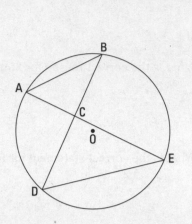

Given: In Circle O, chords $\overline{AE}$ and $\overline{BD}$ intersect at C. $\overline{AB}$ and $\overline{DE}$ are chords.

Prove: $AB \times DC = ED \times BC$

Chapter 7

The Right Triangle

. . . .

The right triangle has many applications in the real world. The Pythagorean theorem shows the special relationship among the sides of a right triangle. You can also discover interesting patterns and concepts from the right triangle. Some of those concepts include Pythagorean triplets and the 45°-45°-90° and 30°-60°-90° special right triangles. The right triangle is also the foundation of trigonometry, giving you the trigonometric ratios *sine*, *cosine*, and *tangent*.

The Problems You'll Work On

In this chapter, you see a variety of geometry problems:

- ✔ Using the Pythagorean theorem to find the length of a side of a triangle
- ✔ Using the Pythagorean theorem to prove that a triangle is a right triangle
- ✔ Understanding proportions of similar right triangles formed by drawing an altitude to the hypotenuse
- ✔ Finding the length of a side of a 30°-60°-90° or 45°-45°-90° triangle
- ✔ Using trigonometric ratios to solve for missing angles and sides of a right triangle

What to Watch Out For

Don't let common mistakes trip you up. Some of the following suggestions may be helpful:

- ✔ Remember that the hypotenuse of a right triangle must be the largest side of the triangle. This side is represented by c in the Pythagorean theorem, $a^2 + b^2 = c^2$.
- ✔ To help you decide which of the three trigonometric ratios to use (sine, cosine, or tangent), label the sides of the triangle as adjacent or opposite. This labeling is dependent on the given angle in the right triangle. The hypotenuse is always the side opposite the right angle.
- ✔ Solving for an angle of a right triangle requires the use of the inverse trigonometric functions.
- ✔ Make sure you understand what the question is asking. Sometimes you simply need to solve for a variable. Other times, you have to find the value of the variable and then plug it in to find the measure of a side.

Pythagorean Theorem

316–325 Use the Pythagorean theorem to solve.

316. In $\triangle ABC$, $\angle C$ is a right angle. If $AC = 6$ and $BC = 8$, find the length of $\overline{AB}$.

317. In $\triangle RST$, $m\angle S = 90°$. If $RS = 5$ and $RT = 13$, find the length of $\overline{ST}$.

318. In $\triangle BAG$, $m\angle A = 90°$. If $AB = 4$ and $AG = 2$, find the length of $\overline{BG}$.

319. In $\triangle ABC$, $\angle C$ is a right angle. If side $a = 24$ and side $c = 25$, what is the length of side b?

320. In $\triangle BAG$, $\angle G$ is a right angle. If side $b = 3$ and side $a = 5$, what is the length of side g?

321. In $\triangle DOG$, $m\angle G = 90°$. If $OG = x$, $DG = x - 7$, and $DO = 17$, solve for x.

322. The diagonal of a rectangle measures 10 inches. If one side of the rectangle measures 8 inches, what is the perimeter of the rectangle?

323. The diagonals of a rhombus measure 10 and 14 inches. What is the length of a side of the rhombus?

324. In isosceles $\triangle ABC$, $AB = BC = 13$. If $AC = 10$, what is the length of the altitude drawn to $\overline{AC}$?

325. True or False? A triangle whose sides measure 10, 20, and 25 is a right triangle.

Right Triangle Proportions

326–330 Use the following figure to answer each question.

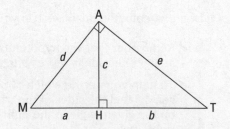

326. △MHA ~ △MAT ~ △AHT

Is this statement always true, sometimes true, or never true?

327. $a^2 + \underline{\quad}^2 = d^2$

328. $\dfrac{d}{a+b} = \dfrac{\quad}{d}$

329. $\dfrac{e}{a+b} = \dfrac{b}{\quad}$

330. $b^2 + c^2 = \underline{\quad}$

331–340 Use the following figure and given information to solve.

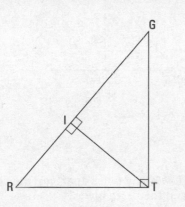

331. If $GI = 9$ and $RI = 4$, find the length of $\overline{TI}$.

332. If $GI = 6$ and $RI = 2$, find the length of $\overline{RT}$.

333. If $RT = 8$ and $GI = 12$, find the length of $\overline{RI}$.

334. If $RI = 10$ and $GI = 14$, find the length of $\overline{IT}$ in simplest radical form.

335. If $GI = 9$ and $IT = 3$, find the length of $\overline{RI}$.

336. If $RI = 9$ and $GT = 20$, find the length of $\overline{GI}$.

337. If $RI = 4$ and $GI = 8$, find the length of $\overline{RT}$ in simplest radical form.

338. If GI is 21 more than RI and $RT = 9$, find the length of $\overline{GI}$.

339. If *GI* is 21 more than *RI* and *IT* = 10, find the length of $\overline{GI}$.

340. If *GI* is 2 more than *IT* and *GT* = 10, find the length of $\overline{IT}$.

Word Problems Involving Right-Triangle Proportions

341–347 Use the proportions in the right triangle to solve for the missing side.

341. △*ABC* is a right triangle, and $\overline{CD}$ is the altitude drawn to hypotenuse $\overline{AB}$. If *AD* = 9 and *BD* = 16, find the length of $\overline{CD}$.

342. In right △*HLS*, $\overline{LR}$ is the altitude drawn to hypotenuse $\overline{HS}$. If *HL* = 6 and *HR* = 4, find $\overline{HS}$.

343. The altitude of a right triangle bisects the hypotenuse. If the altitude equals 25, find the length of the hypotenuse.

344. △*WJD* is a right triangle, and altitude $\overline{JC}$ is drawn to hypotenuse $\overline{WD}$. If *JC* = 3 and $\overline{WC}$ is 8 more than $\overline{DC}$, find the length of $\overline{WC}$.

345. △*WJD* is a right triangle, and altitude $\overline{JC}$ is drawn to hypotenuse $\overline{WD}$. If *JC* = 12 and *DC* is 10 more than *WC*, find the length of $\overline{DC}$.

346. In right △*ABC*, $\overline{CD}$ is the altitude drawn to hypotenuse $\overline{AB}$. If *AC* = $\sqrt{27}$ and $\overline{BD}$ is 3 more than *AD*, find the length of $\overline{AD}$.

347. In △*ABC*, $\overline{CD}$ is the altitude drawn to hypotenuse $\overline{AB}$. If *AC* = 4 and *BD* is 4 more than *AD*, find the length of $\overline{BD}$.

348–350 Use the diagram and the given information to answer each question.

Gavin is in his office, deciding where to go to lunch. The distance from the deli to the coffee shop is 18 miles. The distance from Gavin's office to the coffee shop is 24 miles.

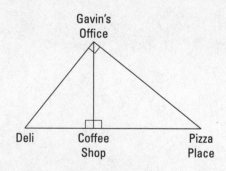

348. How many miles apart are the coffee shop and the pizza place?

349. How many miles is the deli from Gavin's office?

350. How many miles is the pizza place from Gavin's office?

Working with Special Right Triangles

351–354 △FOR is an isosceles right triangle with the right angle at F. Use the given information to answer each question in simplest radical form.

351. If $FR = 10$, find the length of $\overline{OR}$.

352. If $OR = 18\sqrt{2}$, find the length of $\overline{FO}$.

353. If $RF = 12\sqrt{2}$, find the length of $\overline{RO}$.

354. If $RO = 20$, find the length of $\overline{RF}$.

355–359 △SIX is a right triangle with $m\angle X = 60°$ and $m\angle I = 90°$. Use the given information to answer each question.

355. If $IX = 7$, find the length of $\overline{SI}$.

356. If $IX = 5$, find the length of $\overline{SX}$.

357. If $SI = 15\sqrt{3}$, find the length of $\overline{SX}$.

358. If $SX = 26$, find the length of $\overline{SI}$.

359. If $SI = 15$, find the length of $\overline{SX}$.

Application of Special Right Triangles

360–369 Use your knowledge of special right triangles to answer each question.

360. In rhombus $ABCD$, $m\angle B = 60°$. If $AB = 10$, find the length of diagonal $\overline{AC}$.

361. Find the length of the diagonal of a square that has a perimeter of 36.

362. The length of the altitude of an equilateral triangle is $4\sqrt{3}$. Find the length of a side of the triangle.

363. The length of the altitude of an equilateral triangle is 21. Find the perimeter of the triangle.

364. A rhombus contains a 120° angle. The length of each side is 14. Find the length of the longer diagonal.

365. The lengths of the bases of an isosceles trapezoid are 6 feet and 12 feet. Each leg makes an angle of 45° with the longer base. Find the length of the leg of the trapezoid.

366. In isosceles right $\triangle SPE$, $m\angle S = 90°$ and $SE = 12\sqrt{2}$. Rectangle $PLCE$ with diagonal $\overline{LE}$ drawn forms a 30° angle at $\angle LEP$. Find the length of $\overline{PL}$.

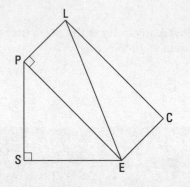

367. In isosceles right $\triangle SPE$, $m\angle S = 90°$ and $SE = 12\sqrt{2}$. Rectangle $PLCE$ with diagonal $\overline{LE}$ drawn forms a 30° angle at $\angle LEP$. Find the length of $\overline{LE}$.

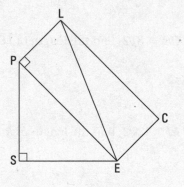

368. In the following figure, △*ABC*, △*ACD*, △*ADE*, △*AEF*, and △*AFG* are all right triangles. If *BC* = 3, find the length of $\overline{AG}$.

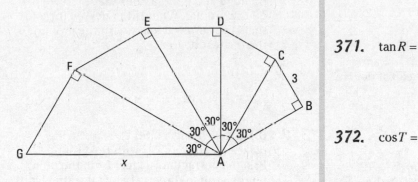

369. In the following diagram, △*ABC*, △*ACD*, and △*ADE* are right triangles. If *AB* = 8, find the length of $\overline{AE}$.

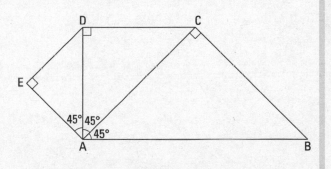

Trigonometric Ratios

370–372 Use the given figure to find the trigonometric ratio.

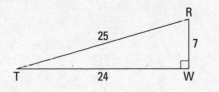

370. $\sin T =$

371. $\tan R =$

372. $\cos T =$

373–377 Use the following diagram and your knowledge of trigonometric ratios to solve each problem.

373. △*CST* is a right triangle with $m\angle T = 30°$. If *ST* = 12, find the length of $\overline{CS}$.

374. △*CST* is a right triangle with $m\angle T = 30°$. If *ST* = 14, find the length of $\overline{CT}$ to the nearest tenth.

375. $\triangle CST$ is a right triangle with $CT = 25$ and $SC = 34$. Find $m\angle STC$ to the nearest degree.

376. $\triangle CST$ is a right triangle with $SC = 5$ and $ST = 12$. Find $m\angle CST$ to the nearest degree.

377. $\triangle CST$ is a right triangle with $m\angle S = 50°$. If $CT = 28$, find the length of $\overline{ST}$ to the nearest tenth.

Applying the Trigonometric Ratios to Word Problems

378–380 Use your knowledge of the trigonometric ratios to solve each word problem.

378. A ladder that is 36 feet long leans against a store and makes an angle of 44° with the ground. Find the distance between the foot of the ladder and the store. Round the answer to the nearest foot.

379. A school is building a ramp for wheelchair access into the school. The ramp needs to be built from a point 20 feet away from the building and up to a door that is 10 feet off the ground. What is the degree measure that the ramp will make with the ground to the nearest degree?

380. Brianna is flying a kite. The kite string makes an angle of 74° with the ground. Brianna is standing 84 feet from the point on the ground directly below the kite. Find the length of the kite string to the nearest tenth of a foot.

Chapter 8

Triangle Inequalities

This chapter focuses on the relationship between the sides and angles of a triangle. This relationship can help you locate a triangle's longest and shortest sides. You also work with two important theorems. One of them states that the sum of the two smaller sides of a triangle must be greater than the longest side of the triangle. The other theorem states that the exterior angle of a triangle is equal to the sum of the two nonadjacent interior angles of the triangle.

The Problems You'll Work On

In this chapter, you see a variety of geometry problems:

- Locating the longest and shortest sides of a triangle
- Determining possible values for the third side of a triangle
- Determining the measure of an interior or exterior angle of a triangle
- Completing inequality proofs

What to Watch Out For

To avoid common mistakes, consider the following:

- The three angles of a triangle must add up to 180°. However, the sides of a triangle do not have to add up to 180.
- Two adjacent angles do not necessarily add up to 180° unless they form a linear pair or it's stated that they're supplementary.
- You can use the Pythagorean theorem to find the length of a side of a triangle only if the triangle is a right triangle.
- The sum of the two shorter sides of a triangle is greater than the longest side, never equal to it.
- Make sure you know what the question is asking. After finding the value of a variable, you may have to plug it in to find the measure of an angle.

Relationships between the Sides and Angles of a Triangle

381–386 The longest side of a triangle is always opposite the largest angle. Use this theorem to solve the problem.

381. In △ABC, $m\angle A = 40°$ and $m\angle C = 30°$. Name the longest side of the triangle.

382. In △MNO, $m\angle N = 45°$ and $m\angle O = 80°$. Name the longest side of the triangle.

383. In △BRI, $m\angle B = 105°$ and $m\angle R = 50°$. Which side of the triangle is the shortest?

384. △PIN is a right triangle with hypotenuse $\overline{PI}$. If $PN = 15$ and $PI = 17$, which angle is the smallest?

385. In △ABC, $\angle A$ is represented by $x^2 + 68$, $\angle B$ is represented by $x + 5$, and $\angle C$ is 87°. Which side of the triangle is the longest?

386. In △EFG, the three sides of the triangle are represented by

$$EF = x^3 + 50$$
$$FG = 20x^2$$
$$GE = x + 62$$

If $x = -2$, which angle is the largest?

387–390 Use the given information to arrange the angles or sides from least to greatest.

387. In △KLM, $KL = 8$, $LM = 7$, and $KM = 6$. List the angles from least to greatest.

388. In △XYZ, $XY = \sqrt{27}$, $XZ = \frac{5}{2}\sqrt{12}$, and $YZ = 4\sqrt{3}$. List the angles from least to greatest.

389. In △NFL, $m\angle N = 74°$ and $m\angle F = 57.9°$. List the sides of the triangle from shortest to longest.

390. In △MLB, $m\angle M = 65°$, $m\angle B$ is represented by $x + 5$, and $m\angle L$ is represented by x. List the sides of the triangle from shortest to longest.

Triangle Inequality Theorem

391–395 Use your knowledge of triangle inequalities to determine whether the following statements are true or false.

391. True or False? It's possible to form a triangle with sides that are 2 inches, 3 inches, and 4.995 inches.

392. True or False? It's possible to form a triangle with sides that are $5\sqrt{2}$ feet, $5\sqrt{2}$ feet, and $5\sqrt{2}$ feet.

393. True or False? It's possible to form a triangle with sides that are $\sqrt{5}$ miles, $\sqrt{7}$ miles, and $\sqrt{26}$ miles.

394. True or False? It's possible to form a triangle with sides that are represented by $a+4$, $a+3$, and $2a+7$.

395. True or False? It's possible to form a triangle with sides that are 7 centimeters, 8 centimeters, and $\sqrt{221}$ centimeters.

Finding the Missing Side Length

396–401 Find the length of the third side of the triangle.

396. If two sides of an isosceles triangle are 6 and 12, what must the third side be?

397. If two sides of an isosceles triangle are 14 and 5, what must the third side be?

398. If two sides of a triangle are 2 and 8, write a list of all integer values that can be the third side.

399. Three sides of a triangle are represented by integers, and one side is equal to 12. What is the smallest possible perimeter of the triangle?

400. Write an inequality that can represent all values of the third side of a triangle if the other two sides are 12 and 25.

401. The lengths of two sides of a triangle are 15 and 21. Write an inequality that can represent all possible values for the third side.

Isosceles Triangles

402–405 Find the missing side or angle.

402. △*ABC* is isosceles. If *BC* = 7 and *AC* = 15, find the length of $\overline{AB}$.

403. △*ABC* is isosceles. If ∠*A* = 22.5°, *BC* = 3.9, and *AC* = 10, find the degree measure of ∠*C*.

404. True or False? Suppose △*ABC* is isosceles. If $\overline{AC} \cong \overline{BC}$, *m*∠*C* = 70°, and *AB* = 20, then 10 is a possible length for $\overline{BC}$.

405. True or False? Suppose △*ABC* is isosceles. If $\overline{AC} \cong \overline{BC}$, *m*∠*C* = 70°, and *AB* = 20, then the length of $\overline{BC}$ can be 25.

Using the Exterior Angle Theorem of a Triangle

406–411 Use the figure and the given information to answer the question.

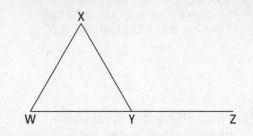

406. In △*WXY*, $\overline{WY}$ is extended to $\overline{WYZ}$. If *m*∠*XYZ* = 120° and *m*∠*W* = 70°, find the degree measure of ∠*X*.

407. In △*WXY*, $\overline{WY}$ is extended to $\overline{WYZ}$. If *m*∠*XYZ* = 110° and $\overline{WY} \cong \overline{XY}$, find the degree measure of ∠*W*.

408. In △*WXY*, $\overline{WY}$ is extended to $\overline{WYZ}$. If *m*∠*W* is represented by *b* + 30, *m*∠*X* is represented by 2*b*, and *m*∠*WYX* is represented by 3*b*, find the degree measure of ∠*XYZ*.

409. In △*WXY*, $\overline{WY}$ is extended to $\overline{WYZ}$. If *m*∠*W* = *x* + 50, *m*∠*X* = 2*x*, and *m*∠*XYZ* = 4*x* + 30, find the degree measure of ∠*WYX*.

410. In △*WXY*, $\overline{WY}$ is extended to $\overline{WYZ}$. If $m\angle W = 4a - 50$, $m\angle X = a + 10$, and $m\angle XYZ = 3a$, find the degree measure of $\angle XYZ$.

411. In △*WXY*, $\overline{WY}$ is extended to $\overline{WYZ}$. Each angle is represented by the following:

$m\angle W = 2a$

$m\angle X = b - 50$

$m\angle WYX = 3a - 15$

$m\angle XYZ = b$

Solve for the variable *b*.

412–418 *Use the following diagram and given information to answer each question.*

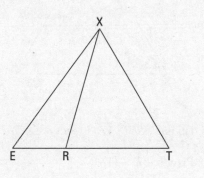

412. $m\angle XRE$ is represented by $3c - 35$, $m\angle RXT$ is represented by c, and $m\angle T$ is represented by $c + 15$. Find the value of *c*.

413. Name the angle that makes the following equation always true:

$m\angle E + m\angle EXR = m$ ____

414. Name the angle that makes the following equation always true:

$m\angle T + m\angle RXT = m$ ____

415. Is the following inequality always, sometimes, or never true? $m\angle XRE > m\angle XRT$

416. Is the following inequality always, sometimes, or never true? $m\angle XRT > m\angle E$

417. Is the following inequality always, sometimes, or never true? $m\angle XRT > m\angle RXT$

418. Is the following inequality always, sometimes, or never true? $m\angle T > m\angle ERX$

Geometric Proofs Involving Triangle Inequality Theorems

419–423 *Complete the following proof by giving the statement or reason.*

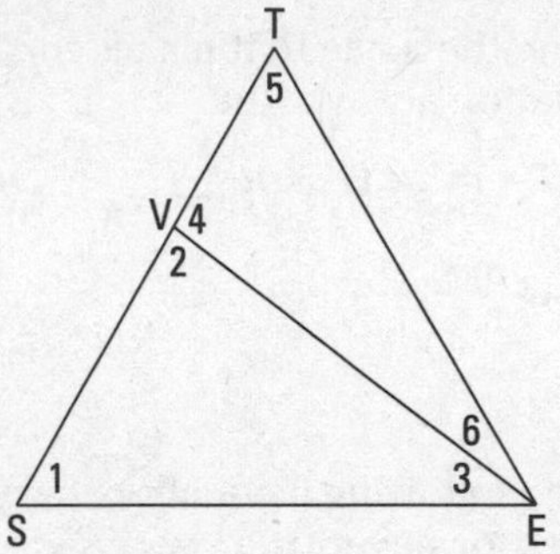

Given: $\triangle SET$, $m\angle 4 > m\angle 2$

Prove: $TE > VE$

Statements	*Reasons*
1. $\triangle SET, m\angle 4 > m\angle 2$	1.
2. $m\angle 2 = m\angle 5 + m$____	2. The exterior angle of a triangle is equal to the sum of the two nonadjacent interior angles.
3. m____ $> m\angle 5$	3. The whole is greater than its parts.
4. $m\angle 4 > m\angle 5$	4.
5. $TE > VE$	5.

419. What is the reason for Statement 1?

420. What is the missing angle in Statement 2?

421. What is the missing angle in Statement 3?

422. What is the reason for Statement 4?

423. What is the reason for Statement 5?

424–428 *Complete the following proof by giving the statement or reason.*

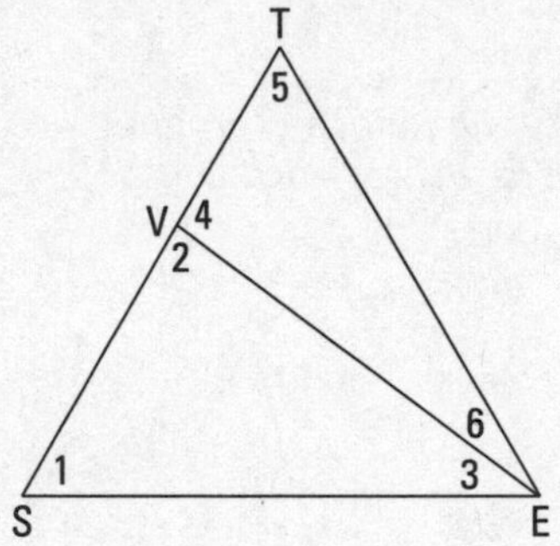

Given: $\triangle SET$, $\overline{VE}$ bisects $\angle SET$

Prove: $ET > TV$

Statements	*Reasons*
1. $\triangle SET, \overline{VE}$ bisects $\angle SET$	1. Given
2.	2. An angle bisector divides an angle into two congruent parts.
3. $m\angle 4 = m\angle 1 + m\angle 3$	3.
4. m____ $> m\angle 3$	4. The whole is greater than its parts.
5. $m\angle 4 > m\angle 6$	5.
6. $ET > TV$	6.

424. What is the statement for Reason 2?

425. What is the reason for Statement 3?

426. What is the missing angle in Statement 4?

427. What is the reason for Statement 5?

428. What is the reason for Statement 6?

429–430 *Complete the following proofs.*

429.

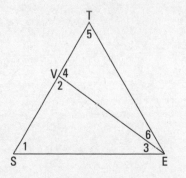

Given: △*SET*, $m\angle 2 > m\angle 4$

Prove: $m\angle 2 > m\angle 1$

430.

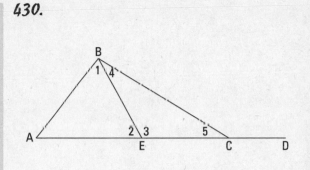

Given: △*ABC* with exterior $\angle BCD$; $\overline{AECD}$

Prove: $m\angle BCD > m\angle A$

Chapter 9

Polygons

• •

A *regular polygon* is a polygon in which all the angles are equal in measure and all the sides are equal in length. This means that the polygon is both equiangular and equilateral. A polygon can have three or more sides. The more sides a regular polygon has, the more the polygon looks like a circle. The name of the polygon depends on the number of sides. In this chapter, you become familiar with types of polygons and properties of their angles.

The Problems You'll Work On

In this chapter, you see a variety of geometry problems:

- ✔ Assigning a name to a polygon
- ✔ Finding the sum of the interior or exterior angles of a polygon
- ✔ Finding the measure of an interior angle of a polygon
- ✔ Using an interior angle to determine how many sides a polygon has
- ✔ Using the apothem to find the area of a regular polygon

What to Watch Out For

Some of the following points may be helpful as you work through this chapter:

- ✔ The sum of the exterior angles of any polygon is 360°.
- ✔ The formula $180°(n-2)$ tells you the sum of the interior angles of a polygon, where n represents the number of sides.
- ✔ The formula for the area of a polygon is $\frac{1}{2}$(apothem)(perimeter). The *apothem* is the line segment that connects the center of the polygon to the midpoint of any side of the polygon.
- ✔ Make sure you know whether you're solving for a variable or for the measure of a side or angle.

Naming Polygons

431–435 Name the polygon when given the number of sides.

431. What is the name of a polygon that has five sides?

432. What is the name of a polygon that has six sides?

433. What is the name of a polygon that has seven sides?

434. What is the name of a polygon that has twelve sides?

435. What is the name of a polygon that has nine sides?

Understanding Angles of a Polygon

436–450 Practice solving problems with interior and exterior angles of a polygon.

436. What is the sum of the interior angles of a pentagon?

437. What is the sum of all the interior and exterior angles of an octagon?

438. What is the sum of the exterior angles of heptagon?

439. What is the sum of the interior angles of a polygon with 22 sides?

440. What is the sum of the exterior angles of a hexagon?

441. Name a polygon whose interior and exterior angles add up to 1,620°.

442. Find the degree measure of each interior angle of a regular octagon.

443. Find the degree measure of each interior angle of a regular dodecagon.

444. Find the degree measure of each exterior angle of a regular pentagon.

445. Find the degree measure of each exterior angle of a regular decagon.

446. If each exterior angle of a regular polygon measures 18°, how many sides does the polygon have?

447. If each exterior angle of a polygon measures 12°, how many sides does the polygon have?

448. If each interior angle of a regular polygon measures 160°, how many sides does the polygon have?

449. If each interior angle of a regular polygon has a degree measure of 168°, how many sides does the polygon have?

450. If each interior angle of a regular polygon is 156°, find the degree measure of each exterior angle.

The Sum of the Interior and Exterior Angles of a Polygon

451–456 Use your knowledge of the sums of the interior and exterior angles of a polygon to answer each question.

451. Solve for x.

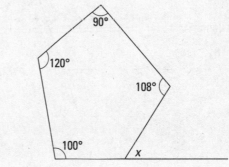

452. Solve for *x*.

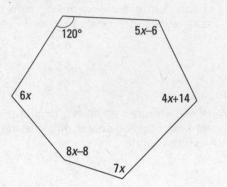

453. Solve for *x*.

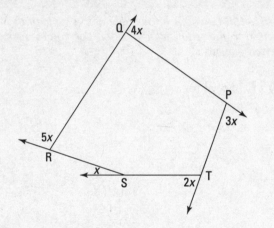

454. The interior angles of a heptagon are represented by $5x + 5$, $2x + 22$, $6x - 6$, $2x + 12$, $4x + 20$, $6x - 3$, and $3x + 10$. What is the value of *x*?

455. Solve for *x*.

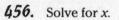

456. Solve for *x*.

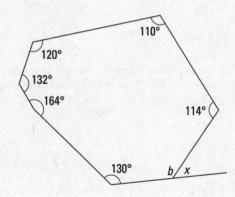

Finding the Area of Regular Polygons

457–461 Use the apothem to find the area of a regular polygon.

457. What is the formula for the area of a regular polygon?

458. Find the area of a regular pentagon whose perimeter is 40 units and whose apothem is 5 units.

459. Find the area of a regular octagon whose side is 12 feet and whose apothem is 14.5 feet.

460. Find the exact area of a regular hexagon that has a perimeter of 60 units.

461. Find the perimeter of a regular hexagon whose area is $54\sqrt{3}$ units^2.

Chapter 10

Properties of Parallel Lines

• •

*P*arallel lines are two lines that lie on the same plane and never intersect. When two parallel lines are intersected by a third line, a *transversal*, congruent pairs of angles are formed. These angles are called alternate interior angles, alternate exterior angles, and corresponding angles. In this chapter, you apply your knowledge of these angles toward algebraic problems and geometric proofs.

The Problems You'll Work On

In this chapter, you see a variety of geometry problems:

✔ Understanding the properties of alternate interior and alternate exterior angles

✔ Classifying triangles as acute, obtuse, equiangular, or right

✔ Understanding the properties of corresponding angles, adjacent angles, and vertical angles

✔ Understanding the relationship between two angles that form a linear pair

✔ Completing geometric proofs involving parallel lines cut by a transversal

What to Watch Out For

Don't let common mistakes trip you up. Some of the following suggestions may be helpful:

✔ When completing geometric proofs in which the given information involves parallel lines, look for corresponding angles, alternate interior angles, and alternate exterior angles in the figure. These will give you pairs of congruent angles.

✔ Trace the two parallel lines and the transversal to form the letter *Z*. The alternate interior angles are located at the corners of the *Z*.

✔ Trace the two parallel lines and the transversal to form the letter *F*. The corresponding angles are located at the corners of the *F*.

✔ Make sure you know what the question is asking for, whether it's the value of a variable or the measure of an angle.

Alternate Interior and Alternate Exterior Angles

462–471 Use the diagram and the given information to solve the problem.

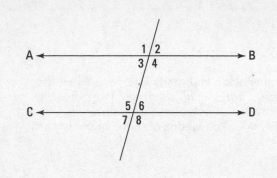

462. Parallel lines $\overline{AB}$ and $\overline{CD}$ are cut by a transversal. $\angle 3 \cong \angle$ __ because they're alternate interior angles.

463. Parallel lines $\overline{AB}$ and $\overline{CD}$ are cut by a transversal. $\angle 2 \cong \angle$ __ because they're alternate exterior angles.

464. Parallel lines $\overline{AB}$ and $\overline{CD}$ are cut by a transversal. If $m\angle 4 = 125°$, find the degree measure of $\angle 5$.

465. Parallel lines $\overline{AB}$ and $\overline{CD}$ are cut by a transversal. If $m\angle 1 = 111.5°$, find the degree measure of $\angle 8$.

466. Parallel lines $\overline{AB}$ and $\overline{CD}$ are cut by a transversal. If $m\angle 4$ is represented by $5x - 80$ and $m\angle 5$ is represented by $2x - 5$, solve for x.

467. Parallel lines $\overline{AB}$ and $\overline{CD}$ are cut by a transversal. If $m\angle 3$ is represented by $\frac{1}{2}x + 25$ and $m\angle 6$ is represented by $4x - 45$, find the value of x.

468. Parallel lines $\overline{AB}$ and $\overline{CD}$ are cut by a transversal. If $m\angle 2$ and $m\angle 7$ are represented by $3x - 40$ and $x + 20$, respectively, find the degree measure of $\angle 7$.

469. Parallel lines $\overline{AB}$ and $\overline{CD}$ are cut by a transversal. If $m\angle 2$ and $m\angle 7$ are represented by $2x - 40$ and $\frac{1}{2}x + 20$, respectively, find the degree measure of $\angle 5$.

470. Parallel lines $\overline{AB}$ and $\overline{CD}$ are cut by a transversal. If $m\angle 4$ and $m\angle 5$ are represented by x^2 and $8x + 20$, respectively, find the positive value of x.

471. Parallel lines $\overline{AB}$ and $\overline{CD}$ are cut by a transversal. If $m\angle 3$ and $m\angle 6$ are represented by $x^2 - 2$ and $3x + 52$, respectively, find the positive value of x.

Classifying Triangles by Their Angle Measurements

472–475 *Given:* △*MNO*, △*PRO*, $\overline{MP}$ intersects $\overline{NR}$ at *O*, and $\overline{MN} \parallel \overline{PR}$. Classify △*MNO* as *acute, obtuse, equiangular, right,* and/or *isosceles*. Note that the figure may not be drawn to scale.

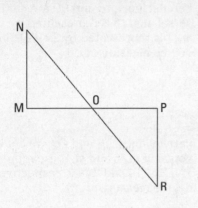

472. What kind of triangle is △*MNO* if $m\angle R = 65°$ and $m\angle MON = 72°$?

473. What kind of triangle is △*MNO* if $m\angle P = 80°$ and $m\angle R = 20°$?

474. What kind of triangle is △*MNO* if $m\angle R = 55°$ and $m\angle NOM = 35°$?

475. What kind of triangle is △*MNO* if $m\angle N = 60°$ and $m\angle P = 60°$?

Finding Angle Measures Involving Parallel Lines

476–479 *Use the following figure and given information to solve the problem.*

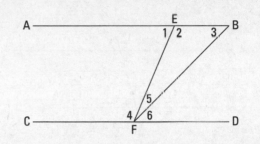

476. Given: $\overline{AB} \parallel \overline{CD}$, $m\angle 1 = 70°$, and $m\angle 5 = 45°$. Find the measure of $\angle 2$.

477. Given: $\overline{AB} \parallel \overline{CD}$, $m\angle 1 = 70°$, and $m\angle 5 = 45°$. Find the measure of $\angle 3$.

478. Given: $\overline{AB} \parallel \overline{CD}$, $m\angle 1 = 70°$, and $m\angle 5 = 45°$. Find the measure of $\angle 4$.

479. Given: $\overline{AB} \parallel \overline{CD}$, $m\angle 1 = 70°$, and $m\angle 3 = 25°$. Find the measure of $\angle 6$.

Reviewing Corresponding, Adjacent, and Vertical Angles

480–488 Use the following figure to answer each question.

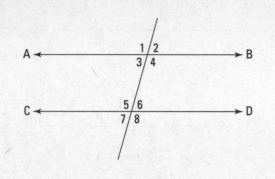

480. Parallel lines $\overline{AB}$ and $\overline{CD}$ are cut by a transversal. Name the angle that is corresponding to $\angle 1$.

481. Parallel lines $\overline{AB}$ and $\overline{CD}$ are cut by a transversal. $\angle 5 \cong$ ___ because vertical angles are congruent.

482. Parallel lines $\overline{AB}$ and $\overline{CD}$ are cut by a transversal. If $m\angle 1$ and $m\angle 2$ are represented by $2x + 20$ and $x - 20$, respectively, find the value of x.

483. Parallel lines $\overline{AB}$ and $\overline{CD}$ are cut by a transversal. If $m\angle 3$ and $m\angle 5$ are represented by $x + 28$ and $2x + 32$, respectively, find the degree measure of $\angle 5$.

484. Parallel lines $\overline{AB}$ and $\overline{CD}$ are cut by a transversal. If $m\angle 4$ is represented by $3x - 50$ and $m\angle 8$ is represented by $2x + 10$, find the degree measure of $\angle 4$.

485. Parallel lines $\overline{AB}$ and $\overline{CD}$ are cut by a transversal. If $m\angle 6$ is represented by $x - 18$ and $m\angle 7$ is represented by $2x - 70$, find the degree measure of $\angle 6$.

486. Parallel lines $\overline{AB}$ and $\overline{CD}$ are cut by a transversal. If $m\angle 5$ and $m\angle 7$ are represented by $x^2 - 2x - 14$ and $2x - 2$, respectively, find the degree measure of $\angle 5$.

487. Parallel lines $\overline{AB}$ and $\overline{CD}$ are cut by a transversal. If $m\angle 4$ is represented by x^3 and $m\angle 6 = 55°$, find the value of x.

488. Parallel lines $\overline{AB}$ and $\overline{CD}$ are cut by a transversal. If $m\angle 2$ and $m\angle 5$ are represented by $x^2 + 15x + 70$ and $x^2 - 3x + 30$, respectively, solve for the positive value of x.

More Practice with Angles Involving Parallel Lines

489–494 Given: △COR, △ESP, $\overline{CO}$ is parallel to $\overline{ES}$, and $\overline{PS}$ is parallel to $\overline{RO}$. Use the following figure and the given information to find the missing angle.

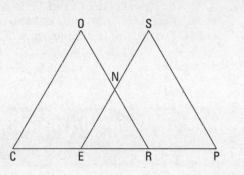

489. If $m\angle COR = 60°$ and $m\angle P = 40°$, find $m\angle C$.

490. If $m\angle PES = 70°$ and $m\angle P = 55°$, find $m\angle O$.

491. If $m\angle RCO = 85°$ and $m\angle S = 75°$, find the degree measure of $\angle PRO$.

492. If $m\angle C = 58°$ and $m\angle P = 62°$, find the degree measure of $\angle RNE$.

493. If $m\angle O = 58°$, find $m\angle RNE$.

494. If $m\angle C$ and $m\angle PES$ are represented by $x^2 - 20$ and x, respectively, find the degree measure of $\angle PES$.

495–498 Use the following figure and the given information to solve for the missing angle.

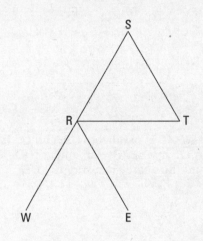

495. $\overline{RE}$ is parallel to $\overline{ST}$, $\overline{WRS}$, and RE bisects $\angle WRT$. If $m\angle WRT = 120°$, find the degree measure of $\angle S$.

496. $\overline{RE}$ is parallel to $\overline{ST}$, $\overline{WRS}$, and $\overline{RE}$ bisects $\angle WRT$. If $m\angle T = 72°$, find $m\angle WRT$.

497. $\overline{RE}$ is parallel to $\overline{ST}$, $\overline{WRS}$, and $\overline{RE}$ bisects $\angle WRT$. If $m\angle SRT = 80°$, find the degree measure of $\angle WRE$.

498. $\overline{RE}$ is parallel to $\overline{ST}$, $\overline{WRS}$, and $\overline{RE}$ bisects ∠*WRT*. If *m*∠*SRT* = 56°, find *m*∠*T*.

499–504 Use the following figure and the given information to solve for the missing angle.

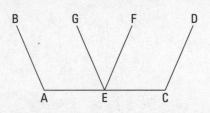

499. $\overline{AEC}$, $\overline{AB}$ is parallel to $\overline{GE}$, and $\overline{DC}$ is parallel to $\overline{FE}$. If *m*∠*BAE* = 95° and *m*∠*DCE* = 120°, find the degree measure of ∠*GEF*.

500. $\overline{AEC}$, $\overline{AB}$ is parallel to $\overline{GE}$, and $\overline{DC}$ is parallel to $\overline{FE}$. If *m*∠*GEF* = 25° and *m*∠*GEA* = 75°, find *m*∠*DCE*.

501. $\overline{AEC}$, $\overline{AB}$ is parallel to $\overline{GE}$, and $\overline{DC}$ is parallel to $\overline{FE}$. If *m*∠*GEF* = 45° and *m*∠*FEC* = 60°, find *m*∠*BAE*.

502. $\overline{AEC}$, $\overline{AB}$ is parallel to $\overline{GE}$, and $\overline{DC}$ is parallel to $\overline{FE}$. If *m*∠*AEF* = 135°, find the degree measure of ∠*DCE*.

503. $\overline{AEC}$, $\overline{AB}$ is parallel to $\overline{GE}$, and $\overline{DC}$ is parallel to $\overline{FE}$. If *m*∠*AEF* = 140° and $\overline{GE}$ bisects ∠*AEF*, find *m*∠*BAE*.

504. $\overline{AEC}$, $\overline{AB}$ is parallel to $\overline{GE}$, and $\overline{DC}$ is parallel to $\overline{FE}$. If *m*∠*BAE* = 120° and the ratio of *m*∠*AEG* to *m*∠*FEC* is 3:2, find *m*∠*GEF*.

Geometric Proof Incorporating Parallel Lines

505 *Complete the geometric proof:*

Given: $\overline{AR} \parallel \overline{PL}$ with transversal $\overline{PR}$, and $\overline{AR} \cong \overline{PL}$

Prove: △*PAR* ≅ △*RLP*

Geometric Proof Incorporating Parallel Lines

506–511 Complete the following geometric proof by giving the statement or reason.

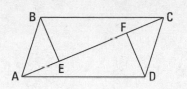

Given: $\overline{BC} \parallel \overline{AD}$, $\overline{AEFC}$, $\overline{BC} \cong \overline{AD}$, and $\overline{AE} \cong \overline{FC}$

Prove: $\overline{BE} \cong \overline{DF}$

Statements	Reasons
1. $\overline{BC} \parallel \overline{AD}$, $\overline{AEFC}$, $\overline{BC} \cong \overline{AD}$, and $\overline{AE} \cong \overline{FC}$	1. Given
2. $\angle BCA$ and _____ are alternate interior angles.	2. Parallel lines cut by a transversal form alternate interior angles.
3. $\angle BCA \cong \angle DAC$	3.
4. $\overline{EF} \cong \overline{EF}$	4.
5. $\overline{AE} + \overline{EF} \cong \overline{FC} + \overline{EF}$ or $\overline{AF} \cong \overline{CE}$	5.
6.	6. SAS
7. $\overline{BE} \cong \overline{DF}$	7.

506. What is the missing angle in Statement 2?

507. What is the reason for Statement 3?

508. What is the reason for Statement 4?

509. What is the reason for Statement 5?

510. What is the statement for Reason 6?

511. What is the reason for Statement 7?

Chapter 11

Properties of Quadrilaterals

• •

*I*n two-dimensional geometry, a four-sided figure is called a *quadrilateral*. A *parallelogram* is a quadrilateral with two pairs of congruent sides and two pairs of congruent angles. A *rectangle* is a parallelogram with four right angles. A *rhombus* is a parallelogram in which all sides are congruent. A *square* has the properties of both a rhombus and a rectangle, which means it has four congruent sides and four right angles. A *trapezoid* is a quadrilateral with only one pair of parallel sides. An *isosceles trapezoid* is a trapezoid in which the nonparallel sides are congruent. This chapter focuses on the similarities and differences between these quadrilaterals and applies them toward algebraic problems.

The Problems You'll Work On

In this chapter, you see a variety of geometry problems:

- ✔ Understanding the properties of a parallelogram
- ✔ Understanding the properties of a rectangle
- ✔ Understanding the properties of a rhombus
- ✔ Understanding the properties of a square
- ✔ Understanding the properties of a trapezoid

What to Watch Out For

To avoid common mistakes, try the following tips:

- ✔ Be careful when naming a quadrilateral by its vertices. The letters must go in consecutive order.
- ✔ Because these quadrilaterals have parallel sides, keep an eye out for alternate interior angles formed when the diagonal acts as the transversal.
- ✔ Remember that the diagonals of a parallelogram do *not* bisect the angles of the parallelogram.
- ✔ Make sure you understand what the question is asking you to solve for.

Properties of Parallelograms

512–525 In parallelogram MATH, diagonals $\overline{MT}$ and $\overline{AH}$ intersect at E. Use the figure and the given information to solve.

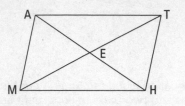

512. If $AT = 8x + 2$ and $MH = 5x + 8$, find the length of $\overline{MH}$.

513. If $AM = 40$ and $TH = 2x - 10$, find the value of x.

514. If $m\angle AMH = 80°$ and $m\angle HTA$ is represented by $x + 50$, find the value of x.

515. If $m\angle AMH$ is represented by $x + 40$ and $m\angle MAT = 110°$, find the measure of $\angle AMH$.

516. If $m\angle ATH$ and $m\angle MHT$ are represented by $2x + 25$ and $3x + 5$, respectively, find the measure of $\angle MHT$.

517. If $ME = 8$, find the length of $\overline{TE}$.

518. If $TE = 5x - 20$ and $ME = x + 20$, find the length of $\overline{TE}$.

519. If $TE = x + 8$ and $MT = 4x - 8$, find the length of $\overline{MT}$.

520. If $AH = 3x - 8$ and $AE = x + 2$, find the length of $\overline{AE}$.

521. If MT and ME are represented by $x^2 + 3x$ and $x + 28$, respectively, find the positive value of x.

522. If $AT = 2a + 3b$, $MH = 28$, $AM = 4a + b$, and $TH = 26$, find the sum of $a + b$.

523. If $m\angle TEA = 2x + 61$ and $m\angle MEH = 3x + 41$, find the degree measure of $\angle AEM$.

524. If $m\angle AMT$ and $m\angle HTM$ are represented by $9x-10$ and $7x+10$, respectively, find the degree measure of $\angle HTM$.

525. If $m\angle MAT$ is represented by $7x+5$ and $m\angle HTA$ is represented by $x+15$, find the degree measure of $\angle MHT$.

Word Problems with Parallelograms

526–531 Use your knowledge of parallelograms to solve each problem.

526. In parallelogram $ABCD$, $m\angle A$ is twice the $m\angle B$. Find the degree measure of $\angle A$.

527. In parallelogram $DREW$, the length of side $\overline{DR}$ is represented by $9x-5$ and the length of side $\overline{WE}$ is represented by $3x+7$. Solve for x.

528. In parallelogram $JONA$, $m\angle O$ and $m\angle A$ are represented by $6x+30$ and $8x+15$, respectively. Find the degree measure of $\angle J$.

529. In parallelogram $ABCD$, $m\angle C$ is represented by $3a-10$, and $m\angle D = a+80$. Find the degree measure of $\angle D$.

530. In parallelogram $SHER$, SH is represented by x^2+4x and ER is represented by x^3+x^2. Find the length of $\overline{ER}$.

531. In parallelogram $LIND$, the four angles are represented by $m\angle L = x+y$, $m\angle I = 3x+2y$, and $m\angle N = 50$. Solve for y.

Properties of Rectangles

532–536 Use your knowledge of rectangles to solve for y in each problem.

532. In rectangle $RSTW$, $RS = y+25$ and $WT = 40$. Solve for y.

533. In rectangle $RSTW$, $m\angle R$ is represented by $10y+50$. Solve for y.

534. In rectangle $RSTW$, $ST = 8y-6$ and $RW = 5y-3$. Solve for y.

535. In rectangle $RSTW$, $m\angle T$ is represented by $3y+45$. Solve for y.

536. In rectangle $RSTW$, $ST = 7y + 10$ and $RW = -y + 34$. Solve for y.

Finding the Diagonal of a Rectangle

537–541 Find the length of the diagonal of rectangle RSTW.

537. In rectangle $RSTW$, $SR = 5$ and $RW = 12$. Find the length of $\overline{SW}$.

538. In rectangle $RSTW$, $TW = 15$ and $ST = 20$. Find the length of $\overline{SW}$.

539. In rectangle $RSTW$, $SR = 7$ and $RW = 24$. Find the length of $\overline{SW}$.

540. In rectangle $RSTW$, $TW = 4$ and $ST = 5$. Find the length of $\overline{RT}$.

541. In rectangle $RSTW$, the length of $\overline{RW}$ is 7 more than the length of $\overline{SR}$, and the length of $\overline{RT}$ is 8 more than the length of $\overline{SR}$. Find the length of $\overline{SW}$.

542–547 Use the given information to solve for the variable y.

542. In rectangle $RECT$, diagonals $\overline{RC}$ and $\overline{TE}$ intersect at A. If $RC = 12y - 8$ and $RA = 4y + 16$, solve for y.

543. In rectangle $RECT$, diagonals $\overline{RC}$ and $\overline{TE}$ intersect at A. If $RC = y + 20$ and $TE = 2y + 4$, solve for y.

544. In rectangle $RECT$, diagonals $\overline{RC}$ and $\overline{TE}$ intersect at A. If $RA = 7y - 3$ and $AE = 2y + 2$, solve for y.

545. In rectangle $RECT$, diagonals $\overline{RC}$ and $\overline{TE}$ intersect at A. If $TA = y + 8$ and $RC = 5y - 20$, solve for y.

546. In rectangle $RECT$, diagonals $\overline{RC}$ and $\overline{TE}$ intersect at A. If $RC = y^2$ and $ET = y + 42$, solve for the positive value of y.

547. In rectangle $RECT$, diagonals $\overline{RC}$ and $\overline{TE}$ intersect at A. If $RC = y^2$ and $AC = 3y + 8$, solve for the positive value of y.

Reviewing the Properties of a Rhombus

548–553 Use the properties of a rhombus to solve for x.

548. In rhombus *LMNO*, diagonals $\overline{LN}$ and $\overline{MO}$ intersect at *P*. If $m\angle NPO$ is represented by $2x + 70$, solve for *x*.

549. In rhombus *LMNO*, diagonals $\overline{LN}$ and $\overline{MO}$ intersect at *P*. If $m\angle LOP$ and $m\angle NOP$ are represented by $3x$ and $2x + 25$, respectively, find the value of *x*.

550. In rhombus *LMNO*, diagonals $\overline{LN}$ and $\overline{MO}$ intersect at *P*. If $m\angle MNP$ is represented by $4x - 12$ and $m\angle ONP$ is represented by $x + 18$, solve for *x*.

551. In rhombus *LMNO*, diagonals $\overline{LN}$ and $\overline{MO}$ intersect at *P*. If $m\angle LMN$ and $m\angle NMO$ are represented by $x^2 - 16$ and $3x$, respectively, find the positive value of *x*.

552. In rhombus *LMNO*, diagonals $\overline{LN}$ and $\overline{MO}$ intersect at *P*. If $m\angle MNO$ is represented by $4x$ and $m\angle MNP$ is represented by $x + 10$, find the value of *x*.

553. In rhombus *LMNO*, diagonals $\overline{LN}$ and $\overline{MO}$ intersect at *P*. If $m\angle MPL$ is represented by $x^2 + 3x + 50$, find the positive value of *x*.

Diagonal Properties of a Rhombus

554–558 The diagonals of a rhombus are perpendicular to each other. Use this property to find the perimeter of each rhombus.

554. In rhombus *DEFG*, diagonals $\overline{DF}$ and $\overline{GE}$ intersect at *R*. If $GR = 3$ and $DR = 4$, find the perimeter of rhombus *DEFG*.

555. In rhombus *DEFG*, diagonals $\overline{DF}$ and $\overline{GE}$ intersect at *R*. If $EG = 16$ and $DF = 30$, find the perimeter of rhombus *DEFG*.

556. In rhombus *DEFG*, diagonals $\overline{DF}$ and $\overline{GE}$ intersect at *R*. If $ER = 15$ and $DF = 72$, find the perimeter of rhombus *DEFG*.

557. In rhombus *DEFG*, diagonals $\overline{DF}$ and $\overline{GE}$ intersect at *R*. If $ER = x$, $DR = x + 7$, and $ED = x + 8$, find the perimeter of rhombus *DEFG*.

558. In rhombus *DEFG*, diagonals $\overline{DF}$ and $\overline{GE}$ intersect at *R*. If $DF = 16$, $ER = x$, and $EF = x + 2$, find the perimeter of rhombus *DEFG*.

559–564 *Apply the properties of a rhombus to solve each problem.*

559. In rhombus *DREW*, side *DR* is represented by $x + 8$ and *DW* is represented by $3x - 4$. Find the length of $\overline{DW}$.

560. In rhombus *DREW*, $m\angle R$ is 4 more than 3 times $m\angle D$. Find $m\angle R$.

561. In rhombus *DREW*, $m\angle W = 140°$. Find the degree measure of $\angle D$.

562. In rhombus *DREW*, diagonal $RW = 20$ and diagonal $DE = 48$. Find the length of $\overline{DR}$.

563. In rhombus *DREW*, $m\angle REW = 120°$ and diagonal $DE = 20$. Find the perimeter of rhombus *DREW*.

564. In rhombus *DREW*, $m\angle REW = 120°$ and $DR = 10$. Find the length of diagonal $\overline{DE}$.

Properties of a Square

565–580 *Apply the properties of a square to solve the problem.*

565. In square *KING*, $KG = 2x + 34$ and $IN = 7x + 4$. Find the length of $\overline{KG}$.

566. In square *KING*, $NG = x^2 + 4x$ and $NI = 12$. Solve for the negative value of *x*.

567. In square *KING*, diagonal $KN = x^2 - 3x$ and diagonal $IG = x + 5$. Find the positive value of *x*.

568. If the perimeter of a square is 400 centimeters, find the length of a side.

569. If the perimeter of a square is 100 feet, find the length of the diagonal.

570. Find the length of the side of a square if its area is 400 square units.

571. In square *RSTW*, diagonals $\overline{RT}$ and $\overline{SW}$ intersect at *E*. If $m\angle SRW$ is represented by $2a+14$, find the value of *a*.

572. In square *RSTW*, diagonals $\overline{RT}$ and $\overline{SW}$ Intersect at *E*. If $m\angle REW$ is represented by $8x-6$, find the value of *x*.

573. In square *RSTW*, diagonals $\overline{RT}$ and $\overline{SW}$ intersect at *E*. If $SE=2x-1$ and $WE=x+8$, find the length of $\overline{WE}$.

574. In square *RSTW*, diagonals $\overline{RT}$ and $\overline{SW}$ intersect at *E*. Find the degree measure of $\angle RSW$.

575. In square *RSTW*, diagonals $\overline{RT}$ and $\overline{SW}$ intersect at *E*. If $RT=7x-22$ and $RE=3x+5$, find the value of *x*.

576. In square *RSTW*, diagonals $\overline{RT}$ and $\overline{SW}$ intersect at *E*. If $m\angle STW=2x-11$, find the value of *x*.

577. In square *JONA*, $JO=7x+2$ and $ON=8x+1$. Find the perimeter of the square.

578. In square *JONA*, diagonal $JN=10\sqrt{2}$. Find the length of a side of the square.

579. In square *JONA*, diagonals $\overline{JN}$ and $\overline{AO}$ intersect at *T*. If $JN=4x-8$ and $AT=x+2$, find the value of *x*.

580. In square *JONA*, diagonals $\overline{JN}$ and $\overline{AO}$ intersect at *T*. If $m\angle JOA$ is represented by x^2-4x, find the positive value of *x*.

Properties of a Trapezoid

*581–591 Use the following figure and given
information to solve each problem.*

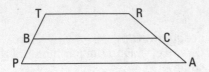

581. In trapezoid *TRAP*, $\overline{BC}$ is a median. If
$TR = 20$ and $PA = 32$, find the length of $\overline{BC}$.

582. In trapezoid *TRAP*, $\overline{BC}$ is a median. If $TR = 17$
and $PA = 22$, find the length of $\overline{BC}$.

583. In trapezoid *TRAP*, $\overline{BC}$ is a median. If $PA = 50$
and $BC = 38$, find the length of $\overline{TR}$.

584. In trapezoid *TRAP*, $\overline{BC}$ is a median. If
$TR = 2x + 8$, $PA = 6x - 2$, and $BC = 51$,
find the value of x.

585. In trapezoid *TRAP*, $\overline{BC}$ is a median. If
$TR = 4x - 1$, $PA = 5x$, and $BC = 2x + 2$, find
the length of $\overline{PA}$.

586. In trapezoid *TRAP*, $m\angle T = 115°$. Find $m\angle P$.

587. In trapezoid *TRAP*, $m\angle R = 2x + 4$ and
$m\angle A = x + 11$. Find the value of x.

588. In trapezoid *TRAP*, $m\angle T = 4x + 3$ and
$m\angle P = 2x - 9$. Find the degree measure
of $\angle P$.

589. If trapezoid *TRAP* is isosceles, diagonal
$TA = 5x$, and diagonal $RP = 95$, find the
value of x.

590. If trapezoid *TRAP* is isosceles, diagonal
$TA = 7x - 21$, and diagonal $RP = 2x - 3$,
find the value of x.

591. If trapezoid *TRAP* is isosceles, diagonal
$TA = x^2 + 3x$, and diagonal $RP = 9x$, find the
positive value of x.

Chapter 12

Coordinate Geometry

. . . .

Coordinate geometry is the study of geometry using ordered pairs that can be plotted on a Cartesian plane. Numerical coordinates are plotted using the *x*- and *y*-axes and are then used along with the distance, midpoint, and slope formulas to show properties of geometric figures.

The Problems You'll Work On

In this chapter, you see a variety of geometry problems:

✔ Using the distance, midpoint, and slope formulas

✔ Understanding the slopes of parallel and perpendicular lines

✔ Writing the equation of a line in slope-intercept form

✔ Completing coordinate geometry proofs

What to Watch Out For

Don't let common mistakes trip you up. Some of the following suggestions may be helpful:

✔ Remember that in the slope formula, the difference between the *y* coordinates — not the *x* coordinates — goes in the numerator of the fraction. Also be sure to keep the order of the coordinates consistent.

✔ The slope formula tells you whether lines are parallel, perpendicular, or neither. Parallel lines have equal slopes, and perpendicular lines have negative reciprocal slopes.

✔ The distance formula tells you the length of a segment. The midpoint formula tells you the point that lands in the middle of a segment.

✔ Coordinate geometry proofs require an understanding of the properties of triangles and quadrilaterals.

Determining Distance

592–599 Use the distance formula to solve each problem. Write your answer in simplest radical form.

592. Find the distance between the following two points: $(-3, 2)$ and $(-3, -5)$.

593. What is the distance from the origin to the point $(6, -4)$ in simplest radical form?

594. Given Point A $(-4, 3)$ and Point C $(2, 11)$, find the length of $\overline{AC}$.

595. Find the distance between the two points that are represented by (a, c) and $(a - b, c)$.

596. The endpoints of the diameter of a circle are represented by $(4a, b)$ and $(3a, 0)$. Write an expression for the length of the diameter.

597. $\triangle JKL$ has coordinates J $(5, 0)$, K $(4, 4)$, and L $(6, 4)$. Classify the triangle as isosceles, equilateral, right, and/or scalene.

598. $\triangle DIS$ has coordinates D $(4, 0)$, I $(-6, 0)$, and S $\left(-1, -5\sqrt{3}\right)$. Classify the triangle as isosceles, equilateral, right, and/or scalene.

599. The distance between two points is 5. If one point is represented by $(2, 6)$ and the other point is represented by $(5, y)$, find the possible values for y.

Using the Midpoint Formula

600–607 Use the midpoint formula to solve the problem.

600. Given M $(14, 2)$ and N $(8, 10)$, find the midpoint of $\overline{MN}$.

601. Find the midpoint of $(-2, 21)$ and $(4, 6)$.

602. Rectangle *DIAG* is represented by D $(1, 2)$, I $(10, 5)$, A $(9, 8)$, and G $(0, 5)$. At what point do the diagonals of the rectangle bisect each other?

603. The endpoints of the diameter of a circle are $(10, -4)$ and $(-6, 8)$. Find the center of the circle.

604. Write an expression for the midpoint of $\overline{MA}$ if Point M is at $(4a+b,\ 3a+b)$ and Point A is at $(2a+3b,\ a-b)$.

605. The midpoint of $\overline{AK}$ is $(10,-3)$. If the coordinates of A are $(4,1)$, find the coordinates of K.

606. The midpoint of $\overline{MP}$ is $(-2,-8)$. If the coordinates of M are $(4,-2)$, find the coordinates of P.

607. $\overline{AC}$ is the diameter of a circle with center O. Point A is represented by $(-2,4)$, and Point O is represented by $(-4,7)$. Find the coordinates of C.

Using the Slope Formula

608–616 Use the slope formula to solve the problem.

608. The coordinates of two points are represented by $(9,8)$ and $(5,0)$. Calculate the slope of the line containing these two points.

609. Find the slope of $\overline{SP}$ with the following coordinates: $S\ (-2,8)$ and $P\ (-4,-10)$.

610. Rectangle *DIAG* is represented by $D\ (1,2)$, $I\ (10,5)$, $A\ (9,8)$, and $G\ (0,5)$. Find the slope of diagonal $\overline{DA}$.

611. True or False? The following points are collinear: $A\ (-2,3)$, $B\ (3,5)$, $C\ (8,7)$.

612. The slope of a line that goes through the points $(4,y)$ and $(2,10)$ is -2. Find the value of y.

613. Find the value of x so that the line passing through points $(4,-2)$ and $(x,3)$ has a slope that is undefined.

614. Given the vertices $R\ (0,3)$, $S\ (5,4)$, $T\ (-2,8)$, and $V\ (3,9)$, state whether $\overline{RS}$ is parallel or perpendicular to $\overline{TV}$.

615. Given the vertices M (3,1), A (10,0), T (7,4), and H (6,–3), state whether $\overline{MA}$ is parallel or perpendicular to $\overline{TH}$.

616. The coordinates of $\triangle ABC$ are A (2,0), B (4,8), and C (8,2). The midpoint of $\overline{AB}$ is M, and the midpoint of $\overline{BC}$ is N. Find the slope of $\overline{MN}$.

Parallel and Perpendicular Lines

617–626 Apply your knowledge of slope to answer each question.

617. What is the slope of the line that is parallel to the line $y = -\frac{3}{2}x + 7$?

618. Two lines are perpendicular, and the slope of one line is $\frac{5}{3}$. What is the slope of the other line?

619. What is the slope of a line that is parallel to the line $y = \frac{1}{2}x + 4$?

620. What is the slope of a line that is perpendicular to the line $y = \frac{1}{2}x + 4$?

621. State whether the following two lines are parallel, perpendicular, or neither: $2y + 3 = 4x$ and $4y + 2x = 12$.

622. State whether the following two lines are parallel, perpendicular, or neither: $y = \frac{2}{3}x - 1$ and $6y = 4x + 3$.

623. State whether the following two lines are parallel, perpendicular, or neither: $y = 5x + 8$ and $10y - 2x = 3$.

624. State whether the following two lines are parallel, perpendicular, or neither: $x = 3$ and $y = -4$.

625. State whether the following two lines are parallel, perpendicular, or neither.

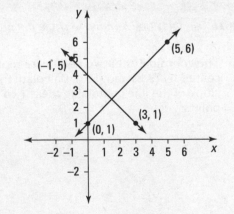

626. State whether the following two lines are parallel, perpendicular, or neither.

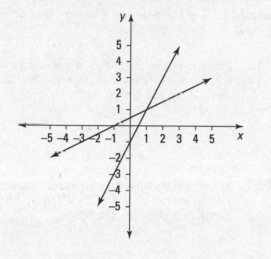

Writing the Equation of a Line in Slope-Intercept Form

627–641 *Express the answer to each question in the form* $y = mx + b$.

627. Write the equation of a line that has a slope of 3 and a y-intercept of –4.

628. Write the equation of a line that has a slope of $-\frac{1}{4}$ and passes through the origin.

629. Write the equation of a line that has a slope of $\frac{3}{5}$ and passes through (5, 7).

630. Write the equation of a line that passes through the points (4, 12) and (6, 12).

631. Write the equation of a line that passes through the points (–1, 5) and (–1, 8).

632. Write the equation of a line that passes through the points (–3, 16) and (–2, 17).

633. Write the equation of a line that passes through the points (–1, 8) and (1, 5).

634. Write the equation of a line that is parallel to $y = 3x + 4$ and has a y intercept of 10.

635. Write the equation of a line that is parallel to $2y = 4x - 2$ and passes through the point $\left(\frac{1}{2}, 6\right)$.

636. Write the equation of a line that passes through the point $(8,2)$ and is perpendicular to the x-axis.

637. Write the equation of a line that passes through the point $(-2,5)$ and is parallel to the x-axis.

638. Write the equation of a line that is perpendicular to $y = 3x + 2$ and passes through the point $(9,-5)$.

639. Write the equation of a line that is perpendicular to $4y - 3x = 5$ and passes through the point $(12,-11)$.

640. Write the equation of a line that represents the perpendicular bisector of $\overline{AB}$ if the coordinates are represented by A $(2,1)$ and B $(10,5)$.

641. Write the equation of a line that represents the perpendicular bisector of $\overline{AB}$ if the coordinates are represented by A $(2,1)$ and B $(10,17)$.

Coordinate Geometry Proofs

642–656 Apply the distance, midpoint, and/or slope formulas to the following proofs.

642. Prove that the diagonals of a rectangle whose vertices are R $(1,2)$, E $(10,5)$, C $(9,8)$, and T $(0,5)$ are congruent.

643. Show that the diagonals of a rectangle whose vertices are R $(1,2)$, E $(10,5)$, C $(9,8)$, and T $(0,5)$ bisect each other.

644. Show that the diagonals of a rectangle whose vertices are R $(1,2)$, E $(10,5)$, C $(9,8)$, and T $(0,5)$ are *not* perpendicular to each other.

645. The points R $(4,1)$, H $(8,3)$, O $(10,7)$, and M $(6,5)$ are the vertices of a rhombus. Show that the diagonals of a rhombus are perpendicular to each other.

646. The points R $(4,1)$, H $(8,3)$, O $(10,7)$, and M $(6,5)$ are the vertices of a rhombus. Show that the diagonals of a rhombus are *not* congruent to each other.

647. The points R (4, 1), H (8, 3), O (10, 7), and M (6, 5) are the vertices of a rhombus. Prove that all four sides of a rhombus are congruent to each other.

648. The points R (4, 1), H (8, 3), O (10, 7), and M (6, 5) are the vertices of a rhombus. Prove that the diagonals of a rhombus bisect each other.

649. The points P (−2, 5), A (7, −1), R (10, −10), and L (1, −4) are vertices of a parallelogram. Show that the diagonals of a parallelogram bisect each other.

650. The points P (−2, 5), A (7, −1), R (10, −10), and L (1, −4) are vertices of a parallelogram. Show that the diagonals of a parallelogram are *not* perpendicular to each other.

651. The points P (−2, 5), A (7, −1), R (10, −10), and L (1, −4) are vertices of a parallelogram. Show that the diagonals of a parallelogram are *not* congruent to each other.

652. Given the following vertices, use the distance formula to classify the triangle as isosceles, equilateral, scalene, and/or right: J (0, −6), K (7, −3), and L (0, 0).

653. Given the following vertices, use the distance formula to classify the triangle as isosceles, equilateral, scalene, and/or right: T (4, 0), R (−6, 0), $S\left(-1,\ 5\sqrt{3}\right)$.

654. Given the following vertices, use the distance formula to classify the triangle as isosceles, equilateral, scalene, and/or right: Q (0, −2), R (6, −4), and S (1, 1).

655. Show that the diagonals of an isosceles trapezoid whose vertices are T (−5, 2), R (4, 5), A (4, 10), and P (−8, 6) are congruent to each other.

656. Show that bases $\overline{TR}$ and $\overline{AP}$ of the trapezoid whose vertices are T (−5, 2), R (4, 5), A (4, 10), and P (−8, 6) are parallel to each other.

Chapter 13

Transformational Geometry

· ·

*I*n geometry, transformations change the size, location, or appearance of geometric figures. Reflections, rotations, translations, and dilations are all transformations that this chapter covers. Whether you're doing one transformation or a composition of multiple transformations, you'll see how each transformation affects a point or figure.

The Problems You'll Work On

In this chapter, you see a variety of geometry problems:

✔ Understanding rigid motion by observing the changes in a geometric figure

✔ Practicing reflecting points over a line and writing the equation of a line of reflection

✔ Understanding point symmetry and point reflections

✔ Translating, dilating, and rotating points

✔ Working with the composition of two or more transformations

✔ Understanding glide reflections and direct and opposite isometries

✔ Completing constructions involving transformations

What to Watch Out For

Some of the following suggestions may be helpful:

✔ Remember that rigid motion means changing the location of an object without changing its shape or size.

✔ An *isometry* is a transformation that preserves distance. A *direct isometry* is a transformation that preserves both distance and orientation.

✔ When performing a rotation, the sign of the degree of rotation tells you the direction in which you're rotating. A positive degree measurement means you're rotating counter-clockwise, whereas a negative degree measurement means you're rotating clockwise.

Rigid Motion

657–670 *Determine the rigid motion that will map one triangle onto another.*

657. △*ABC* ≅ △*A′B′C′*. What rigid motion would map one triangle onto the other?

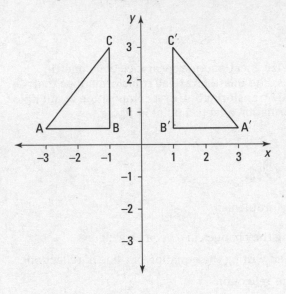

658. △*ABC* ≅ △*A′B′C′*. What rigid motion would map one triangle onto the other?

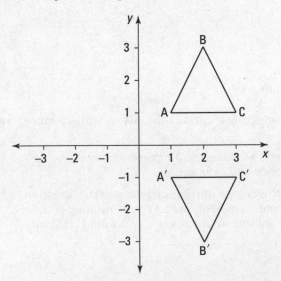

659. △*ABC* ≅ △*A′B′C′*. What rigid motion would map one triangle onto the other?

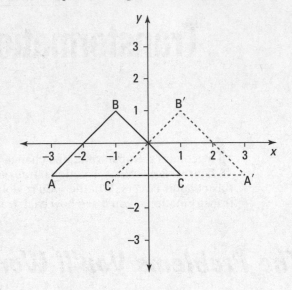

660. △*ABC* ≅ △*A′B′C′*. What rigid motion would map one triangle onto the other?

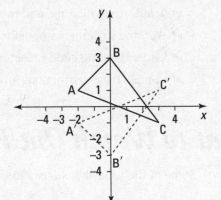

661. $\triangle ABC \cong \triangle A'B'C'$. What rigid motion would map one triangle onto the other?

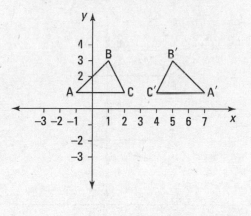

662. $\triangle ABC \cong \triangle A'B'C'$. What rigid motion would map one triangle onto the other?

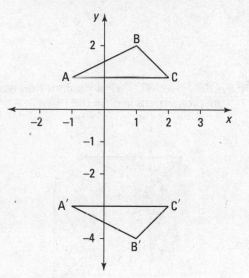

663. $\triangle ABC \cong \triangle A'B'C'$. What rigid motion would map one triangle onto the other?

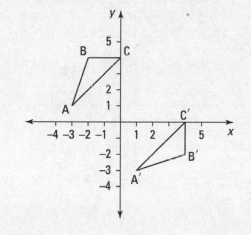

664. $\triangle ABC \cong \triangle A'B'C'$. What rigid motion would map one triangle onto the other?

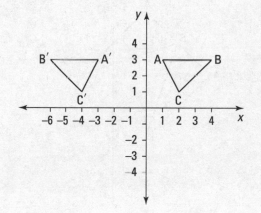

665. △*ABC* ≅ △*A′B′C′*. What rigid motion would map one triangle onto the other?

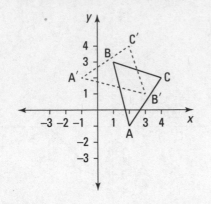

666. △*ABC* ≅ △*A′B′C′*. What rigid motion would map one triangle onto the other?

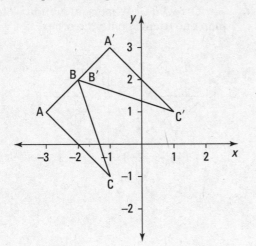

667. △*ABC* ≅ △*A′B′C′*. What rigid motion would map one triangle onto the other?

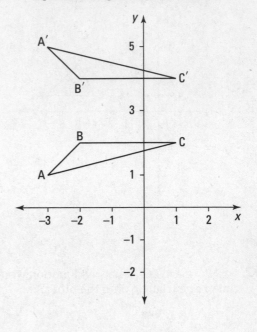

668. △*ABC* ≅ △*A′B′C′*. What rigid motion would map one triangle onto the other?

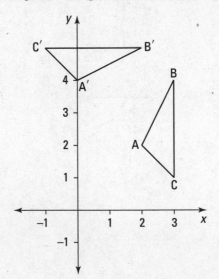

669. $\triangle ABC \cong \triangle A'B'C'$. What rigid motion would map one triangle onto the other?

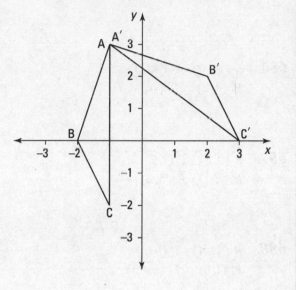

671–675 *Find the image of the point after a reflection over the x-axis.*

671. $(1, 3)$

672. $(-2, 4)$

673. $(5, -3)$

674. $(-4, -10)$

675. (a, b)

670. $\triangle ABC \cong \triangle A'B'C'$. Name the two line reflections that would map one triangle onto the other.

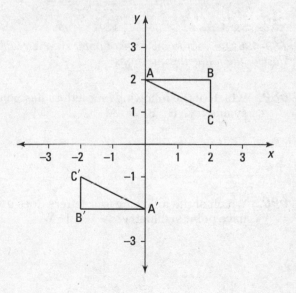

676–680 *Find the image of the point after a reflection over the y-axis.*

676. $(4, 5)$

677. $(-2, 8)$

678. (3, –10)

679. (–1, –2)

680. (a, b)

Writing Equations for Lines of Reflection

681–688 *The image of $\overline{RT}$ is $\overline{R'T'}$. Use the coordinates to write an equation for the line of reflection.*

681. R (0, 1)   T (3, 2)
    R' (1, 0)   T' (2, 3)

682. R (–2, 4)   T (0, –1)
    R' (–4, 2)   T' (1, 0)

683. R (1, 1)   T (3, 5)
    R' (5, 1)   T' (3, 5)

684. R (3, 4)   T (9, 4)
    R' (3, –1)   T' (9, –1)

685. R (3, 1)   T (7, 2)
    R' (–1, 1)   T' (–5, 2)

686. R (4, –2)   T (–3, 5)
    R' (4, 2)   T' (–3, –5)

687. R (0, 3)   T (1, 6)
    R' (2, 1)   T' (5, 2)

688. R (–1, 4)   T (–1, 9)
    R' (1, 0)   T' (5, –3)

Understanding Point Symmetry

689–696 *Use your knowledge of point symmetry to answer the following questions.*

689. Which of the following four letters has point symmetry? H B R M

690. Which of the following four letters does not have point symmetry? N O I W

691. Find the image of point (2, –5) under a reflection in the origin.

692. Under a reflection in the origin, find the image of $P(-6, 3)$.

693. A reflection in the origin is equivalent to a rotation of how many degrees about the origin?

694. Determine the reflection that maps $\triangle ABC$ to $\triangle A'B'C'$, given the following coordinates:

$A(2, 3) \qquad B(2, 5) \qquad C(4, 1)$
$A'(-2, -3) \quad B'(-2, -5) \quad C'(-4, -1)$

695. Reflection, dilation, translation, and rotation are types of transformations. Which one of these is not an example of rigid motion?

696. What is the image of $P(3, 5)$ under a reflection in the point $(5, 2)$?

Triangle Translations

697–701 *A translation is a rigid motion. Find the rigid motion that maps one congruent triangle onto the other.*

697.

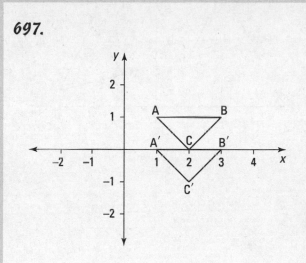

698.

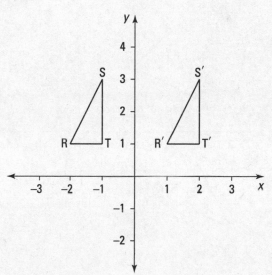

699.

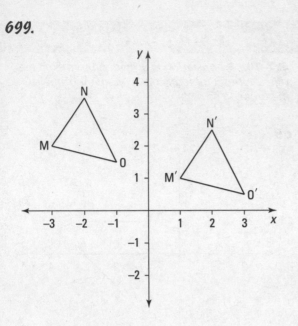

700.

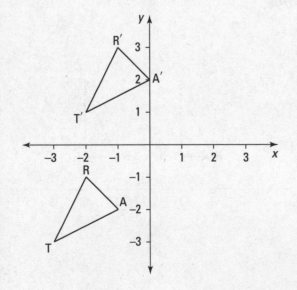

701.

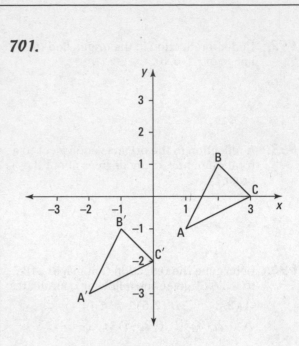

Translating Points

702–705 Find P', the image of P, under the translation $(x, y) \to (x+4, y-3)$.

702. $P(2,5)$

703. $P(-2,0)$

704. $P(7,-3)$

705. $P(-1, 15)$

Finding Translation Rules

706–709 *Find the rule for the translation that maps T to T'.*

706. $T(4, 6) \to T'(5, 5)$

707. $T(-2, 10) \to T'(6, 12)$

708. $T(5, -1) \to T'(-1, -2)$

709. $T(-8, -3) \to T'(8, 4)$

710–712 $\triangle WXY$ *has vertices W (–1, 2), X (1, 5), and Y (6, 0). Find the requested image.*

710. Find W', the image of W after a rigid motion of $(x+1, y+4)$.

711. Find X', the image of X after a rigid motion of $T_{10,-2}$.

712. Find Y', the image of Y after a rigid motion of $(x-4, y+3)$.

713–715 *Use your knowledge of translations to answer the following questions.*

713. What single translation is equivalent to $T_{-2,3} \circ T_{6,12}$?

714. A translation moves $P(6, -2)$ to $P'(11, -5)$. What would the image of $T(4, 1)$ be under the same translation?

715. A translation moves $B(3, 0)$ to $B'(3, -5)$. What would the image of $G(-2, 4)$ be under the same translation?

Doing Dilations

716–724 *Use your knowledge of dilations to solve the following problems. The center of dilation is the origin.*

716. Find the image of $A(12, 25)$ under D_2.

717. Find the image of $B(27, -3)$ under $D_{-\frac{1}{3}}$.

718. Find the image of P $(-2, -11)$ under D_{-3}.

719. Find D', the image of D $(-2, 14)$ under $D_{\frac{1}{2}}$.

720. If Point A' $(9, 6)$ is the image of Point A $(3, 2)$ under a dilation with respect to the origin, what is the constant of dilation?

721. If Point B' $(-8, 4)$ is the image of point B $(-12, 6)$ under a dilation with respect to the origin, what is the constant of dilation?

722. A dilation maps A $(4, 2)$ to A' $(16, 8)$. What would the image of B $(-2, 3)$ be under the same dilation?

723. A dilation maps J $(6, 3)$ to J' $(3, 1.5)$. What would the image of K $(-2, 10)$ be under the same dilation?

724. True or False? Under a dilation, the midpoint, angle measure, and distance are preserved.

Practicing with Rotations

725–728 *The figure shows a regular octagon with l, m, and p as lines of symmetry. Find the image of the given point after the composition of rigid motions.*

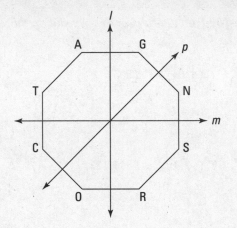

725. $r_p \circ R_{90°}(S)$

726. $R_{135°} \circ R_{-180°}(T)$

727. $R_{-135°} \circ r_l \circ R_{405°}(A)$

728. $R_{315°} \circ r_m \circ R_{-225°} \circ R_{45°}(R)$

Understanding the Rules for Rotations

729–735 Use your knowledge of rotations to answer the following questions.

729. What are the coordinates of R', the image of R (2, 1) after a rotation of 90° about the origin?

730. What are the coordinates of P', the image of P (−5, 10) after $R_{180°}$ about the origin?

731. What are the coordinates of A', the image of P (−5, 10) after $R_{90°}$ about the origin?

732. What are the coordinates of the image of (4, −6) after a clockwise rotation of 90° about the origin?

733. What clockwise rotation is equivalent to a 120° counterclockwise rotation about the origin?

734. What single rotation is equivalent to the following composition of rotations about the origin? $R_{−120°} \circ R_{100°} \circ R_{250°}$

735. $\overline{AB}$ is located at A (4, 3) and B (5, 7). In what quadrant is $\overline{A'B'}$ located after $R_{270°}$ about the origin?

Rigid Motion of Triangles

736–740 $\triangle MNO \cong \triangle M'N'O'$. Find the two rigid motions (a line reflection followed by a translation) that map one triangle onto the other.

736. The vertices of $\triangle MNO$ are M (0, 0), N (1, 5), and O (4, 2). The vertices of $\triangle M'N'O'$ are M' (−1, 1), N' (0, −4), and O' (3, −1). Find the two rigid motions that map one triangle onto the other.

737. The vertices of $\triangle MNO$ are M (0, −5), N (3, −2), and O (10, −3). The vertices of $\triangle M'N'O'$ are M' (−2, −3), N' (−5, 0), and O' (−12, −1). Find the two rigid motions that map one triangle onto the other.

738. The vertices of △*MNO* are *M* (3, 2), *N* (5, 1), and *O* (7, 4). The vertices of △*M'N'O'* are *M'*(−3, 4), *N'*(−5, 3), and *O'* (−7, 6). Find the two rigid motions that map one triangle onto the other.

Compositions of Transformations

741–746 In a composition of transformations, two or more transformations are combined to form a new transformation. Find P', the image of P (5, 1) after a composition of transformations.

741. $r_{y\text{-axis}} \circ R_{90°}(P)$

739.

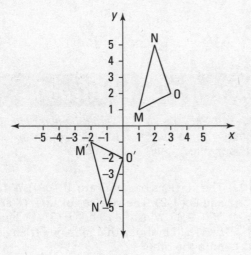

742. $R_{90°} \circ T_{2,-3} \circ D_{2}(P)$

743. $r_{y=x} \circ D_{\frac{1}{3}} \circ T_{4,-1}(P)$

744. $R_{180°} \circ r_{x=2} \circ r_{y=-3}(P)$

740.

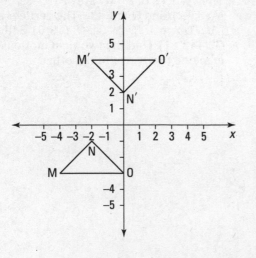

745. $r_{x\text{-axis}} \circ T_{-6,2} \circ r_{y=x} \circ R_{-90°}(P)$

746. $D_{\frac{1}{2}} \circ T_{3,7} \circ r_{y=x}$

Glide Reflections and Direct and Indirect Isometries

747–754 Use your knowledge of glide reflections and isometries to determine whether the following statements are true or false.

747. True or False? A reflection over the x-axis followed by a translation $T_{5,0}$ is an example of a glide reflection.

748. True or False? A reflection over the y-axis followed by a translation of $T_{3,2}$ is an example of a glide reflection.

749. True or False? The following glide reflections are equivalent to one another: $r_{y\text{-axis}} \circ T_{0,8}$ and $T_{0,8} \circ r_{y\text{-axis}}$

750. True or False? A glide reflection is a direct isometry.

751. True or False? A translation is a direct isometry.

752. True or False? The following composition of transformations is an opposite isometry: $r_{y=x} \circ T_{2,5} \circ r_{x=2}$

753. True or False? A rotation of 90° followed by a reflection over the line $y = x$ is an example of a direct isometry.

754. True or False? A rotation of 90° followed by a reflection over the line $y = x$ is an example of an isometry.

Transformations of a Segment

755–757 The vertices of segment $\overline{HA}$ are H (2, 3) and A (6, 7). $\overline{H'A'}$ is the image of HA after the following composition: $r_{y=x} \circ R_{90°}$.

755. True or False? $\overline{HA} \cong \overline{H'A'}$

756. True or False? The coordinates of H' are represented by (–2, 3).

757. True or False? The coordinates of A' are represented by (6, –7).

Trying Rigid Motion Constructions

758–766 Use a compass and straight edge for each construction.

758. Construct $\overline{R'E'}$, the reflection of $\overline{RE}$ over the given line.

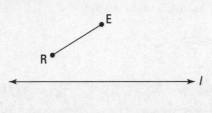

759. Construct △$T'R'A'$, the image of △TRA after a reflection over the given line.

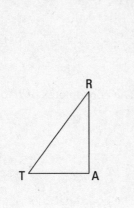

760. Construct the line of reflection for the following figure.

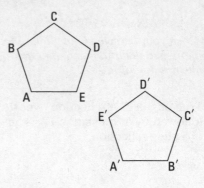

761. In the following figure, there is a rigid motion that transforms △REF to △$R'E'F'$. Construct the line of reflection.

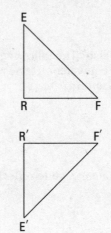

762. Construct △*P'A'L'* by applying a translation to △*PAL* using vector $\overrightarrow{MJ}$.

763. Construct $\overline{R'S'}$, the image of $\overline{RS}$ after it is translated using the given vector $\overrightarrow{VT}$.

764. Construct a vector that defines the translation. (Use Point *V* as the endpoint of the vector.)

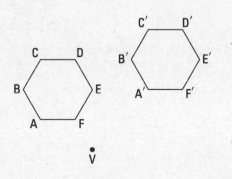

765. Construct the center of rotation, Point *P,* for the following figure.

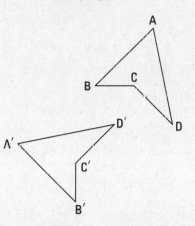

766. Construct the rotation of $\overline{AK}$ using the given angle and Point *P* as the center of rotation.

Chapter 14

Exploring Circles

· ·

Humans seem to have a fascination with circles. After all, circles are all around you, from jewelry to the Olympics. There are so many formulas and theorems regarding circles that we had to divide them into two chapters. This chapter focuses on the many circle formulas, from area and circumference to the equation of a circle.

The Problems You'll Work On

In this chapter, you see a variety of geometry problems:

- ✔ Determining the circumference and area of a circle
- ✔ Finding the area of a sector of a circle
- ✔ Calculating the length of an arc of a circle
- ✔ Understanding the standard form of the equation of a circle
- ✔ Completing the square to determine the center and radius of a circle in general form

What to Watch Out For

Some of the following points may be helpful as you work through these problems:

- ✔ The diameter of a circle is equal to twice the radius of the circle; conversely, the radius of a circle is equal to half the diameter of the circle.
- ✔ To determine the center of a circle whose equation is written in standard form, you can just negate the numbers in the parentheses of the equation.
- ✔ When determining the length of an arc of a circle or the area of a sector of a circle, you need to know the central angle of the circle in radians.
- ✔ When calculating the area or circumference of a circle, always leave your answer in terms of π unless the question asks otherwise.

Working with the Circumference of a Circle

767–773 Use the formula for the circumference of a circle to solve the following problems.

767. In terms of π, find the circumference of a circle if the diameter is 8 inches.

768. In terms of π, find the circumference of a circle if the radius is 10 centimeters.

769. The circumference of a circle is 14π units. Find the radius of the circle.

770. The area of a circle is 64π square units. Find the circumference of the circle.

771. Circle *O* is inscribed in square *ABCD*. The area of the square is 144 square inches. Find the circumference of Circle *O*.

772. The following figure consists of a right triangle and two semicircles. Find the perimeter of the figure in terms of π.

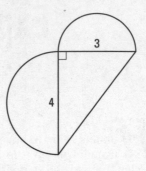

773. Square *ABCD* is inscribed in Circle *E*. If the circumference of the circle is 8π units, find the perimeter of the square in simplest radical form.

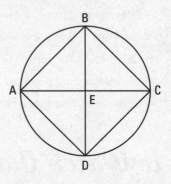

Understanding the Area of a Circle

774–780 Use the formula for the area of a circle to solve the following problems.

774. In terms of π, find the area of a circle if the radius is 10 units.

775. In terms of π, find the area of a circle if the diameter is 28 feet.

776. The circumference of a circle is 50π units. Find the area of this circle.

777. The area of a circle is 289π square units. Find the diameter of the circle.

778. A circle is inscribed in a square. The perimeter of the square is 80 units. Find the area of the circle.

779. A square is inscribed in Circle O. If the area of the square is 100 square units, find the area of Circle O in terms of π.

780. The radii of the two circles are 8 and 14. Find the area of the shaded region in terms of π.

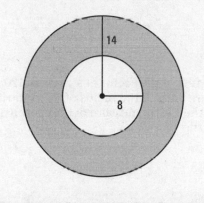

Working with Sectors

781–785 *The formula for the area of a sector of a circle is* $A = \frac{1}{2}r^2\theta$, *where* θ *is measured in radians.*

Use this formula to solve the following problems.

781. Find the area of a sector of a circle that has a radius of 8 feet and a central angle of 2.5 radians.

782. Find the area of a sector of a circle that has a diameter of 12 feet and a central angle of 2 radians.

783. The circumference of a circle is 48π. Find the area of a sector that has a central angle of 6 radians.

784. The area of the sector of a circle is 147π square units. The central angle measures $\frac{3\pi}{2}$ radians. Find the radius.

785. The area of the sector of a circle is 120π square units. If the radius of the circle is 12, find the radian measure of the central angle in terms of π.

Arc Length

786–792 Use the formula s = rθ, where s is the length of the intercepted arc, r is the radius, and θ is the central angle of the circle in radians.

786. Find the area of a sector of a circle that has an arc length of 20 centimeters and a radius of 8 centimeters.

787. In terms of π, find the length of the arc of a circle whose radius is 16 and whose central angle is $\frac{3\pi}{4}$ radians.

788. The radius of a circle is 10 kilometers. To the nearest hundredth, find the length of the arc intercepted by a central angle of $\frac{2\pi}{3}$ radians.

789. The length of an arc of a circle is 8 centimeters. The arc is intercepted by a central angle of 45°. Find the circle's radius to the nearest hundredth of a centimeter.

790. A circle has a radius of 2 meters. In terms of π, find the length of the arc intercepted by a central angle of 120°.

791. Circle O has a radius of 22 centimeters and an arc length of 36 centimeters. To the nearest tenth, find the radian measure of the central angle.

792. The central angle of a circle whose radius is 3 feet intercepts an arc measuring 54 inches. Find the radian measure of the central angle of the circle.

The Equation of a Circle in Standard Form

793–797 Given the equation of a circle in standard form, find the center and the radius of each circle.

793. $(x+4)^2 + y^2 = 25$

794. $(x-3)^2 + \left(y+\frac{2}{3}\right)^2 = 50$

795. $x^2 + (y-6)^2 = 144$

796. $(x-1)^2 + (y+10)^2 = 45$

797. $(x-h)^2 + (y-k)^2 = r^2$

798–804 *Write the equation of a circle in standard form using the given information.*

798. A circle with its center at the origin has a radius of 7.

799. A circle with its center at (9, –8) has a radius of $\frac{2}{3}$.

800. The center of a circle is (–7, 13), and the diameter is 10.

801. The point (5, 6) lies on a circle whose center is at (3, –4).

802. The endpoints of the diameter of a circle are (2, 8) and (6, 8).

803. The endpoints of the diameter of a circle are (–11, 5) and (–3, 7).

804. The endpoints of the diameter of a circle are (4, 12) and (–6, 8).

805–808 *The equation of a circle is written in general form. Find the center and radius of the circle by completing the square.*

805. $x^2 + y^2 + 2x - 15 = 0$

806. $x^2 + y^2 + 4x + 6y + 4 = 0$

807. $x^2 + y^2 - 10x + 2y - 74 = 0$

808. $2x^2 + 2y^2 + 12x - 28y + 104 = 0$

809–811 The coordinates of three points on a circle are (0, 6), (6, 4) and (–8, 2).

809. Write the equation of the circle in standard form.

810. Write the equation of the circle in general form.

811. What are the center and radius of this circle?

Chapter 15

Circle Theorems

· ·

This chapter focuses on the theorems associated with circles. You get a better understanding of theorems regarding the angles and arcs of circles as well as the lengths of segments within a circle.

The Problems You'll Work On

In this chapter, you see a variety of geometry problems:

- ✔ Understanding the relationship between central, inscribed, vertical, or exterior angles and the arcs they intercept
- ✔ Finding the lengths of two chords intersecting in a circle
- ✔ Finding the lengths of tangents and/or secants intersecting a circle
- ✔ Completing geometric proofs involving circles

What to Watch Out For

Don't let common mistakes trip you up. Some of the following points may be helpful:

- ✔ A central angle is equal to the arc it intercepts. An inscribed angle is equal to half of the intercepted arc.
- ✔ The sum of all the arcs around a circle is 360°.
- ✔ A *secant* is a line that goes through a circle at two locations, whereas a *tangent* is a line that touches the circle only once.
- ✔ Make sure you understand what the question is asking you to solve for. After finding the value of the variable, you may have to plug it in to find the measure of a segment or angle.

Central Angles and Arcs

812–817 In the following figure, $\overline{DF}$ and $\overline{EF}$ are chords in Circle O and $\angle DOE$ is a central angle. Use the given information to solve for the missing angle or arc.

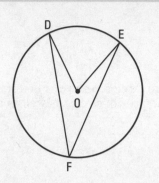

812. If $m\angle DOE = 50°$, find $m\widehat{DE}$.

813. If $m\widehat{DE} = 64°$, find $m\angle DOE$.

814. If $m\angle DOE = 75°$, find $m\widehat{DFE}$.

815. If $m\widehat{DFE} = 300°$, find $m\angle DOE$.

816. If $m\widehat{DF} = 150°$ and $m\widehat{EF} = 140°$, find $m\angle DOE$.

817. If $\widehat{DF} \cong \widehat{EF}$ and $m\angle DOE = 40°$, find $m\widehat{DF}$.

Inscribed Angles and Arcs

818–819 Equilateral $\triangle EQU$ is inscribed in Circle O. Find the missing arc.

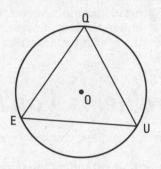

818. Find $m\widehat{EQ}$.

819. Find $m\widehat{QUE}$.

820–823 In Circle O, $\triangle RTI$ is a right triangle with diameter $\overline{IR}$ as the hypotenuse. If $m\widehat{IT} = 120°$ and $OR = 20$, find the missing angle or arc.

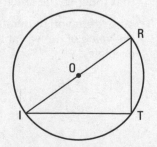

820. Find $m\angle IRT$.

821. Find $m\angle RIT$.

822. Find the length of $\overline{RT}$.

823. Find the length of $\overline{IT}$.

Angles Formed by Intersecting Chords of a Circle

824–826 In Circle O, chords $\overline{IX}$ and $\overline{MX}$ are congruent. If $m\overparen{IX}$ is 9 less than 3 times $m\overparen{IM}$, find the missing angle or arc.

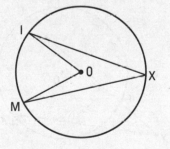

824. Find $m\overparen{IM}$.

825. Find $m\angle MOI$.

826. Find $m\angle X$.

827–830 In Circle O, chords $\overline{UP}$ and $\overline{LS}$ intersect at E. Use the given information to solve for the missing angle or arc.

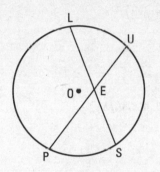

827. If $m\overparen{LU} = 43°$ and $m\overparen{PS} = 57°$, find $m\angle LEU$.

828. If $m\overparen{PL} = 114°$ and $m\overparen{US} = 76°$, find $m\angle LEP$.

829. If $m\overparen{PL} = 130°$ and $m\angle SEU = 100°$, find $m\overparen{US}$.

830. If $m\angle LEP = 132°$ and $m\overparen{LU} = 46°$, find $m\overparen{PS}$.

831–833 In Circle O, chords $\overline{AC}$ and $\overline{BD}$ intersect at E. If $m\widehat{AB} = 65°$, $m\widehat{BC} = 85°$, and $m\widehat{CD} = 150°$, find the missing angle or arc.

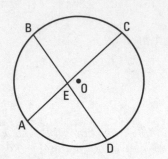

831. Find $m\widehat{AD}$.

832. Find $m\angle AED$.

833. Find $m\angle CED$.

834–837 In Circle O, chords $\overline{AH}$ and $\overline{MT}$ intersect at E. If $m\widehat{AM} = 64°$, $m\widehat{AT} = 59°$, and $m\widehat{MH}$ is twice $m\widehat{TH}$, find the missing angle or arc.

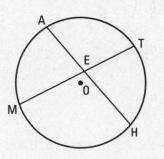

834. Find $m\widehat{TH}$.

835. Find $m\widehat{MH}$.

836. Find $m\angle AEM$.

837. Find $m\angle AET$.

Angles Formed by Secants and Tangents.

838–843 Circle O has secants $\overline{GEY}$ and $\overline{MTY}$ meeting at Point Y. Use the given information to solve for the missing angle or arc.

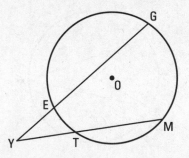

838. If $m\widehat{GM} = 100°$ and $m\widehat{ET} = 20°$, find $m\angle Y$.

839. If $m\widehat{GM} = 155°$ and $m\widehat{ET} = 43°$, find $m\angle Y$.

840. If $m\widehat{GM} = 99°$ and $m\angle Y = 39°$, find $m\widehat{ET}$.

841. If $m\widehat{GM} = 142°$ and $m\angle Y = 58.5°$, find $m\widehat{ET}$.

842. If $m\widehat{ET} = 70°$ and $m\angle Y = 52°$, find $m\widehat{GM}$.

843. If $m\widehat{ET} = 37°$ and $m\angle Y = 73.25°$, find $m\widehat{GM}$.

844–848 *In Circle O, tangents $\overline{TA}$ and $\overline{TN}$ meet at an external Point T. Use the following information to solve for the missing angle or arc.*

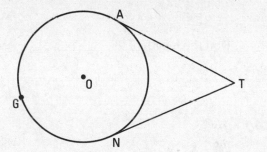

844. If $m\widehat{AGN} = 300°$, find $m\angle T$.

845. If $m\widehat{AGN} = 250°$, find $m\angle T$.

846. If $m\widehat{AN} = 105°$, find $m\angle T$.

847. If $m\angle T = 60°$, find $m\widehat{AGN}$.

848. If $m\angle T = 81°$, find $m\widehat{AN}$.

849–852 *In Circle O, $\overline{AOC}$ is a diameter, $\overline{ADB}$ is a secant, and $\overline{BC}$ is a tangent. If $m\widehat{DC}$ is 3 less than twice $m\widehat{AD}$, find the missing angle or arc.*

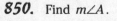

849. Find $m\widehat{DC}$.

850. Find $m\angle A$.

851. Find $m\angle B$.

852. Find $m\angle C$.

The Intersecting Chord Theorem

853–861 In Circle O, chords $\overline{CI}$ and $\overline{RL}$ intersect at E. Use the given information to find the missing segment.

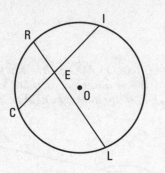

853. $\overline{RE} \times \overline{LE} = \overline{CE} \times \underline{\hspace{0.5cm}}$

854. $\overline{CE} \times \overline{IE} = \underline{\hspace{0.5cm}} \times \overline{LE}$

855. $RE = 3$
$LE = 8$
$IE = 6$
$CE = x$
Find the length of $\overline{CE}$.

856. $RE = x$
$LE = 12$
$IE = 8$
$CE = 6$
Find the length of $\overline{RE}$.

857. $RE = 5$
$LE = x$
$IE = 12.5$
$CE = 8$
Find the length of $\overline{LE}$.

858. $RE = 3$
$LE = 10$
$IE = x + 1$
$CE = x$
Find the length of $\overline{CE}$.

859. $RE = 4$
$LE = 16$
$\overline{IE} \cong \overline{CE}$
Find the length of $\overline{IE}$.

860. $RE = x$
$LE = x + 10$
$IE = 8$
$CE = 7$
Find the length of $\overline{RE}$.

861. $RE = x + 5$
$LE = 8$
$IE = x + 2$
$CE = x$

Find the length of $\overline{CE}$.

862–867 *In each of the following figures, the diameter of Circle O is perpendicular to the given chord.*

862. Solve for x.

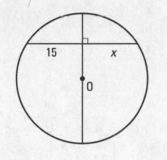

863. Solve for x.

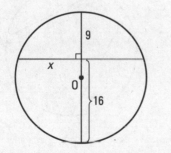

864. Solve for x.

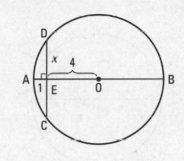

865. Solve for x.

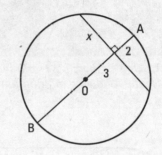

866. Solve for x.

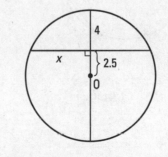

867. Find the length of the diameter of the circle.

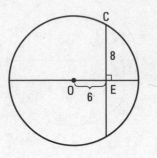

Lengths of Tangents and Secants

868–872 *In Circle O, tangent $\overline{AT}$ and secant $\overline{SET}$ meet at T. Use the following information to solve for x.*

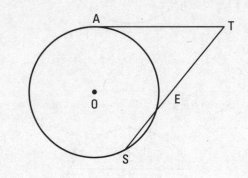

868. $(TE)(TS)=(TA)(x)$

869. $TS = 12$

$TE = 3$

$TA = x$

870. $TS = 16$

$TE = x$

$TA = 8$

871. $TE = 9$

$TA = 12$

$SE = x$

872. $TE = x$

$TA = 10$

$SE = x + 10$

873–879 *In Circle O, secants $\overline{ABE}$ and $\overline{CDE}$ meet at external Point E. Use the following information to solve for x.*

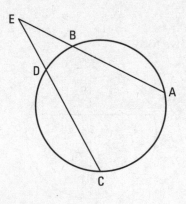

873. $(x)(EA)=(ED)(EC)$

874. $(x)(CE) = (BE)(AE)$

875. $AE = 16$
$BE = 6$
$CE = 24$
$DE = x$

876. $AE = 30$
$BE = 10$
$CE = 20$
$DE = x$

877. $EB = 2$
$BA = 10$
$ED = x$
$DC = 5$

878. $EB = 4$
$BA = 8$
$ED = 3$
$DC = x$

879. $EB = x$
$BA = 10$
$ED = x - 2$
$DC = 22$

880–884 Use your knowledge of circles to solve each problem.

880. $\overline{HA}$ and $\overline{HT}$ are tangents to Circle O. If $HA = 28$, find the length of $\overline{HT}$.

881. $\overline{AB}$ is tangent to Circle O at B. If the diameter of Circle O is 14 and if $AO = 25$, find the length of $\overline{AB}$.

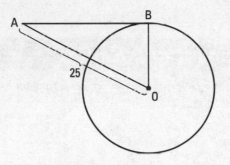

882. Find the perimeter of $\triangle TAN$ if $\overline{TA}$, $\overline{AN}$, and $\overline{NT}$ are all tangent to Circle O.

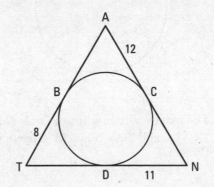

883. The perimeter of △RST is 34. If $\overline{RS}$, $\overline{ST}$, and $\overline{TR}$ are tangent to Circle O at A, B, and C, respectively, find the length of $\overline{RT}$.

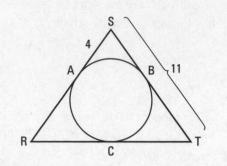

Tangent and Radius

884–887 Determine whether $\overline{TA}$ is tangent to Circle O.

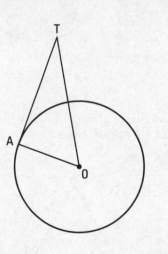

884. The length of the radius of Circle O is 10, $TA = 24$, and $TO = 26$. Is $\overline{TA}$ tangent to Circle O?

885. In Circle O, $AO = 3\sqrt{41}$, $TO = 25$, and $TA = 16$. Is $\overline{TA}$ tangent to Circle O?

886. In Circle O, $AO = 12$, $TA = 15$, and $TO = 3\sqrt{41}$. Is $\overline{TA}$ tangent to Circle O?

887. In Circle O, $AO = 5$, $TA = 12$, and $TO = 15$. Is $\overline{TA}$ tangent to Circle O?

"BIG" Circle Problems

888–892 In Circle O, $\overline{NR}$ is a tangent, $\overline{NEY}$ is a secant, $\overline{RW}$ is a diameter, and $\overline{EY}$ and $\overline{ER}$ are chords. If $m\angle N = 40°$ and the ratio of $m\overset{\frown}{WY}$ to $m\overset{\frown}{YR}$ is 1:9, find the missing angle or arc.

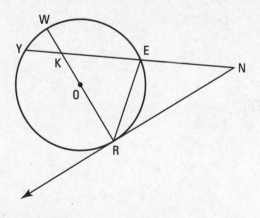

888. Find $m\overset{\frown}{WY}$.

889. Find $m\overset{\frown}{ER}$.

890. Find $m\angle WKY$.

891. Find $m\angle ERW$.

892. Find $m\angle NER$.

893–897 *In Circle O, chord $\overline{IC}$ is parallel to diameter $\overline{BE}$. $\overline{GCL}$ is a secant, and $\overline{GI}$ is a tangent to Circle O. If $m\overset{\frown}{EL} = 74°$ and $\overset{\frown}{IC} \cong \overset{\frown}{CE}$, find the missing angle or arc.*

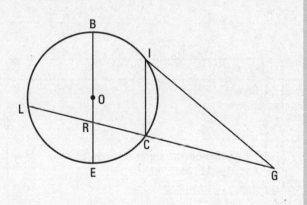

893. Find $m\overset{\frown}{BI}$.

894. Find $m\angle BRL$.

895. Find $m\angle ICL$.

896. Find $m\angle G$.

897. Find $m\angle GIC$.

Circle Proofs

898–902 Complete the proof by giving the statement or reason.

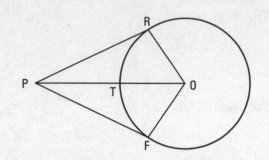

Given: $\overline{PR}$ and $\overline{PF}$ are tangent to Circle O; $\overline{RO}$ and $\overline{FO}$ are radii.

Prove: $\angle POR \cong \angle POF$

898. What is the reason for Statement 2?

899. What is the statement for Reason 3?

900. What is the statement for Reason 4?

901. What is the reason for Statement 5?

902. Because $\triangle POR \cong \triangle POF$, you know that $\angle PRO \cong \angle PFO$. What type of angles are $\angle PRO$ and $\angle PFO$?

Statements	Reasons
1. $\overline{PR}$ and $\overline{PF}$ are tangent to Circle O; $\overline{RO}$ and $\overline{FO}$ are radii.	1. Given
2. $\overline{RO} \cong \overline{FO}$	2.
3.	3. Tangents from the same external point are congruent to each other.
4.	4. Reflexive postulate
5. $\triangle POR \cong \triangle POF$	5.
6. $\angle POR \cong \angle POF$	6. CPCTC

903–905 *Complete the proof by giving the statement or reason.*

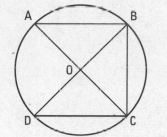

Given: In Circle O, $\overline{AOC}$ and $\overline{BOD}$ are diameters. $\overline{AB}$, $\overline{BC}$, and $\overline{DC}$ are chords.

Prove: $\triangle ABC \cong \triangle DCB$ (by using the triangle congruence theorem HL)

903. What is the reason for Statement 2?

904. What is the reason for Statement 3?

905. What is the reason for Statement 4?

Statements	Reasons
1. In Circle O, $\overline{AOC}$ and $\overline{BOD}$ are diameters. $\overline{AB}$, $\overline{BC}$, and $\overline{DC}$ are chords.	1. Given
2. $\overline{AC} \cong \overline{DB}$	2.
3. $\overline{BC} \cong \overline{BC}$	3.
4. $\angle ABC$ and $\angle DCB$ are right angles.	4.
5. $\angle ABC \cong \angle DCB$	5. All right angles are congruent to each other.
6. $\triangle ABC$ and $\triangle DCB$ are right triangles.	6. If a triangle contains a right angle, then it's a right triangle.
7. $\triangle ABC \cong \triangle DCB$	7. HL

906. In Circle *O*, $\overline{AOC}$ and $\overline{BOD}$ are diameters. $\overline{AB}$, $\overline{BC}$, and $\overline{DC}$ are chords. Why is ∠*BAC* ≅ ∠*CDB*?

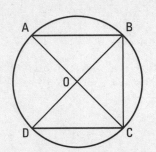

Chapter 16

Three-Dimensional Geometry

- -

A large part of geometry is based on visualizing different shapes. Three-dimensional geometry requires an understanding of cubes, rectangular solids, triangular prisms, cylinders, cones, and spheres. Three-dimensional geometry also lends itself to the discussion of three undefined terms: *point, line,* and *plane.*

The Problems You'll Work On

In this chapter, you see a variety of geometry problems:

- ✔ Understanding the edges and planes of rectangular prisms
- ✔ Making sense of skew lines
- ✔ Understanding points, lines, and planes
- ✔ Finding the surface area and volume of a cube, rectangular solid, cylinder, right circular cone, and sphere
- ✔ Finding the lateral area of a solid cylinder and a right circular cone
- ✔ Rotating two-dimensional figures to create three-dimensional figures

What to Watch Out For

Some of the following points may help you avoid common mistakes:

- ✔ Be sure that all measurements are in the same unit.
- ✔ When answering questions regarding points, lines, and planes, you may find it helpful to model the problem using props such as a pencil to represent a line and a piece of paper to represent a plane.
- ✔ Be careful not to confuse the slant height of a three-dimensional object with the actual height of the three-dimensional object.

Understanding Points, Lines, and Planes

907–915 The following figure is a right rectangular prism. State whether each statement is true or false.

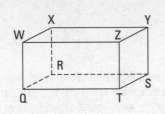

907. True or False? Edge $\overline{QR}$ is parallel to edge $\overline{ZY}$.

908. True or False? Edge $\overline{WX}$ is perpendicular to edge $\overline{ZT}$.

909. True or False? Edge $\overline{RS}$ is parallel to edge $\overline{YS}$.

910. True or False? Plane *XRSY* is parallel to plane *STZY*.

911. True or False? Plane *QRXW* is perpendicular to plane *QRST*.

912. True or False? Edge $\overline{XR}$ is perpendicular to plane *QRST*.

913. True or False? There are only two planes in the rectangular prism that are perpendicular to plane *QRXW*.

914. True or False? $\overline{QW}$ and $\overline{ZY}$ are examples of skew lines.

915. True or False? $\overline{QW}$ and $\overline{SY}$ are coplanar.

916–922 State whether each of the following statements is true or false.

916. True or False? Through a point not on a plane, there is only one line that can be drawn through the point that is perpendicular to the plane.

917. True or False? If two planes are perpendicular to a line, the two planes are perpendicular to each other.

918. True or False? If $\overline{AB}$ is perpendicular to plane P, an infinite number of planes can contain $\overline{AB}$ and also be perpendicular to plane P.

919. True or False? If a line is parallel to one of two perpendicular planes, then it's parallel to the other plane.

920. True or False? If two points lie in a plane, the line joining them lies in the same plane.

921. True or False? Skew lines are straight lines that are not parallel and do not intersect in three dimensional space.

922. True or False? If Point P is on line n, an infinite number of planes are perpendicular to line n and pass through Point P.

Surface Area of Solid Figures

923–938 Find the surface area.

923. The edge of a cube measures 11 inches. Find the surface area of the cube.

924. Calculate the surface area of the following rectangular solid.

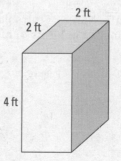

925. Find the surface area of the following rectangular prism.

2 ft

2 ft

4 ft

2 ft

926. Calculate the surface area of a rectangular prism that has a length of 6 centimeters, a width of 5 centimeters, and a height of 3 centimeters.

927. The surface area of a rectangular prism is 2,280 square feet. Find the length of the prism if the width and height are 20 and 15 feet.

928. The radius of the base of a cylinder is 20 inches, and the height is 10 inches. Find the lateral surface area of the solid cylinder. Round your answer to the nearest hundredth.

929. The diameter of the base of a cylinder is 15 inches, and the height is 26 inches. Find the surface area of the solid cylinder. Round your answer to the nearest tenth.

930. Find the surface area of the solid cylinder. Leave the answer in terms of π.

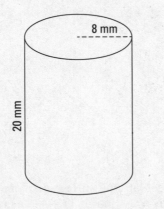

931. Find the lateral area of the solid cylinder to the nearest hundredth.

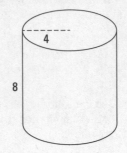

932. Calculate the surface area of the cone to the nearest tenth.

933. Find the lateral area of a right circular cone that has a base with a radius of 5 meters and a slant height of 7 meters. Leave your answer in terms of π.

934. Find the lateral area of a right circular cone that has a base with a diameter of 12 inches and an altitude of 8 inches. Leave your answer in terms of π.

935. In terms of π, find the surface area of a right circular cone that has a base with a radius of 5 feet and an altitude of 12 feet.

936. To the nearest hundredth, find the surface area of a sphere that has a radius of 2 feet.

937. What is the radius of a sphere that has a surface area of 256π square units?

938. In terms of π, find the surface area of a spherical planet with a radius of 1,000 kilometers.

Calculating the Volume of Solid Figures

939–951 Use the appropriate volume formula to solve each problem.

939. The edge of a cube measures 4.5 feet. What is the volume of the cube?

940. Calculate the volume of a rectangular prism that has a length of 8.5 meters, a width of 3.5 meters, and a height of 2.4 meters.

941. Find the volume of the rectangular prism.

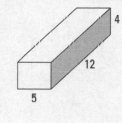

942. Find the volume of the triangular prism.

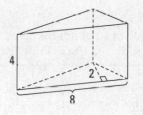

943. A cylinder has a radius of 4 inches and a height of 12 inches. Find the volume of the cylinder, leaving the answer in terms of π.

944. A cylinder has a diameter of 20 inches and a height of 1.5 feet. In terms of π, find the volume of the cylinder in cubic inches.

945. Calculate the volume of the cylinder to the nearest cubic foot.

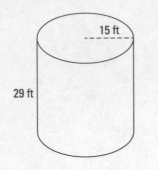

946. The volume of a cylinder is 392π cubic feet. The radius of the cylinder is 7 feet. Find the height of the cylinder.

947. In terms of π, find the volume of a right circular cone with a radius of 8 kilometers and a height of 15 kilometers.

948. The volume of a right circular cone is $2{,}700\pi$ cubic inches, and the height of the cone is 36 inches. Calculate the length of the radius.

949. Calculate the volume of the square pyramid.

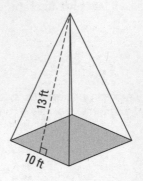

950. Find the volume of a sphere with a radius of 24 inches. Leave your answer in terms of π.

951. Find the volume of a sphere with a diameter of 9 feet. Round your answer to the nearest tenth.

Rotations of Two-Dimensional Figures

952–956 Use your knowledge of two- and three-dimensional figures to answer each question.

952. What three-dimensional figure is created when a triangle is rotated about an axis of rotation that bisects the triangle?

953. What three-dimensional figure is created when a rectangle is rotated about an axis of rotation that bisects the rectangle?

954. What three-dimensional figure is created when a circle is rotated about an axis of rotation that bisects the circle?

955. What is the volume, in terms of π, of the three-dimensional figure created when the following rectangle is rotated about an axis of rotation that bisects the rectangle?

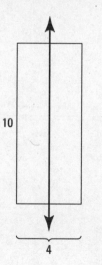

956. What is the volume, in terms of π, of the three-dimensional figure created when a circle whose radius is 3 is rotated about an axis of rotation that bisects the circle?

Chapter 17

Locus Problems

. . . .

*I*n geometry, the word *locus* represents a set of points that satisfy a specific condition. This chapter allows you to practice the five fundamental locus problems and then apply the concepts toward more-complex or compound locus problems. Many conic sections, such as parabolas, ellipses, and hyperbolas, are also defined using loci.

The Problems You'll Work On

In this chapter, you see a variety of geometry problems:

- ✔ Determining the equation for the locus of points equidistant from a given point or line
- ✔ Determining the equation for the locus of points equidistant from two parallel or intersecting lines
- ✔ Determining the equation for the locus of points equidistant from two points
- ✔ Finding the solution for compound locus problems using coordinate geometry
- ✔ Finding the equation of an ellipse

What to Watch Out For

Don't let common mistakes trip you up. Some of the following suggestions may be helpful:

- ✔ Drawing a picture of the situation can make finding the locus of points easier. Try drawing the locus of points using dotted lines and leave the given lines solid so you're able to distinguish between the two.
- ✔ You need to use the equation of a line, $y = mx + b$, and the equation of a circle, $(x-h)^2 + (y-k)^2 = r^2$, to write the equation representing the locus of points.
- ✔ To determine how many solutions are in a compound locus problem, note how many times the two loci intersect. For accuracy, consider using a compass to draw the circles and a straight edge to draw the lines.
- ✔ When finding the locus, first locate one point that satisfies the condition. Then locate a second point. Repeat this process by locating several other points that satisfy the condition. Finally, you can connect these points to describe the locus.

Basic Locus Theorems

957–961 Describe the locus of points that satisfy the following conditions.

957. The locus of points equidistant from Point P

958. The locus of points equidistant from two Points A and B

959. The locus of points equidistant from one line

960. The locus of points equidistant from two parallel lines

961. The locus of points equidistant from a line and a point not on the line

Loci Using Coordinate Geometry

962–966 Are each of the following points on the locus?

962. True or False? The point $(4, -1)$ is on the locus of points that are equidistant from the two points $(1, -1)$, and $(7, -1)$.

963. True or False? The point $(3, 2)$ is on the locus of points that are equidistant from the two lines $x = 8$ and $x = -1$.

964. True or False? The point $(3, 3)$ is on the locus of points that are 3 units from the origin.

965. True or False? The point $(-2, 4)$ is on the locus of points that are 5 units from the point $(2, 1)$.

966. True or False? The point $(18, 2)$ is on the locus of points that are 4 units away from $y = -2$.

The Locus of Points Equidistant from One or Two Lines

967–974 Write the equation of the line that would satisfy the following conditions.

967. The locus of points 4 units to the right of $x = 2$

968. The locus of points 3 units below $y = -1$

969. The locus of points 10 units from the x-axis

970. The locus of points equidistant from $y = 5$ and $y = 11$

971. The locus of points equidistant from $x = -6$ and $x = 14$

972. The locus of points equidistant from the x- and y-axes

973. The locus of points equidistant from $y = 2x + 5$ and $y = 2x + 15$

974. The locus of points equidistant from $y = -\frac{1}{2}x - 4$ and $y = -\frac{1}{2}x + 6$

The Locus of Points Equidistant from Two Points

975–982 Write the equation of the line that would satisfy the following conditions.

975. The locus of points equidistant from $(3, 5)$ and $(3, 11)$

976. The locus of points equidistant from $(10, 20)$ and $(10, 8)$

977. The locus of points equidistant from $(6, -2)$ and $(-14, -2)$

978. The locus of points equidistant from $(-8, 1)$ and $(1, 1)$

979. The locus of points equidistant from $(2, 4)$ and $(6, 12)$; give your answer in slope-intercept form

980. The locus of points equidistant from (0, 8) and (12, 4); give your answer in slope-intercept form

981. The locus of points equidistant from (–2, 2) and (–14, 10); give your answer in slope-intercept form

982. The locus of points equidistant from the origin and (–4, 4); give your answer in slope-intercept form

Writing the Equation of a Circle

983–988 Write the equation of the circle that would satisfy the following conditions.

983. The locus of points 10 units from the origin

984. The locus of points 5 units from the point (–2, 8)

985. The locus of points 6 units from the point (–3, 15)

986. The locus of points 3.5 units from the point (–1, –2)

987. The locus of points equidistant from $x^2 + y^2 = 49$ and $x^2 + y^2 = 121$

988. The locus of points equidistant from $x^2 + y^2 = 81$ and $x^2 + y^2 = 196$

Compound Locus in Coordinate Geometry

989–993 Find the number of points that satisfy both conditions.

989. **Condition 1:** Equidistant from the x-axis and y-axis

Condition 2: Five units from the origin

990. **Condition 1:** Two units from the y-axis

Condition 2: Two units from the origin

991. **Condition 1:** Equidistant from points (3, 1) and (3, 7)

Condition 2: Three units away from (3, 1)

992. **Condition 1:** Equidistant from $x = -4$ and $x = 6$

Condition 2: Four units from a point on the line $x = 6$

993. **Condition 1:** Five units from the line $y = 3$

Condition 2: Equidistant from the x- and y-axes

Compound and Challenging Locus Problems

994–1,001 Use your knowledge of loci to answer the following questions.

994. Points A and B are 10 feet apart. How many points are equidistant from both A and B and are also 5 feet from A?

995. Given two parallel lines k and m, find the number of points that are equidistant from k and m and are also equidistant from two points on line k.

996. Point P is 3 inches from $\overleftrightarrow{GT}$. How many points are 5 inches from $\overleftrightarrow{GT}$ and 8 inches from P?

997. $\overline{MA}$ and $\overline{TH}$ are two parallel line segments that are 14 feet apart, and E is a point on $\overline{MA}$. How many points are equidistant from $\overline{MA}$ and $\overline{TH}$ and 7.5 feet from E?

998. What is the number of points equidistant from two parallel lines and also equidistant from two points on one of these lines?

999. $\overleftrightarrow{LO}$ and $\overleftrightarrow{CU}$ are two parallel lines that are 8 meters apart, and S is a point on $\overleftrightarrow{LO}$. How many points are equidistant from $\overleftrightarrow{LO}$ and $\overleftrightarrow{CU}$ and 3 meters from S?

1,000. Find the number of points that are 4 units from a given line and also 6 units from a given point on that line.

1,001. An *ellipse* is defined as the locus of points such that the sum of the distances from two fixed points is a constant. The two fixed points are called the *foci*. Find the equation of the locus of points whose sum of distances from the fixed points $(-4, 1)$ and $(4, 1)$ is equal to 10.

Part II
The Answers

In this part . . .

Here you get answers and explanations for all 1,001 problems. As you read the solutions, you may realize that you need a little more help. Lucky for you, the *For Dummies* series offers several excellent resources. We recommend checking out the following titles, depending on your needs:

- ✔ *Geometry For Dummies*, by Mark Ryan
- ✔ *Geometry Workbook For Dummies,* by Mark Ryan
- ✔ *Geometry Essentials For Dummies,* by Mark Ryan

Visit www.dummies.com for more information.

Chapter 18

Answers and Explanations

• •

1. $\overline{BM}$

A *midpoint* divides a segment into two congruent segments. Therefore, $\overline{AM} \cong \overline{BM}$.

2. $\overline{ME}$

A *midpoint* divides a segment into two congruent segments. Therefore, $\overline{AE} \cong \overline{ME}$.

3. $\angle ABC$

When two lines are *perpendicular*, they form right angles at their point of intersection.

4. True

You're given that $\overline{RS} \perp \overline{TS}$. Perpendicular lines form right angles. Therefore, $\angle RST$ is a right angle.

5. False

$\angle RSQ$ and $\angle RSW$ form a linear pair but are not congruent.

6. True

$\angle QSR$ and $\angle WSR$ form a linear pair because they're two adjacent angles that form a straight angle.

7. True

You're given that $\overline{QW}$ bisects $\angle RST$. A bisector divides an angle into two congruent angles. Therefore, $\angle WST \cong \angle WSR$.

8. True

$\overline{RS} \perp \overline{TS}$, which means that $\angle RST = 90°$. $\overline{QW}$ bisects $\angle RST$, so $\angle WST \cong \angle WSR = 45°$. $\overline{QSW}$ is a straight angle, which means that $\angle RSQ$ and $\angle RSW$ form a linear pair and are supplementary angles. Therefore, $\angle RSQ = 180° - 45° = 135°$, which is an obtuse angle.

9. **False**

A *midpoint* divides a segment into two congruent segments, so $\overline{QS} \cong \overline{SW}$.

10. $\overline{AF} \cong \overline{CF}$

A *median* of a triangle is a line segment that connects the vertex of a triangle to the midpoint of the opposite side.

11. **∠ADB and ∠CDB are right angles.**

The *altitude* of a triangle is a segment that connects the vertex of a triangle perpendicular to the opposite side.

12. **∠ABE ≅ ∠CBE**

A *bisector* divides an angle into two congruent angles.

13. $\overline{AF} \cong \overline{CF}$

A *midpoint* divides a segment into two congruent segments.

14. **Right**

A *perpendicular bisector* divides a segment into two congruent segments and also forms right angles at the point of intersection.

15. **12.5**

A *midpoint* divides a segment into two congruent segments. Set the two segments equal and solve for x:

$$CE = BE$$
$$2x + 25 = 50$$
$$2x = 25$$
$$x = 12.5$$

16. **80°**

A *bisector* divides an angle into two congruent angles. Set the two angles equal and solve for x:

$$m\angle ADE = m\angle CDE$$
$$3x - 5 = x + 25$$
$$2x - 5 = 25$$
$$2x = 30$$
$$x = 15$$

After finding the value of *x*, plug it in to find the value of the desired angle:

$$m\angle ADE = 3(15) - 5 = 40°$$
$$m\angle ADC = 2(m\angle ADE) = 2(40°) - 80°$$

---

17.   22

Perpendicular lines form right angles. Set the angle equal to 90° and solve for *x*:

$$m\angle BAD = 90°$$
$$5x - 20 = 90$$
$$5x = 110$$
$$x = 22$$

---

18.   21

A *bisector* divides a segment into two congruent segments. This means that $\overline{BE} \cong \overline{CE}$. Find the sum of *BE* and *CE*, set it equal to *BC*, and solve for *x*:

$$BE + CE = BC$$
$$(x + 12) + (x + 12) = 5x - 3$$
$$2x + 24 = 5x - 3$$
$$24 = 3x - 3$$
$$27 = 3x$$
$$x = 9$$

Now plug in the value of *x* to find *BE*:

$$BE = CE = x + 12 = 9 + 12 = 21$$

---

19.   Linear pair

A *linear pair* is a pair of adjacent angles whose sum is a straight angle.

---

20.   Perpendicular

Perpendicular lines are two lines that intersect to form right angles.

---

21.   Acute

An *acute angle* is an angle measuring more than 0° but less than 90°.

---

22.   Isosceles

An *isosceles triangle* is a triangle that has two congruent angles with the sides opposite them congruent as well.

23. **Bisector**

A *bisector* is a line that divides a line segment or angle into two congruent parts.

24. **Obtuse**

An *obtuse angle* is an angle measuring more than 90° but less than 180°.

25. **Median**

A *median* is a segment connecting the vertex of a triangle to the midpoint of the opposite side.

26. **Altitude**

The *altitude* of a triangle is a line segment connecting the vertex of a triangle perpendicular to the opposite side. This represents the height of the triangle.

27. $\overline{DE}$

$\overline{DE}$ is the sum of $\overline{DR}$ and $\overline{RE}$ because Point R is between Points D and E.

28. $\overline{DW}$

$\overline{DW}$ is the sum of $\overline{DR}$ and $\overline{RW}$ because Point R is between Points D and W.

29. $\overline{DE}$

$\overline{DW} - \overline{EW} = \overline{DE}$ because $\overline{DE} + \overline{EW} = \overline{DW}$.

30. $\overline{DR}$

$\overline{DE} - \overline{RE} = \overline{DR}$ because $\overline{DR} + \overline{RE} = \overline{DE}$.

31. **Reflexive property**

The *reflexive property* states that a segment or angle is always congruent to itself.

32. $\overline{RE}$

The reflexive property states that $\overline{RE} \cong \overline{RE}$. Through the addition postulate, you obtain $\overline{DR} + \overline{RE} \cong \overline{WE} + \overline{RE}$.

33. $\overline{DE}$

The reflexive property states that $\overline{RE} \cong \overline{RE}$. Through the subtraction postulate, you obtain $\overline{RW} - \overline{RE} \cong \overline{DE} - \overline{RE}$.

34. **Substitution postulate**

The *substitution postulate* states that if two segments are equal to the same segment, then they're equal to each other.

35. **Reflexive property**

The *reflexive property* states that an angle is congruent to itself.

36. ∠NOA

The reflexive property indicates that ∠MOA ≅ ∠MOA. The addition postulate lets you conclude that ∠JOM + ∠MOA ≅ ∠NOA + ∠MOA.

37. *m*∠NOA

∠JON is the entire angle. When you subtract ∠JOA from it, the only angle remaining is ∠NOA.

38. ∠JOA ≅ ∠NOM

As long as ∠JOA ≅ ∠NOM, the subtraction postulate will state that
∠JOA − ∠MOA ≅ ∠NOM − ∠MOA.

39. ∠JON

A bisector divides an angle into two congruent angles, which makes ∠JOM ≅ ∠MON. Therefore, if you double either ∠JOM or ∠MON, you get ∠JON.

40. ∠JOM or ∠MON

A bisector divides an angle into two congruent angles, which is the same as saying that it cuts the angle in half. Therefore, $\frac{1}{2}(\angle JON)$ is congruent to ∠JOM or ∠MON.

41. ∠MEH

Intersecting lines form vertical angles, and vertical angles are congruent. This means that ∠AET ≅ ∠MEH.

42. **Vertical**

Intersecting lines form vertical angles. This also means that ∠AEM ≅ ∠TEH.

43. **180°**

Angles that form a linear pair combine to form a straight angle. A straight angle measures 180°.

44. **158°**

$\angle AET \cong \angle MEH$ because they are vertical angles. Set the angle measures equal to each other and solve for x:

$$m\angle AET = m\angle MEH$$
$$\frac{1}{2}x + 10 = x - 2$$
$$10 = \frac{1}{2}x - 2$$
$$12 = \frac{1}{2}x$$
$$24 = x$$

$\angle MEH$ and $\angle MEA$ form a linear pair, which means their sum is 180°. Use this info to solve for $m\angle MEA$:

$$m\angle MEH = 24 - 2 = 22°$$

$$m\angle MEH + m\angle MEA = 180°$$
$$22° + m\angle MEA = 180°$$
$$m\angle MEA = 158°$$

45. **20**

Start with the given information:

$$m\angle MEA = 2a$$
$$m\angle AET = 2a + b$$
$$m\angle MEH = 3a - 20$$

$\angle MEH$ and $\angle MEA$ form a linear pair, which means their sum is 180°. Set up the following equation and solve for a:

$$m\angle MEH + m\angle MEA = 180°$$
$$(3a - 20) + 2a = 180$$
$$5a - 20 = 180$$
$$5a = 200$$
$$a = 40$$

Plug in the value of a to find $m\angle MEH$:

$$m\angle MEH = 3a - 20 = 3(40) - 20 = 100°$$

$\angle AET \cong \angle MEH$ because they're vertical angles. Set them equal to each other, plug in the value of a, and solve for b:

$$m\angle AET = m\angle MEH$$
$$2a + b = 100°$$
$$2(40) + b = 100$$
$$80 + b = 100$$
$$b = 20$$

46.

Adjacent

Adjacent angles are angles that have a vertex and a side in common.

47.

124°

Angles that form a linear pair add up to 180°. Set the sum of $m\angle HET$ and $m\angle TEA$ equal to 180° and solve for x:

$$m\angle HET + m\angle TEA = 180°$$
$$2x + (5x - 16) = 180°$$
$$7x - 16 = 180°$$
$$7x = 196°$$
$$x = 28°$$

Now plug in the value of x to solve for $m\angle TEA$:

$$m\angle TEA = 5x - 16 = 5(28) - 16 = 124°$$

48.

40°

Complementary angles are two angles whose sum is 90°:

$$m\angle A + m\angle C = 90°$$
$$50° + m\angle C = 90°$$
$$m\angle C = 40°$$

49.

60°

Supplementary angles are two angles whose sum is 180°:

$$m\angle B + m\angle D = 180°$$
$$120° + m\angle D = 180°$$
$$m\angle D = 60°$$

50. 45°

Complementary angles are two angles whose sum is 90°:

$$x + x = 90°$$
$$2x = 90°$$
$$x = 45°$$

51. Right

Supplementary angles are two angles whose sum is 180°:

$$x + x = 180°$$
$$2x = 180°$$
$$x = 90°$$

52. 108°

Supplementary angles are two angles whose sum is 180°. When you're given angles as a ratio, place an x next to the numbers in the ratio to represent each angle:

$$2x + 3x = 180°$$
$$5x = 180°$$
$$x = 36°$$

Because the larger angle is represented by $3x$, the larger angle is $3(36°) = 108°$.

53. 70°

Supplementary angles are two angles whose sum is 180°. Let x = one angle, and let $x + 40$ = another angle. Then set their sum equal to 180° and solve for x:

$$x + (x + 40) = 180°$$
$$2x + 40 = 180°$$
$$2x = 140°$$
$$x = 70°$$

The smaller angle is 70°.

54. 30°

Complementary angles are two angles whose sum is 90°. Let x = one angle, and let $2x$ = the other angle. Set their sum equal to 90° and solve for x:

$$x + 2x = 90°$$
$$3x = 90°$$
$$x = 30°$$

55. 58°

Complementary angles are two angles whose sum is 90°. Let $x =$ one angle, and let $2x - 6 =$ the other angle. Set their sum equal to 90° and solve for x:

$$x + 2x - 6 = 90°$$
$$3x - 6 = 90°$$
$$3x = 96°$$
$$x = 32°$$

The other angle is

$$2x - 6 = 2(32°) - 6 = 58°$$

56. 180°

Angles that form a linear pair are supplementary angles. Their sum is 180°.

57. 40°

Complementary angles are two angles whose sum is 90°. When you're given angles as a ratio, place an x next to the numbers in the ratio to represent each angle. Then solve for x:

$$5x + 4x = 90°$$
$$9x = 90°$$
$$x = 10°$$

Because $4x$ represents the smaller angle, the value of the smaller angle is $4(10°) = 40°$.

58. 35

You first need to find $m\angle VEI$. The sum of the three angles in $\triangle VIE$ is 180°:

$$50° + 70° + m\angle VEI = 180°$$
$$120° + m\angle VEI = 180°$$
$$m\angle VEI = 60°$$

Intersecting lines form vertical angles, and vertical angles are congruent. This means that

$$m\angle VEI = m\angle RET$$
$$60 = 2x - 10$$
$$70 = 2x$$
$$35 = x$$

59. 120°

You first need to find the value of $m\angle VEI$. The sum of the three angles in $\triangle VIE$ is 180°:

$$50° + 70° + m\angle VEI = 180°$$
$$120° + m\angle VEI = 180°$$
$$m\angle VEI = 60°$$

$\angle VEI$ and $\angle TEI$ form a linear pair, which means their sum is 180°:

$$60° + m\angle TEI = 180°$$
$$m\angle TEI = 120°$$

60. 40

$\angle VEI$ and $\angle RET$ are vertical angles, which means they're congruent:

$$2a + b = a - b$$
$$a + b = -b$$
$$a = -2b$$

$\angle VER$ and $\angle VEI$ form a linear pair, which means their sum is 180°:

$$4a + 2b + 2a + b = 180°$$
$$6a + 3b = 180°$$

Your previous work allows you to substitute –2b in for a:

$$6(-2b) + 3b = 180°$$
$$-12b + 3b = 180°$$
$$-9b = 180°$$
$$b = -20$$

Because $a = -2b$,

$$a = -2(-20) = 40$$

61.

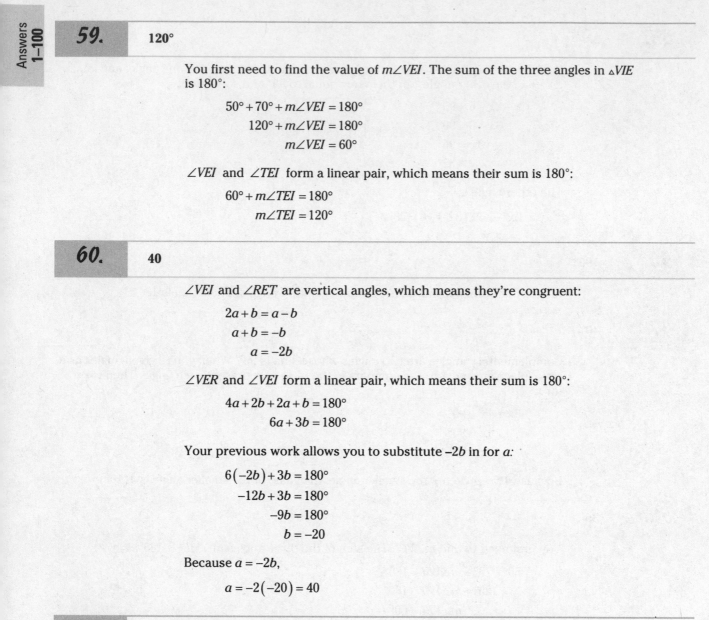

Use a straight edge to draw a ray with endpoint C. Place the compass point at A and measure the length of $\overline{AB}$. Keeping that measurement, place the compass point at C and make an arc. Label that point D.

62.

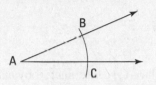

Use a straight edge to draw a ray with endpoint *D*. Place the compass point on *A*, and with any width, draw an arc intersecting the angle at two points.

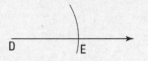

Using the same width, place the compass point at *D* and make an arc.

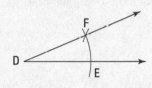

Using the compass, measure the distance between *B* and *C*. Keeping that compass width, place the compass point at *E* and draw an arc. Connect the point where the arcs intersect to *D*.

63.

Place the compass point at *B* and measure the length of $\overline{BC}$. Draw Point *D* on your paper. Keeping the length of $\overline{BC}$, place the compass point on *D* and draw an arc. Place a point on the arc and label it *E*.

• D   • E

Use your compass to measure the length of $\overline{AB}$. Keeping that compass width, place the compass point at *D* and draw an arc where the third vertex would be located.

• D   • E

Use your compass to measure the length of $\overline{AC}$. Keeping that compass width, place the compass point at *E* and draw an arc where the third vertex would be. Name the point where the arcs intersect *F*. Connect the three vertices of the triangle.

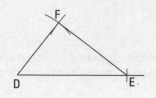

64.

Copy of an angle

A copy of an angle would show a large arc intersected by a small arc that shows the width of the angle.

65.

Place the compass point at *S* and measure the length of $\overline{ST}$. Draw Point *A* on your paper. Keeping the length of $\overline{ST}$, place the compass point on A and draw an arc. Place a point on the arc and label it *B*.

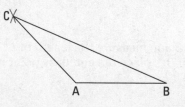

Use your compass to measure the length of $\overline{RS}$. Keeping that compass width, place the compass point at *A* and draw an arc where the third vertex would be located. Use your compass to measure the length of $\overline{RT}$. Keeping that compass width, place the compass point at *B* and draw an arc where the third vertex would be. Name the point where the arcs intersect *C*. Connect the three vertices of the triangle.

66.

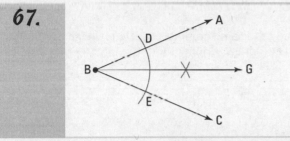

Use a straight edge to draw a ray with endpoint *C*. Place the compass point at *A* and measure the length of $\overline{AB}$. Keeping that measurement, place the compass point at *C* and make an arc. Place the compass point where the arc intersects the ray and draw another arc. Label that point of intersection *D*.

67.

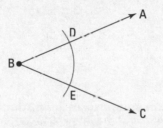

Place the compass point at *B*, and using any width, make an arc through the angle. Label the points of intersection *D* and *E*.

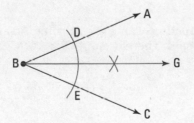

Place the compass point at *D*, and using the same compass width, make an arc in the interior of the angle. Repeat this step at Point *E*. Connect *B* to the intersecting arcs.

68.

Place the compass point at *E*, and using any width, make an arc through the angle.

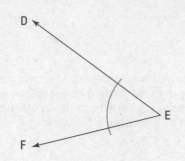

Place the compass point at *A*, and using the same compass width, make an arc in the interior of the angle. Repeat this step at Point *B*. Connect *E* to the intersecting arcs.

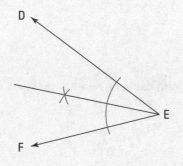

---

69. **Angle bisector**

A construction of an angle bisector shows an arc going through the angle and then two small arcs intersecting in the interior of the angle. The point of intersection connects to the vertex angle.

---

70. **True**

The figure shows the construction of an angle bisector. An angle bisector divides an angle into two congruent angles. Therefore, $\angle ABG \cong \angle GBC$.

71.

Place Point *E* anywhere on $\overline{AB}$. Place the compass point on *E*, and using any width, draw two arcs through $\overline{AB}$.

Place the compass point at both locations where the arcs intersect $\overline{AB}$. At each location, make an arc above E. Label the intersection of those arcs D and connect D to E.

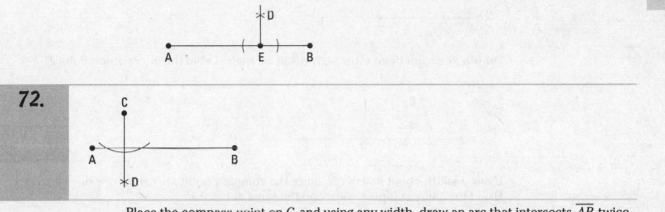

72.

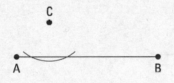

Place the compass point on C, and using any width, draw an arc that intersects $\overline{AB}$ twice.

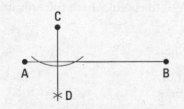

Place the compass point at both locations where the arcs intersect $\overline{AB}$. At each location, make an arc below $\overline{AB}$. Label the intersection of those arcs D and connect D to C.

73.

Place the compass point on A. Open the width of the compass to a little more than halfway through the line. Draw arcs above and below $\overline{AB}$. Using the same compass width, place the compass point at B and draw arcs above and below $\overline{AB}$. Connect the intersection points of both pairs of arcs.

74. **A perpendicular line**

The arcs on and below the line show the construction of a perpendicular line.

75.

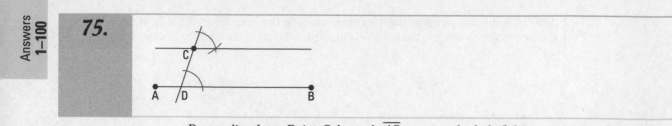

Draw a line from Point *C* through $\overline{AB}$ at an angle. Label the intersection Point *D*.

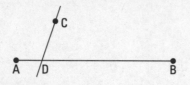

Using a width about half of $\overline{CD}$, place the compass point at *D* and draw one large arc. Using the same compass width, do the same at Point *C*.

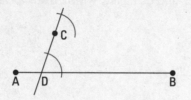

Measure the width of the first arc. Using that same width, place the compass point at the location above point *C* where the arc intersects $\overline{CD}$ and draw an arc intersecting the arc near point *C*. Connect *C* to the point where the arcs intersect.

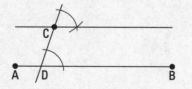

76. **False**

The figure shows a line perpendicular to $\overline{AB}$. Perpendicular lines form right angles, not congruent line segments.

77. **False**

A true construction of a line parallel to $\overline{AB}$ would require a line drawn at an angle through $\overline{AB}$ with the proper arcs drawn, as shown in the following figure.

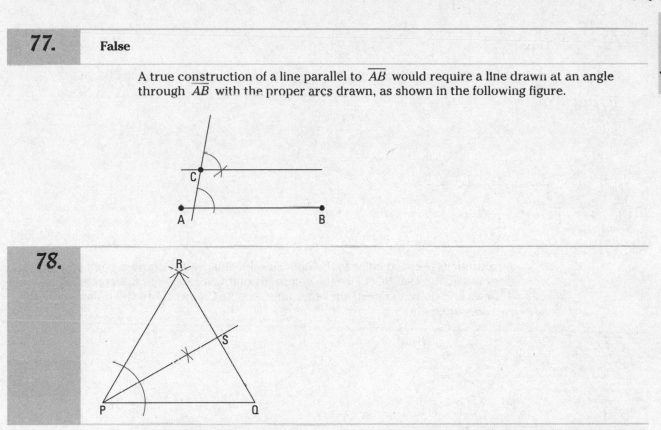

78.

To construct a 30° angle, you first need to construct an equilateral triangle, which will have three 60° angles. To do this, draw a line segment and label it $\overline{PQ}$. Using your compass, measure the length of $\overline{PQ}$. Without changing the width of the compass, place your compass at Point P and draw an arc above $\overline{PQ}$. Repeat this step with your compass at Point Q. The intersection of these arcs is the third vertex of the equilateral triangle. Label that point R and connect the points.

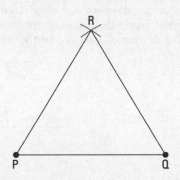

You now have three 60° angles. You can bisect any one of those angles to create a 30° angle. Put your compass point at P and draw an arc through the angle. Place the compass point at both locations where the arc intersects the angle and draw an arc each time. Label the intersection of the arcs S. Connect Point S to Point P. $\angle RPS$ and angle $\angle QPS$ are both 30° angles.

79.   **True**

The first step in constructing a 45° angle is to construct a perpendicular line.

80.

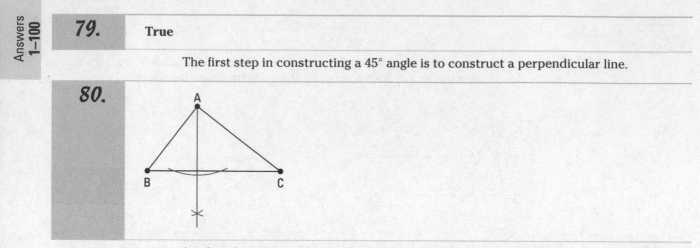

An altitude is perpendicular to the opposite side. Place the compass point at A and draw an arc through $\overline{BC}$. Place the compass point where the arc intersects $\overline{BC}$ and draw an arc; do the same at the other intersection. Connect A to the point where the two arcs intersect.

81.

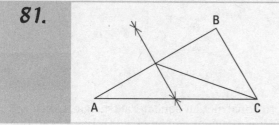

The *median* connects the vertex of a triangle to the midpoint of the opposite side. To construct the median, you need to find the midpoint by constructing the perpendicular bisector of $\overline{AB}$.

Place the compass point on A. Open the width of the compass to a little more than halfway through the line. Draw an arc above and below $\overline{AB}$. Using the same compass width, place the compass point at B and draw an arc above and below $\overline{AB}$. Connect the intersection points of both pairs of arcs. Connect C to the midpoint of $\overline{AB}$.

82.

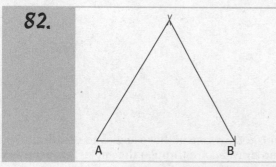

Use the compass to measure the length of $\overline{AB}$. Place the compass at A and draw an arc where the third angle should be; then place the compass at B and draw an arc where the third angle should be. Connect A and B to the intersection of the arcs.

83.

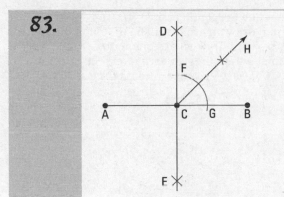

Draw a line segment to be used as a part of the angle; call it $\overline{AB}$. You first need to create a right angle. To do so, draw a perpendicular bisector to $\overline{AB}$. Place the compass point on A and open the width of the compass a little more than halfway through the line. Draw arcs above and below $\overline{AB}$. Using the same compass width, place the compass point at B and draw arcs above and below $\overline{AB}$. Connect the intersection points of both pairs of arcs. Let C be the point where the perpendicular bisector intersects $\overline{AB}$. This means that $\angle DCB$ is a 90° angle.

To create a 45° angle, you need to bisect $\angle DCB$. To do so, place your compass point at C and draw an arc through $\angle DCB$. Using the same compass width, place your compass point at F and draw an arc in the interior of $\angle DCB$. Do the same at Point G. Connect Point C to the intersection of the two arcs. Both $\angle DCH$ and $\angle HCB$ are 45° angles.

84.

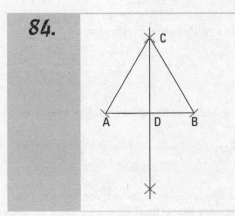

To construct a 30°-60°-90° triangle, you first need to construct an equilateral triangle, where each angle measures 60°.

Draw a segment to be used as a side of the triangle. Call it $\overline{AB}$. Measure the length of $\overline{AB}$ with your compass. Using that compass width, place your compass point at A and draw an arc above $\overline{AB}$. Do the same at Point B. Connect Points A and B to the point where the arcs intersect, which you can call Point C. A, B, and C are now the vertices of an equilateral triangle, which means all three angles of the triangle measure 60°.

To create the 90° angle, you need to draw a line perpendicular to $\overline{AB}$ from Point *C*. To do so, place your compass point at C and draw an arc through $\overline{AB}$. Using the same compass width, place the compass point at *A* and draw an arc below $\overline{AB}$. Do the same thing at Point *B*. Connect Point *C* to the point where the two arcs intersect. △*CDA* is a right triangle where ∠*CDA* is a 90° angle, ∠*CAD* is a 60° angle, and ∠*DCA* is a 30° angle.

85.

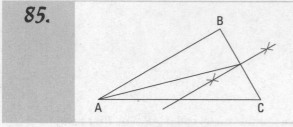

The *median* connects the vertex of a triangle to the midpoint of the opposite side. To construct the median, you need to find the midpoint by constructing the perpendicular bisector of $\overline{BC}$.

Place the compass point on *B*. Open the width of the compass to a little more than halfway through the line. Draw arcs above and below $\overline{BC}$. Using the same compass width, place the compass point at *C* and draw arcs above and below $\overline{BC}$. Connect the intersection points of both pairs of arcs. Connect *A* to the midpoint of $\overline{BC}$.

86. **AAS**

If two angles and the non-included side of one triangle are congruent to two angles and the non-included side of another triangle, then the triangles are congruent by AAS (angle-angle-side).

87. **SSS**

If three sides of one triangle are congruent to three sides of another triangle, then the triangles are congruent by SSS (side-side-side).

88. **ASA**

If two angles and the included side of one triangle are congruent to two angles and the included side of another triangle, then the triangles are congruent by ASA (angle-side-angle).

89. **SAS**

If two sides and the included angle of one triangle are congruent to two sides and the included angle of another triangle, then the triangles are congruent by SAS (side-angle-side).

90. $\overline{AC} \cong \overline{DF}$

If three sides of one triangle are congruent to three sides of another triangle, then the triangles are congruent by SSS (side-side-side).

91. $\angle A \cong \angle D$

If two sides and the included angle of one triangle are congruent to two sides and the included angle of another triangle, then the triangles are congruent by SAS (side-angle-side).

92. $\angle F \cong \angle C$

If two angles and the non-included side of one triangle are congruent to two angles and the non-included side of another triangle, then the triangles are congruent by AAS (angle-angle-side).

93. $\angle B \cong \angle D$

If two angles and the included side of one triangle are congruent to two angles and the included side of another triangle, then the triangles are congruent by ASA (angle-side-angle).

94. $\overline{BE} \cong \overline{CE}$

If three sides of one triangle are congruent to three sides of another triangle, then the triangles are congruent by SSS (side-side-side).

95. $\angle B \cong \angle E$

If two sides and the included angle of one triangle are congruent to two sides and the included angle of another triangle, then the triangles are congruent by SAS (side angle side).

96. **Proving these two triangles congruent isn't possible.**

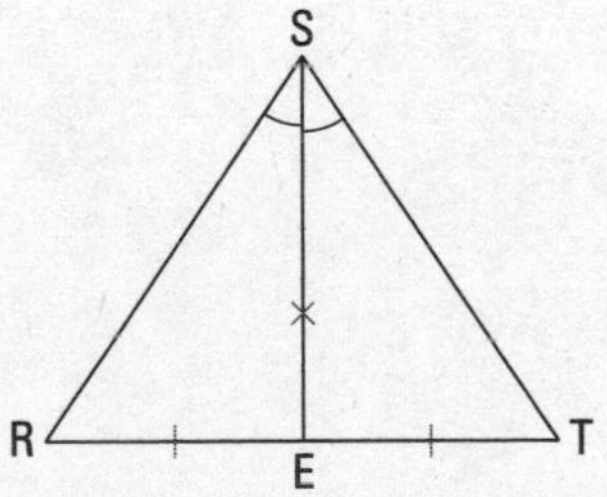

The markings in the figure show SSA (side-side-angle), which is not a method of proving triangle congruence.

97. ASA

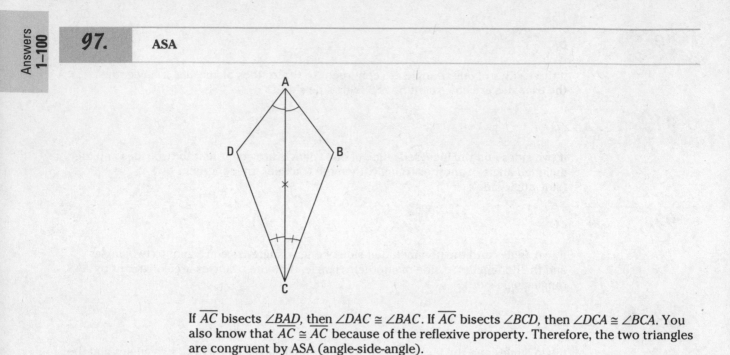

If $\overline{AC}$ bisects $\angle BAD$, then $\angle DAC \cong \angle BAC$. If $\overline{AC}$ bisects $\angle BCD$, then $\angle DCA \cong \angle BCA$. You also know that $\overline{AC} \cong \overline{AC}$ because of the reflexive property. Therefore, the two triangles are congruent by ASA (angle-side-angle).

98. SSS

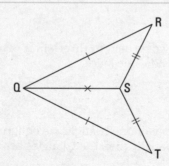

$\overline{QS} \cong \overline{QS}$ by the reflexive property. The two givens show the other two sides of the triangles are congruent to each other. Therefore, the two triangles are congruent by SSS (side-side-side).

99. ASA

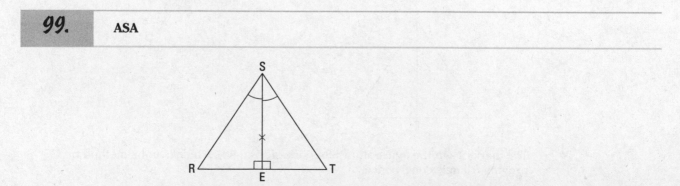

If $\overline{SE}$ is an altitude, then $\angle RES \cong \angle TES$ because they're both right angles. If $\overline{SE}$ bisects $\angle RST$, then $\angle RSE \cong \angle TSE$. $\overline{SE} \cong \overline{SE}$ by the reflexive property. Therefore, the two triangles are congruent by ASA (angle-side-angle).

100.   **ASA or AAS**

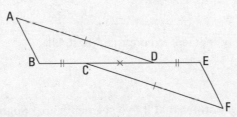

If $\overline{AB} \parallel \overline{DE}$, then $\angle B \cong \angle D$ and $\angle A \cong \angle E$ because alternate interior angles are congruent to each other. Using the given $\overline{AB} \cong \overline{DE}$, the triangles are congruent by ASA (angle-side-angle). However, $\angle ACB \cong \angle ECD$ because vertical angles are congruent. In this case, you can also use AAS (angle-angle-side) to prove that the triangles are congruent to each other.

101.   **SAS**

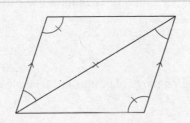

Apply the addition postulate to the second given, which is $\overline{BC} \cong \overline{ED}$. Add $\overline{CD}$ to both segments, and the result is $\overline{BD} \cong \overline{EC}$. Use the other two givens, and the triangles are congruent by SAS (side-angle-side).

102.   **AAS**

If $\overline{AD} \parallel \overline{BC}$, then $\angle ADB \cong \angle CBD$ because alternate interior angles are congruent to each other. $\overline{BD} \cong \overline{BD}$ by the reflexive property. Use the other given, and the triangles are congruent by AAS (angle-angle-side).

103.

A bisector divides a segment into two congruent segments.

A *bisector* divides a segment or angle into two congruent parts, so $\overline{AB} \cong \overline{BD}$ and $\overline{EB} \cong \overline{BC}$.

104.

∠ABE and ∠DBC are vertical angles.

Intersecting lines form vertical angles.

105.

If two angles are vertical angles, then they're congruent.

Vertical angles are congruent, so $\angle ABE \cong \angle DBC$.

106.

SAS

If two sides and the included angle of one triangle are congruent to two sides and the included angle of another triangle, then the triangles are congruent by SAS (side-angle-side). Therefore, $\triangle ABE \cong \triangle DBC$.

107.

CPCTC

Corresponding parts of congruent triangles are congruent to each other, so $\overline{AE} \cong \overline{DC}$.

108.

Reflexive property

$\overline{CD} \cong \overline{CD}$ because any segment is congruent to itself.

109.

Addition postulate

The *addition postulate* states that if two segments are congruent to two other segments, then the sums of the segments are also congruent to each other. Therefore, $\overline{BC} + \overline{CD} \cong \overline{CD} + \overline{DE}$.

110.

SAS

If two sides and the included angle of one triangle are congruent to two sides and the included angle of another triangle, then the triangles are congruent by SAS (side-angle-side). Therefore, $\triangle ABD \cong \triangle FCE$.

111.

CPCTC

$\angle A \cong \angle F$ because corresponding parts of congruent triangles are congruent to each other.

112. $\angle ABD$ and $\angle CDB$ are alternate interior angles.

If two parallel lines are cut by a transversal, then alternate interior angles are formed.

113. **If two angles are alternate interior angles, then they're congruent.**

$\angle ABD \cong \angle CDB$ because alternate interior angles are congruent.

114. $\overline{BD} \cong \overline{BD}$

The reflexive property says that a segment is congruent to itself.

115. **ASA**

If two angles and the included side of one triangle are congruent to two angles and the included side of another triangle, then the triangles are congruent by ASA (angle-side-angle). Therefore, $\triangle ABD \cong \triangle CDB$.

116. **CPCTC**

$\overline{AB} \cong \overline{CD}$ because corresponding parts of congruent triangles are congruent to each other.

117. **Here is one way to complete the proof. Other answers are possible.**

Statements	Reasons
1. $\overline{LP} \parallel \overline{ON}$, and M is the midpoint of $\overline{LO}$.	1. Given.
2. $\angle L$ and $\angle O$, $\angle P$ and $\angle N$ are alternate interior angles.	2. When two parallel lines are cut by a transversal, alternate interior angles are formed.
3. $\angle L \cong \angle O$ and $\angle P \cong \angle N$	3. If two angles are alternate interior angles, then they're congruent.
4. $\overline{LM} \cong \overline{OM}$	4. A midpoint divides a segment into two congruent parts.
5. $\triangle PLM \cong \triangle NOM$	5. AAS
6. $\overline{PM} \cong \overline{NM}$	6. CPCTC

118. **Reflexive property**

An angle is congruent to itself, so $\angle RTL \cong \angle MTS$.

119. ASA

If two angles and the included side of one triangle are congruent to two angles and the included side of another triangle, then the triangles are congruent by ASA (angle-side-angle). Therefore, $\triangle RTL \cong \triangle MTS$.

120. CPCTC

$\overline{TL} \cong \overline{TS}$ because corresponding parts of congruent triangles are congruent to each other.

121. ∠*BAE* and ∠*DEA* are right angles.

Perpendicular lines form right angles.

122. If two angles are right angles, then they're congruent.

$\angle BAE \cong \angle DEA$ because they're both right angles.

123. Reflexive property

A segment is congruent to itself, so $\overline{AE} \cong \overline{AE}$.

124. ASA

If two angles and the included side of one triangle are congruent to two angles and the included side of another triangle, then the triangles are congruent by ASA (angle-side-angle). Therefore, $\triangle ADE \cong \triangle EBA$.

125. CPCTC

$\angle ABE \cong \angle EDA$ because corresponding parts of congruent triangles are congruent to each other.

126. Reflexive property

$\overline{ED} \cong \overline{ED}$ because a segment is congruent to itself.

127. Addition postulate

The addition postulate states that if two segments are congruent to two other segments, then the sums of the segments are also congruent to each other. Therefore, $\overline{AE} + \overline{ED} \cong \overline{CD} + \overline{ED}$, or $\overline{AD} \cong \overline{CE}$.

128. If two sides of a triangle are congruent, the angles opposite those sides are also congruent.

$\angle BDE \cong \angle BED$ because they're the angles opposite the congruent sides $\overline{BE}$ and $\overline{BD}$.

129. **SAS**

If two sides and the included angle of one triangle are congruent to two sides and the included angle of another triangle, then the triangles are congruent by SAS (side-angle-side). Therefore, $\triangle BDA \cong \triangle BEC$.

130. **CPCTC**

$\overline{AB} \cong \overline{CB}$ because corresponding parts of congruent triangles are congruent to each other.

131. **Reflexive property**

$\overline{SN} \cong \overline{SN}$ because a segment is congruent to itself.

132. **Addition postulate**

The addition postulate states that if two segments are congruent to two other segments, then the sums of the segments are also congruent to each other. Therefore, $\overline{LN} + \overline{NS} \cong \overline{RS} + \overline{SN}$, or $\overline{LS} \cong \overline{RN}$.

133. **If two angles form a linear pair, then they're supplementary.**

$\angle LNE$ and $\angle ENR$ form a linear pair, as do $\angle TSR$ and $\angle TSL$.

134. **If two angles are congruent, their supplements are congruent.**

$\angle TSL \cong \angle ENR$ because their supplements are congruent.

135. **ASA**

If two angles and the included side of one triangle are congruent to two angles and the included side of another triangle, then the triangles are congruent by ASA (angle-side-angle). Therefore, $\triangle TSL \cong \triangle ENR$.

136. **CPCTC**

$\overline{TL} \cong \overline{ER}$ because corresponding parts of congruent triangles are congruent to each other.

137. **$\overline{QS}$ and $\overline{PT}$ bisect each other.**

In an indirect proof, assume the opposite of what needs to be proven is true.

138. **A bisector divides a segment into two congruent segments.**

139. **∠*PRQ* and ∠*SRT* are vertical angles.**

Intersecting lines form vertical angles.

140. **If two angles are vertical angles, then they're congruent.**

141. **SAS**

If two sides and the included angle of one triangle are congruent to two sides and the included angle of another triangle, then the triangles are congruent by SAS (side-angle-side).

142. **CPCTC**

Corresponding parts of congruent triangles are congruent to each other.

143. **Contradiction**

Step 7 contradicts Step 1; therefore, the opposite must be true.

144. **∠*DCB* ≅ ∠*AEB***

In an indirect proof, assume that the opposite of what needs to be proven is true.

145. **Perpendicular lines form right angles.**

146. **If two angles are right angles, then they're congruent.**

147. **ASA**

If two angles and the included side of one triangle are congruent to two angles and the included side of another triangle, then the triangles are congruent by ASA (angle-side-angle).

148. **CPCTC**

Corresponding parts of congruent triangles are congruent to each other.

149. **Contradiction**

Statements 1 and 6 lead to a contradiction; therefore, the opposite must be true.

150.

Statements	Reasons
1. $\overline{AC} \perp \overline{DB}$, $\overline{AB} \not\cong \overline{AD}$	1. Given
2. $\overline{BC} \cong \overline{DC}$	2. Assumption
3. $\angle DCA$ and $\angle BCA$ are right angles.	3. Perpendicular lines form right angles.
4. $\angle DCA \cong \angle BCA$	4. All right angles are congruent.
5. $\overline{AC} \cong \overline{AC}$	5. Reflexive property
6. $\triangle DCA \cong \triangle BCA$	6. SAS
7. $\overline{AB} \cong \overline{AD}$	7. CPCTC
8. $\overline{BC} \not\cong \overline{DC}$	8. Contradiction in Statements 1 and 7; therefore, the opposite of the assumption is true.

151. **Isosceles**

An *isosceles triangle* is a triangle with two congruent sides.

152. **Scalene**

A *scalene triangle* is a triangle with all three sides measuring a different length.

153. **Equilateral**

An *equilateral triangle* is a triangle that has all three sides congruent.

154. **Isosceles right**

The triangle is isosceles because it has two congruent sides. The triangle is a right triangle because its sides satisfy the Pythagorean theorem:

$$a^2 + b^2 = c^2$$
$$6^2 + 6^2 = \left(6\sqrt{2}\right)^2$$
$$36 + 36 = 72$$
$$72 = 72$$

155. **Scalene right**

The triangle is a right triangle because it satisfies the Pythagorean theorem:

$$a^2 + b^2 = c^2$$
$$7^2 + \left(7\sqrt{3}\right)^2 = 14^2$$
$$49 + 147 = 196$$
$$196 = 196$$

It is also scalene because each side measures a different length.

156. **Equilateral**

An *equilateral triangle* is a triangle that has all three sides congruent.

157. **Isosceles**

An *isosceles triangle* is a triangle that has two sides congruent. Although it doesn't look like two sides of this triangle are congruent, $\sqrt[3]{8} = 2$. Therefore, two sides measure 2 units in length, making the triangle isosceles.

158. **Equilateral**

The given perimeter means that all three sides of the triangle add up to 108:

$$(5x + 11) + (7x + 1) + (8x - 4) = 108$$
$$5x + 11 + 7x + 1 + 8x - 4 = 108$$
$$20x + 8 = 108$$
$$20x = 100$$
$$x = 5$$

Find the length of each side of the triangle by plugging in 5 for x:

$$PE = 5(5) + 11 = 36$$
$$ER = 7(5) + 1 = 36$$
$$RP = 8(5) - 4 = 36$$

Because all three sides of the triangle are equal, the triangle is equilateral.

159. **Isosceles**

The given perimeter means that all three sides of the triangle add up to 210:

$$(x + 80) + (2x + 60) + (x - 10) = 210$$
$$x + 80 + 2x + 60 + x - 10 = 210$$
$$4x + 130 = 210$$
$$4x = 80$$
$$x = 20$$

Find the length of each side of the triangle by plugging in 20 for x:

$$PE = 20 + 80 = 100$$
$$ER = 2(20) + 60 = 100$$
$$RP = 20 - 10 = 10$$

Because two sides of the triangle are congruent, the triangle is isosceles.

160. Scalene right

The given perimeter means that all three sides of the triangle add up to 60:

$$(2a+8)+(2a-6)+(3a+2)=60$$
$$2a+8+2a-6+3a+2=60$$
$$7a+4=60$$
$$7a=56$$
$$a=8$$

Find the length of each side of the triangle by plugging in 8 for a:

$$PE=2(8)+8=24$$
$$ER=2(8)-6=10$$
$$RP=3(8)+2=26$$

The three sides of the triangle are all different in length, so the triangle is scalene.

Because the three sides of the triangle satisfy the Pythagorean theorem, the triangle is also a right triangle:

$$a^2+b^2=c^2$$
$$10^2+24^2=26^2$$
$$100+576=676$$
$$676=676$$

161. Scalene right

The given perimeter means that all three sides of the triangle add up to 24:

$$(x)+(x+2)+(2x-2)=24$$
$$x+x+2+2x-2=24$$
$$4x=24$$
$$x=6$$

Find the length of each side of the triangle by plugging in 6 for x:

$$PE=6$$
$$ER=6+2=8$$
$$RP=2(6)-2=10$$

The three sides of the triangle are all different in length, so the triangle is scalene.

Because the three sides of the triangle satisfy the Pythagorean theorem, the triangle is also a right triangle:

$$a^2+b^2=c^2$$
$$6^2+8^2=10^2$$
$$36+64=100$$
$$100=100$$

162. **Scalene**

The given perimeter means that all three sides of the triangle add up to 34:

$$(x)+(4x)+(3x+2)=34$$
$$x+4x+3x+2=34$$
$$8x+2=34$$
$$8x=32$$
$$x=4$$

Find the length of each side of the triangle by plugging in 4 for x:

$$PE=4$$
$$ER=4(4)=16$$
$$RP=3(4)+2=14$$

Because no sides of the triangle are equal in length, the triangle is scalene.

163. **Isosceles**

The given perimeter means that all three sides of the triangle add up to $19\sqrt{2}$:

$$PE=\sqrt{50}=\sqrt{25}\sqrt{2}=5\sqrt{2}$$
$$ER=9\sqrt{2}$$

$$5\sqrt{2}+9\sqrt{2}+x=19\sqrt{2}$$
$$14\sqrt{2}+x=19\sqrt{2}$$
$$x=5\sqrt{2}$$

Because two sides of the triangle are equal in length, the triangle is isosceles.

164. **Isosceles**

The given perimeter means that all three sides of the triangle add up to 42:

$$(x^2)+(4x)+(3x-2)=42$$
$$x^2+4x+3x-2=42$$
$$x^2+7x-2=42$$
$$x^2+7x-44=0$$
$$(x+11)(x-4)=0$$
$$\cancel{x=-11} \text{ or } x=4$$

Find the length of each side of the triangle by plugging in 4 for x:

$$PE=4^2=16$$
$$ER=4(4)=16$$
$$RP=3(4)-2=10$$

Because two sides of the triangle are equal in length, the triangle is isosceles.

165. 14

An *isosceles triangle* is a triangle that has two congruent sides. In this isosceles triangle, because E is the vertex, $DE = FE$. Set the two sides equal to each other to determine the value of x:

$$2x + 10 = 3x - 4$$
$$10 = x - 4$$
$$14 = x$$

166. 46

An *equilateral triangle* is a triangle that has all three sides congruent. Set any two sides of the triangle equal to each other to determine the value of x:

$$2.5x - 14 = x + 22$$
$$1.5x - 14 = 22$$
$$1.5x = 36$$
$$x = 24$$

To find the length of $\overline{LI}$, plug 24 in for x in either of the expressions:

$$LN = 24 + 22 = 46$$

167. 9

Using the given relationships, let

$$HP = x$$
$$PY = x + 3$$
$$HY = 2x - 3$$

A right triangle is a triangle whose sides satisfy the Pythagorean theorem. Set up the Pythagorean theorem and solve for x:

$$a^2 + b^2 = c^2$$
$$(x)^2 + (x+3)^2 = (2x-3)^2$$
$$x^2 + x^2 + 6x + 9 = 4x^2 - 12x + 9$$
$$6x + 9 = 2x^2 - 12x + 9$$
$$6x = 2x^2 - 12x$$
$$0 = 2x^2 - 18x$$
$$0 = 2x(x - 9)$$
$$\cancel{x = 0} \text{ or } x = 9$$

Therefore, $HP = x = 9$.

168. 4

An *isosceles triangle* is a triangle that has two congruent sides. In this isosceles triangle, because H is the vertex, $SH = HE$. Set the two sides equal to each other to determine the value of x:

$$x^2 = x + 12$$
$$x^2 - x - 12 = 0$$
$$(x+3)(x-4) = 0$$
$$\cancel{x = -3} \text{ or } x = 4$$

169. 4

An *equilateral triangle* is a triangle that has all three sides congruent. Set any two sides of the triangle equal to each other to determine the value of x:

$$x^3 = 64$$
$$\sqrt[3]{x^3} = \sqrt[3]{64}$$
$$x = 4$$

170. Right

The sum of the three angles in a triangle is 180°:

$$40° + 50° + x = 180°$$
$$90° + x = 180°$$
$$x = 90°$$

Because the largest angle in the triangle is 90°, the triangle is a right triangle.

171. Equiangular

The sum of the three angles in a triangle is 180°:

$$60° + 60° + x = 180°$$
$$120° + x = 180°$$
$$x = 60°$$

Because all three angles are equal, the triangle is equiangular.

172. Acute

The sum of the three angles in a triangle is 180°:

$$50° + 70° + x = 180°$$
$$120° + x = 180°$$
$$x = 60°$$

Because the largest angle in the triangle is less than 90° and no angles are equal, the triangle is acute.

173. **Obtuse**

The sum of the three angles in a triangle is 180°:

$$20° + 60° + x = 180°$$
$$80° + x = 180°$$
$$x = 100°$$

Because the largest angle is greater than 90°, the triangle is obtuse.

174. **Right**

The sum of the three angles in a triangle is 180°:

$$10° + 80° + x = 180°$$
$$90° + x = 180°$$
$$x = 90°$$

Because the largest angle in the triangle is 90°, the triangle is right.

175. **Equiangular**

The sum of the three angles in a triangle is 180°.

$$(2x) + (x + 30) + (3x - 30) = 180°$$
$$2x + x + 30 + 3x - 30 = 180°$$
$$6x = 180°$$
$$x = 30°$$

To find the value of each angle, plug 30° in for x:

$$2(30°) = 60°$$
$$(30°) + 30° = 60°$$
$$3(30°) - 30° = 60°$$

Because all three angles of the triangle are equal, the triangle is equiangular.

176. **Right**

The sum of the three angles in a triangle is 180°:

$$(a) + (b) + (a + b) = 180°$$
$$2a + 2b = 180°$$
$$2(a + b) = 180°$$
$$a + b = 90°$$

Because $a + b$ is an angle in the triangle and $a + b = 90°$, the triangle is a right triangle.

177. **Acute**

Write down the given angle measures:

$$m\angle XWY = 50°$$
$$m\angle WYX = x$$
$$m\angle WXY = x + 20°$$

The sum of the three angles of a triangle is 180°:

$$(50) + (x) + (x + 20) = 180°$$
$$50 + x + x + 20 = 180°$$
$$2x + 70 = 180°$$
$$2x = 110°$$
$$x = 55°$$

Find the measure of each angle by plugging in 55° for x:

$$m\angle XWY = 50°$$
$$m\angle WYX = x = 55°$$
$$m\angle WXY = x + 20° = 55° + 20° = 75°$$

Because the largest angle of this triangle is less than 90°, this triangle is acute.

178. **Equiangular**

Write down the given angle measures:

$$m\angle WXY = 60°$$
$$m\angle XWY = 2x$$
$$m\angle XYW = x + 30$$

The sum of the three angles of a triangle is 180°:

$$(60) + (2x) + (x + 30) = 180°$$
$$60 + 2x + x + 30 = 180°$$
$$3x + 90 = 180°$$
$$3x = 90°$$
$$x = 30°$$

Find the measure of each angle by plugging in 30° for x:

$$m\angle WXY = 60°$$
$$m\angle XWY = 2x = 2(30°) = 60°$$
$$m\angle XYW = x + 30° = 30° + 30° = 60°$$

Because all three angles in this triangle equal 60°, this triangle is equiangular.

179.   **Cannot be determined**

Write down the given angle measures:

$$m\angle XYW = x$$
$$m\angle XYZ = 2x$$

$\angle XYZ$ and $m\angle XYW$ form a linear pair, which means together they add up to 180°:

$$x + 2x = 180°$$
$$3x = 180°$$
$$x = 60°$$

The sum of the angles of a triangle is 180°. Because $m\angle XYW = 60°$, the other two angles of the triangle must total 120°. This can occur with one angle being acute and the other being obtuse or with both angles being acute. This means that you can't determine whether the triangle is acute, obtuse, equiangular, or right.

180.   **Acute**

The sum of the three angles of a triangle is 180°:

$$(x + 40) + (x + 10) + (2x - 10) = 180°$$
$$x + 40 + x + 10 + 2x - 10 = 180°$$
$$4x + 40 = 180°$$
$$4x = 140°$$
$$x = 35°$$

To find the measure of each angle, plug 35° in for x:

$$m\angle X = x + 40° = 35° + 40° = 75°$$
$$m\angle W = x + 10° = 35° + 10° = 45°$$
$$m\angle WYX = 2x - 10° = 2(35°) - 10° = 60°$$

Because the largest angle of the triangle is less than 90°, the triangle is acute.

181.   **Right**

Let $m\angle WYX = x$ and let $m\angle XYZ = 2x + 30$. These two angles form a linear pair, so they add up to 180°. Use these relationships to solve for x and find $m\angle WYX$:

$$x + (2x + 30) = 180°$$
$$x + 2x + 30 = 180°$$
$$3x + 30 = 180°$$
$$3x = 150°$$
$$x = 50°$$

Therefore, $m\angle WYX = x = 50°$. With this info and the given $m\angle W = 40°$, find the measure of the remaining angle in the triangle:

$$50° + 40° + m\angle X = 180°$$
$$90° + m\angle X = 180°$$
$$m\angle X = 90°$$

If one angle in the triangle is 90°, then the triangle is a right triangle.

182.   Right

The sum of the three angles of a triangle is 180°:

$$x + 2x + 3x = 180°$$
$$6x = 180°$$
$$x = 30°$$

To find the measure of each angle, plug 30° in for x:

$$m\angle X = x = 30°$$
$$m\angle W = 2x = 2(30°) = 60°$$
$$m\angle WYX = 3x = 3(30°) = 90°$$

If one angle in the triangle is 90°, then the triangle is a right triangle.

183.   40°

In an isosceles triangle, the base angles are congruent. Because $\angle E$ is the vertex angle, that would make $\angle D$ and $\angle F$ the base angles. Set them equal to each other and solve for x:

$$4x - 40 = x + 20$$
$$3x - 40 = 20$$
$$3x = 60$$
$$x = 20$$

To find the measure of $\angle F$, plug 20 in for x:

$$m\angle F = x + 20° = 20° + 20° = 40°$$

184.   30

An equiangular triangle is a triangle where all three angles are congruent; therefore all three angles have to equal 60°. To solve for x, set any angle equal to 60°:

$$x + 30 = 60$$
$$x = 30$$

185. 50°

The sum of the three angles of a triangle is 180°:

$$90 + (x + 10) + (2x - 10) = 180°$$
$$90 + x + 10 + 2x - 10 = 180°$$
$$3x + 90 = 180°$$
$$3x = 90°$$
$$x = 30°$$

To find the measure of $\angle C$, plug 30 in for x:

$$m\angle C = 2x - 10 = 2(30) - 10 = 50°$$

186. 65°

The sum of the three angles of a triangle is 180°:

$$90° + x^2 + 13x = 180°$$
$$x^2 + 13x + 90 = 180°$$
$$x^2 + 13x - 90 = 0$$
$$(x + 18)(x - 5) = 0$$
$$\cancel{x = -18} \text{ or } x = 5$$

To find the measure of $\angle C$, plug 5 in for x:

$$m\angle C = 13(5) = 65°$$

187. True

An *equilateral* triangle has all three angles congruent. This means that all three angles must equal 60°. Because the angles are congruent, the triangles are similar.

188. True

An angle bisector divides an angle into two congruent angles, a median divides a segment into two congruent segments, and an altitude forms right angles. In an isosceles triangle, the altitude, angle bisector, and median all coincide.

189. True

The angle bisector of an equilateral triangle is always perpendicular to the opposite side.

190. **False**

The sum of the three angles of a triangle must equal 180°. An obtuse angle is an angle measuring between 90° and 180°. It would be impossible to have two obtuse angles in a triangle because the sum of the angles would be greater than 180°.

191. $\angle A \cong \angle C$

Because you're given that $\triangle ABC$ is an isosceles triangle with vertex B, you know that $\angle A$ and $\angle C$ must be the two base angles. In an isosceles triangle, the base angles are congruent.

192. $\overline{BA} \cong \overline{BC}$

The legs of an isosceles triangle are the two sides that join at the vertex angle. The two legs of an isosceles triangle are congruent.

193. **Subtraction postulate**

If congruent segments are subtracted from congruent segments, their difference is congruent.

194. **ASA**

If two angles and the included side of a triangle are congruent to two angles and the included side of another triangle, then the triangles are congruent by ASA (angle-side-angle).

195. **A midpoint divides a segment into two congruent segments.**

$\overline{TM} \cong \overline{NM}$ because midpoint M divides $\overline{TN}$ into two congruent segments.

196. $\angle TRM \cong \angle NAM$

If two angles are right angles, then they're congruent.

197. **AAS**

If two angles and a non-included side of one triangle are congruent to two angles and a non-included side of another triangle, the triangles are congruent by AAS (angle-angle-side).

198. $\angle T \cong \angle N$

Statement 9 involves the triangle being isosceles, so Statement 8 needs to show either that the two legs are congruent or that the two base angles are congruent. You can conclude only that the two base angles are congruent from CPCTC, which states that corresponding parts of congruent triangles are congruent.

199. **If two angles of a triangle are congruent, then the triangle is isosceles.**

200.

Statements	Reasons
1. △*NBA* is isosceles with vertex *B*; △*NMA* is isosceles with vertex *M*	1. Given
2. ∠*BNA* ≅ ∠*BAN*; ∠*MNA* ≅ ∠*MAN*	2. In an isosceles triangle, the two base angles are congruent.
3. $\overline{NA} \cong \overline{NA}$	3. Reflexive property
4. △*PNA* ≅ △*RAN*	4. ASA

201. **Perpendicular lines form right angles.**

202. $\overline{RI} \cong \overline{RH}$

You're given that △*RHI* is isosceles with vertex *R*, so $\overline{RI}$ and $\overline{RH}$ are the legs of the triangle. The legs of isosceles triangles are congruent.

203. $\overline{RG} \cong \overline{RG}$

Both triangles have this side in common.

204. **If a triangle has a right angle, then the triangle is a right triangle.**

205. **HL**

When the hypotenuse and leg of a right triangle are congruent to the corresponding hypotenuse and leg of another right triangle, the two triangles are congruent.

206.

Statements	Reasons
1. $\overline{AY} \cong \overline{WC}$, $\overline{XY} \perp \overline{AW}$, $\overline{BC} \perp \overline{AW}$, and $\overline{AB} \cong \overline{WX}$	1. Given
2. ∠*XYW* and ∠*BCA* are right angles.	2. Perpendicular lines form right angles.
3. ∠*XYW* ≅ ∠*BCA*	3. If two angles are right angles, then they're congruent.
4. $\overline{YC} \cong \overline{YC}$	4. Reflexive property
5. $\overline{AY} + \overline{YC} \cong \overline{WC} + \overline{YC}$ or $\overline{AC} \cong \overline{WY}$	5. Addition postulate
6. △*BCA* and △*XYW* are right triangles	6. A triangle that contains a right angle is a right triangle.
7. △*BCA* ≅ △*XYW*	7. HL
8. $\overline{BC} \cong \overline{XY}$	8. CPCTC

207. $\overline{AB} \cong \overline{CB}$

In order to prove two triangles congruent using hypotenuse-leg theorem, you must have a leg and the hypotenuse from one triangle congruent to a leg and the hypotenuse of the other triangle. You're given $\overline{AD} \cong \overline{CD}$, which are legs of the right triangles. $\overline{AB}$ and $\overline{CB}$ are hypotenuses in the right triangles.

208. $\overline{AD} \cong \overline{CD}$

If $\triangle ABC$ is isosceles with vertex B, then $\overline{AB} \cong \overline{BC}$ and $\angle A \cong \angle C$ because an isosceles triangle has congruent legs and congruent base angles. This gives you a side and an angle. In order to use SAS, you need to obtain a pair of congruent sides that follow the congruent angles.

209. $\angle BDA \cong \angle BDC$

If $\triangle ABC$ is isosceles with vertex B, then $\overline{AB} \cong \overline{BC}$ and $\angle A \cong \angle C$ because an isosceles triangle has congruent legs and congruent base angles. This gives you a side and an angle. In order to use AAS, you need to obtain a pair of congruent angles that follow another pair of congruent angles.

210. HL

When the hypotenuse and leg of a right triangle are congruent to the corresponding hypotenuse and leg of another right triangle, the two triangles are congruent.

211. 38°

The *incenter* of a triangle is the point where the bisectors of each angle of the triangle intersect. A *bisector* divides an angle into two congruent angles, so $\angle ICN \cong \angle ICE$.

212. 26°

The *incenter* of a triangle is the point where the bisectors of each angle of the triangle intersect. A *bisector* divides an angle into two congruent angles.

Find the measure of the third angle of $\triangle CEN$ and then cut the angle in half:

$$60 + 68 + x = 180°$$
$$128 + x = 180°$$
$$x = 52°$$

$$m\angle CNI = \frac{52°}{2} = 26°$$

213.   4

The *incenter* of a triangle is the point where the bisectors of each angle of the triangle intersect. A *bisector* divides an angle into two congruent angles. If $m\angle ENI = 40°$, then

$$m\angle CNI = 40°$$
$$2c = 40$$
$$c = 20$$

If $m\angle CEI = 28°$, then

$$m\angle NEI = 28°$$
$$2b + c = 28$$
$$2b + 20 = 28$$
$$2b = 8$$
$$b = 4$$

214.   115"

The *incenter* of a triangle is the point where the bisectors of each angle of the triangle intersect. A bisector divides an angle into two congruent angles. $m\angle ENC = 70°$, which means that $m\angle CNI = 35°$. $m\angle ECN = 60°$, which means that $m\angle NCI = 30°$.

The sum of the angles of a triangle is 180°. Therefore,

$$30 + 35 + m\angle CIN = 180°$$
$$65 + m\angle CIN = 180°$$
$$m\angle CIN = 115°$$

215.   42°

The *incenter* of a triangle is the point where the bisectors of each angle of the triangle intersect. A *bisector* divides an angle into two congruent angles. $m\angle CEN = 76°$, which means that $m\angle CEI = 38°$.

The sum of the angles of a triangle is 180°. Therefore,

$$38 + 100 + m\angle ECI = 180°$$
$$138 + m\angle ECI = 180°$$
$$m\angle ECI = 42°$$

216.   70°

The *incenter* of a triangle is the point where the bisectors of each angle of the triangle intersect. A *bisector* divides an angle into two congruent angles. $m\angle CNE = 50°$, which means that $m\angle ENI = 25°$.

The sum of the angles of a triangle is 180°. Therefore,

$$25 + 120 + m\angle NEI = 180°$$
$$145 + m\angle NEI = 180°$$
$$m\angle NEI = 35°$$

$\angle CEI \cong \angle NEI$ because $\overline{EI}$ is an angle bisector, so multiply $m\angle NEI$ by 2:

$$m\angle CEN = 2(35°) = 70°$$

217.   **40°**

The *incenter* of a triangle is the point where the bisectors of each angle of the triangle intersect. A *bisector* divides an angle into two congruent angles.

The sum of the angles around a point is 360°:

$$110° + 120° + m\angle NIE = 360°$$
$$230° + m\angle NIE = 360°$$
$$m\angle NIE = 130°$$

The sum of the angles of a triangle is 180°:

$$130° + 30° + m\angle ENI = 180°$$
$$160° + m\angle ENI = 180°$$
$$m\angle ENI = 20°$$

$\angle ENI \cong \angle CNI$ because $\overline{IN}$ is an angle bisector, so multiply $m\angle ENI$ by 2:

$$m\angle ENC = 2(20) = 40°$$

218.   **40°**

The *incenter* of a triangle is the point where the bisectors of each angle of the triangle intersect. A *bisector* divides an angle into two congruent angles. $m\angle NIT = 25°$; therefore, $m\angle CIT = 25°$ as well. This means that $m\angle NIC = 25 + 25 = 50°$. Because $\angle NIC \cong \angle NCI$, $m\angle NCI = 50°$.

The sum of the angles of a triangle is 180°, so

$$50 + 50 + m\angle CNI = 180°$$
$$100 + m\angle CNI = 180°$$
$$m\angle CNI = 80°$$

$m\angle CNT = \dfrac{1}{2}(80) = 40°$ because $\overline{NT}$ is an angle bisector.

219. (–8, –6)

The *orthocenter* of a triangle is the point where the three altitudes of the triangle intersect. An altitude of a triangle is perpendicular to the opposite side. Because perpendicular lines have negative reciprocal slopes, you need to know the slope of the opposite side. Here's the slope of $\overline{BC}$:

$$\text{slope}_{\overline{BC}} = \frac{y_2 - y_1}{x_2 - x_1} = \frac{0-6}{4-(-2)} = -1$$

This means that the slope of the altitude to $\overline{BC}$ needs to be 1.

The point-slope formula of a line is $y - y_1 = m(x - x_1)$, where m is the slope and (x_1, y_1) are the coordinates of a point on the line. To find the altitude formed when you connect Point A to $\overline{BC}$, plug in $m = -1$ and the coordinates of Point A, $(0, 2)$:

$$y - y_1 = m(x - x_1)$$
$$y - 2 = 1(x - 0)$$
$$y - 2 = x$$
$$y = x + 2$$

The slope of $\overline{AB}$ is

$$\text{slope}_{\overline{AB}} = \frac{6-2}{-2-0} = \frac{4}{-2} = -2$$

This means that the slope of the altitude to $\overline{AB}$ is $\frac{1}{2}$.

The altitude formed when you connect Point C, $(4, 0)$, to $\overline{AB}$ is

$$y - 0 = \frac{1}{2}(x - 4)$$
$$y = \frac{1}{2}x - 2$$

To find the orthocenter, you need to find where two altitudes intersect. Set them equal and solve for x:

$$x + 2 = \frac{1}{2}x - 2$$
$$\frac{1}{2}x = -4$$
$$x = -8$$

Now plug the x value into one of the altitude formulas and solve for y:

$$y = x + 2$$
$$y = -8 + 2$$
$$y = -6$$

Therefore, the altitudes cross at (–8, –6).

220. $(-2, -4)$

The *orthocenter* of a triangle is the point where the three altitudes of the triangle intersect. An altitude of a triangle is perpendicular to the opposite side. Because perpendicular lines have negative reciprocal slopes, you need to know the slope of the opposite side. Here's the slope of $\overline{RT}$:

$$\text{slope}_{\overline{RT}} = \frac{y_2 - y_1}{x_2 - x_1} = \frac{0-4}{6-(-2)} = \frac{-4}{8} = \frac{-1}{2}$$

This means that the slope of the altitude to $\overline{RT}$ needs to be 2.

The point-slope formula of a line is $y - y_1 = m(x - x_1)$, where m is the slope and (x_1, y_1) are the coordinates of a point on the line. To find the altitude formed when you connect Point O to $\overline{RT}$, plug in $m = 2$ and the coordinates of O, $(0, 0)$:

$$y - y_1 = m(x - x_1)$$
$$y - 0 = 2(x - 0)$$
$$y = 2x$$

Now find the equation for the altitude to $\overline{OT}$: The slope of $\overline{OT}$ is

$$\text{slope}_{\overline{OT}} = \frac{0-4}{0-(-2)} = \frac{-4}{2} = -2$$

This means that the slope of the altitude to $\overline{OT}$ is $\frac{1}{2}$.

The altitude formed when you connect Point R, $(6, 0)$, to $\overline{OT}$ is

$$y - 0 = \frac{1}{2}(x - 6)$$
$$y = \frac{1}{2}x - 3$$

To find the orthocenter, you need to find where the two altitudes intersect. Set them equal and solve for x:

$$2x = \frac{1}{2}x - 3$$
$$\frac{3}{2}x = -3$$
$$x = -2$$

Now plug the x value into one of the altitude formulas and solve for y:

$$y = 2x$$
$$y = 2(-2)$$
$$y = -4$$

Therefore, the altitudes cross at $(-2, -4)$.

221. $(-2, -2)$

The *orthocenter* of a triangle is the point where the three altitudes of the triangle intersect. An altitude of a triangle is perpendicular to the opposite side. Because perpendicular lines have negative reciprocal slopes, you need to know the slope of the opposite side. Here's the slope of $\overline{DN}$:

$$\text{slope}_{\overline{DN}} = \frac{y_2 - y_1}{x_2 - x_1} = \frac{8 - 0}{-2 - 0} = -1$$

This means that the slope of the altitude to $\overline{DN}$ needs to be 1.

The point-slope formula of a line is $y - y_1 = m(x - x_1)$, where m is the slope and (x_1, y_1) are the coordinates of a point on the line. To find the altitude formed when you connect Point A to $\overline{DN}$, plug in $m = 1$ and the coordinates of A, $(0, 0)$:

$$y - y_1 = m(x - x_1)$$
$$y - 0 = 1(x - 0)$$
$$y = x$$

Now find the equation for the altitude to $\overline{AD}$. The slope of $\overline{AD}$ is

$$\text{slope}_{\overline{AD}} = \frac{0 - 8}{0 - (-2)} = \frac{-8}{2} = -4$$

This means that the slope of the altitude to $\overline{AD}$ is $\frac{1}{4}$.

The altitude formed when you connect Point N, $(6, 0)$, to $\overline{AD}$ is

$$y - 0 = \frac{1}{4}(x - 6)$$
$$y = \frac{1}{4}x - \frac{3}{2}$$

To find the orthocenter, you need to find where the two altitudes intersect. Set them equal and solve for x:

$$x = \frac{1}{4}x - \frac{3}{2}$$
$$\frac{3}{4}x = \frac{-3}{2}$$
$$x = -2$$

Now plug the x value into one of the altitude formulas and solve for y:

$$y = x$$
$$y = -2$$

Therefore, the altitudes cross at $(-2, -2)$.

222. **Right**

The orthocenter of a right triangle is found on the vertex angle, which is the right angle.

223. **Obtuse**

The orthocenter of an obtuse triangle is found outside the triangle.

224. **2**

The *centroid* of a triangle divides each median of the triangle into segments with a 2:1 ratio. Set up a proportion and solve for *DC:*

$$\frac{SC}{DC} = \frac{2}{1}$$
$$\frac{4}{DC} = \frac{2}{1}$$
$$4 = 2(DC)$$
$$2 = DC$$

225. **4**

The *centroid* of a triangle divides each median of the triangle into segments with a 2:1 ratio. Set up a proportion to find *SC:*

$$\frac{SC}{DC} = \frac{2}{1}$$
$$\frac{SC}{2} = \frac{2}{1}$$
$$SC = 4$$

226. **14**

The *centroid* of a triangle divides each median of the triangle into segments with a 2:1 ratio. You don't know the length of either segment of the median, so you'll use an *x* in the ratio to represent the shorter length.

$$SC = 2x$$
$$DC = x$$

You're given that $SD = 21$; therefore,

$$SC + DC = 21$$
$$2x + x = 21$$
$$3x = 21$$
$$x = 7$$

$$SC = 2x = 2(7) = 14$$

Answers
201–300

227.   8

The *centroid* of a triangle divides each median of the triangle into segments with a 2:1 ratio. You don't know the length of either segment of the median, so you'll use an x in the ratio to represent the shorter length.

$$RC = 2x$$
$$CE = x$$

You're given that $RE = 24$; therefore,

$$RC + CE = 24$$
$$2x + x = 24$$
$$3x = 24$$
$$x = 8$$

$$CE = x = 8$$

228.   18

The *centroid* of a triangle divides each median of the triangle into segments with a 2:1 ratio. You don't know the length of either segment of the median, so you'll use an x in the ratio to represent the shorter length.

$$SC = 2x$$
$$DC = x$$

You're given that $SD = 54$; therefore,

$$SC + DC = 54$$
$$2x + x = 54$$
$$3x = 54$$
$$x = 18$$

$$DC = x = 18$$

229.   18

The *centroid* of a triangle divides each median of the triangle into segments with a 2:1 ratio. You don't know the length of either segment of the median, so you'll use an x in the ratio to represent the shorter length.

$$RC = 2x$$
$$CE = x$$

You're given that $RE = 27$; therefore,

$$RC + CE = 27$$
$$2x + x = 27$$
$$3x = 27$$
$$x = 9$$

$$RC = 2x = 2(9) = 18$$

230.   45

The *centroid* of a triangle divides each median of the triangle into segments with a 2:1 ratio. Set up a proportion to find *SC*:

$$\frac{SC}{DC} = \frac{2}{1}$$
$$\frac{SC}{15} = \frac{2}{1}$$
$$SC = 30$$

Therefore, $DS = DC + SC = 15 + 30 = 45$.

231.   16.5

The centroid of a triangle divides each median of the triangle into segments with a 2:1 ratio. Set up a proportion to find *SC*:

$$\frac{SC}{DC} = \frac{2}{1}$$
$$\frac{SC}{5.5} = \frac{2}{1}$$
$$SC = 11$$

Therefore, $DS = DC + SC = 5.5 + 11 = 16.5$.

232.   (2, 2)

You find the centroid of a triangle by averaging the x coordinates and the y coordinates of all three vertices of the triangle. The average of the x coordinates is

$$\frac{0 + 4 + 2}{3} = 2$$

The average of the y coordinates is

$$\frac{-2 + 0 + 8}{3} = 2$$

Therefore, the centroid of the triangle is $(2, 2)$.

233. (3, 3)

You find the centroid of a triangle by averaging the x coordinates and the y coordinates of all three vertices of the triangle. The average of the x coordinates is

$$\frac{0+0+9}{3} = 3$$

The average of the y coordinates is

$$\frac{0+9+0}{3} = 3$$

Therefore, the centroid of the triangle is $(3,3)$.

234. $\left(\frac{-4}{3}, \frac{14}{3}\right)$

You find the centroid of a triangle by averaging the x coordinates and the y coordinates of all three vertices of the triangle. The average of the x coordinates is

$$\frac{-6+-2+4}{3} = \frac{-4}{3}$$

The average of the y coordinates is

$$\frac{2+8+4}{3} = \frac{14}{3}$$

Therefore, the centroid of the triangle is $\left(\frac{-4}{3}, \frac{14}{3}\right)$.

235. 5

You find the centroid of a triangle by averaging the x coordinates and the y coordinates of all three vertices of the triangle. The average of the x coordinates is

$$\frac{1+6+4}{3} = \frac{11}{3}$$

The average of the y coordinates is

$$\frac{1+0+3}{3} = \frac{4}{3}$$

Therefore, the centroid of the triangle is $\left(\frac{11}{3}, \frac{4}{3}\right)$. The sum of the x and y coordinates of the centroid is $\frac{11}{3} + \frac{4}{3} = 5$.

236. 6

The *circumcenter* of a triangle is the point where all three perpendicular bisectors of the triangle intersect. A *perpendicular bisector* divides a segment into two congruent segments. If one part is 6, the other part is also 6.

237. 4

The *circumcenter* of a triangle is the point where all three perpendicular bisectors of the triangle intersect. A *perpendicular* bisector divides a segment into two congruent segments. If one part is 4, the other part is also 4.

238. 5

The *circumcenter* of a triangle is the point where all three perpendicular bisectors of the triangle intersect. A *perpendicular bisector* divides a segment into two congruent segments. If one part is 5, the other part is also 5.

239. 30

The *circumcenter* of a triangle is the point where all three perpendicular bisectors of the triangle intersect. A *perpendicular bisector* divides a segment into two congruent segments.

The perimeter is the sum of all three sides of the triangle:

$$4+4+5+5+6+6=30$$

240. 90°

You're given that Point C is the circumcenter of the triangle. The *circumcenter* of a triangle is the point where all three perpendicular bisectors of the triangle intersect. A perpendicular bisector forms right angles with the side that it intersects. Therefore, b is a right angle.

241. 90°

You're given that Point C is the circumcenter of the triangle. The *circumcenter* of a triangle is the point where all three perpendicular bisectors of the triangle intersect. A perpendicular bisector forms right angles with the side that it intersects. Therefore, d is a right angle.

242. **Circumcenter**

The *circumcenter* of a triangle is the point where all three perpendicular bisectors of the triangle intersect. You know that the segments intersecting at O are perpendicular bisectors because they form right angles and because when one meets the opposite side, it divides the side into two equal parts.

243. **Incenter**

The *incenter* of a triangle is the point where the bisectors of each angle of the triangle intersect. The figure shows that at all three vertices of the triangle, a segment divides each angle into two congruent angles. This is the definition of an *angle bisector*.

244. **Orthocenter**

The *orthocenter* of a triangle is the point where the three altitudes of the triangle intersect. You can tell that the three segments drawn are altitudes because when each segment meets the opposite side, it forms a right angle.

245. **Centroid**

The *centroid* of a triangle is the point where the three medians of the triangle intersect. The figure shows that a segment connects each vertex to the midpoint of the opposite side. This is the definition of a *median*.

246. **Orthocenter**

The orthocenter of a right triangle is found on the vertex angle, which is the right angle.

247. **Incenter**

The *incenter* of a triangle is the point where the bisectors of each angle of the triangle intersect.

248. **Circumcenter**

The *circumcenter* of a triangle is the point where all three perpendicular bisectors of the triangle intersect.

249. **Orthocenter**

The *orthocenter* of a triangle is the point where all three of the altitudes of the triangle intersect.

250. **Centroid**

The *centroid* of a triangle is the point where all three medians of the triangle intersect.

251. **Incenter**

The *incenter* of a triangle is the point where the bisectors of each angle of the triangle intersect. The incenter of a triangle is also the center of the triangle's incircle. This means that it's the center of the circle inscribed in the triangle.

252. **Circumcenter**

The *circumcenter* of a triangle is the point where the three perpendicular bisectors of the triangle intersect. The figure shows the construction of perpendicular bisectors.

253. **Incenter**

The *incenter* of a triangle is the point where the three angle bisectors of the triangle intersect. The figure shows the construction of angle bisectors.

254. **Centroid**

The *centroid* of a triangle is the point where the three medians of the triangle intersect. The figure shows the construction of medians.

255. **Orthocenter**

The *orthocenter* of a triangle is the point where the three altitudes of the triangle intersect. The figure shows the construction of altitudes.

256.

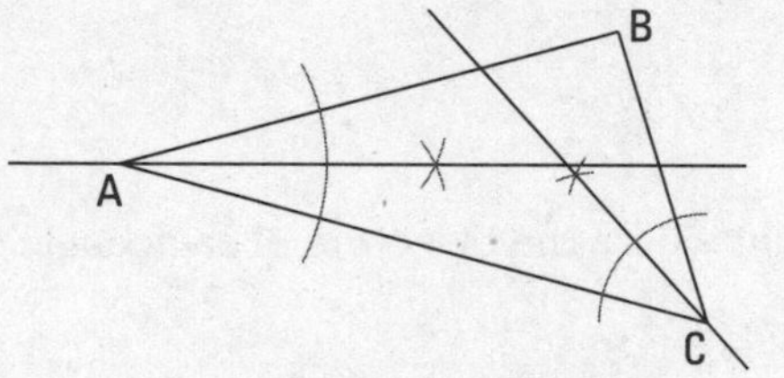

The incenter of a triangle is the point where the three angle bisectors of the triangle intersect. Construct angle bisectors for the angles of the triangle. Place the compass point at *A,* and using any width, make an arc through the angle.

Using the same compass width, place the compass point at both places where the arc intersects the angle. The new arcs should intersect. Make a line by connecting Point *A* to the point of intersection.

Follow the same steps for Point *C* on the triangle. The point where the two lines intersect is the incenter of the triangle.

257. **Centroid, circumcenter, and orthocenter**

The *Euler line* is the line that contains the centroid, circumcenter, and orthocenter of a triangle. You need to know any two of these points to determine the line.

258. **2.75**

The circumcenter is on the Euler line. Find the *y* value by plugging *x* into the equation of the Euler line:

$$y = \frac{3}{4}(5) - 1 = 2.75$$

259. $y = \frac{1}{7}x + \frac{8}{7}$

The *Euler line* is the line that contains the centroid, circumcenter, and orthocenter of a triangle. To find the equation of the line, you first need to determine the slope of the line that contains the two given points:

$$\text{slope} = \frac{\frac{8}{3} - 2}{\frac{32}{3} - 6} = \frac{\frac{2}{3}}{\frac{14}{3}} = \frac{1}{7}$$

Then use the point-slope formula to find the equation of the line:

$$y - y_1 = m(x - x_1)$$
$$y - 2 = \frac{1}{7}(x - 6)$$
$$y - 2 = \frac{1}{7}x - \frac{6}{7}$$
$$y = \frac{1}{7}x + \frac{8}{7}$$

260. $y = \frac{1}{3}x - \frac{4}{3}$

The *Euler line* is the line that contains the centroid, circumcenter, and orthocenter of the triangle. Find the equation of a line going through any two of those points. Here's the slope if you're using the centroid and the orthocenter:

$$m = \frac{0 - \frac{-2}{3}}{4 - 2} = \frac{\frac{2}{3}}{2} = \frac{2}{3} \cdot \frac{1}{2} = \frac{1}{3}$$

Now use the slope and one of the points (here, the centroid) to find the equation of the line:

$$y - y_1 = m(x - x_1)$$
$$y - \frac{-2}{3} = \frac{1}{3}(x - 2)$$
$$y + \frac{2}{3} = \frac{1}{3}x - \frac{2}{3}$$
$$y = \frac{1}{3}x - \frac{4}{3}$$

261. 3

If two triangles are similar, their sides are in proportion:

$$\frac{AB}{DE} = \frac{AC}{DF}$$
$$\frac{4}{12} = \frac{AC}{9}$$
$$12(AC) = 36$$
$$AC = 3$$

262.   20

If two triangles are similar, their sides are in proportion:

$$\frac{AB}{DE} = \frac{AC}{DF}$$

$$\frac{10}{DE} = \frac{8}{16}$$

$$8(DE) = 160$$

$$DE = 20$$

263.   52

If two triangles are similar, the ratio of the sides of the triangles is equal to the ratio of the perimeters of the triangles. The perimeter of the smaller triangle is $8 + 12 + 6 = 26$, so set up the proportion and solve:

$$\frac{AB}{DE} = \frac{\text{perimeter } \triangle ABC}{\text{perimeter } \triangle DEF}$$

$$\frac{8}{16} = \frac{26}{\text{perimeter } \triangle DEF}$$

$$8(\text{perimeter } \triangle DEF) = 416$$

$$\text{perimeter } \triangle DEF = 52$$

264.   6

If two triangles are similar, the ratio of the sides of the triangles is equal to the ratio of the perimeters of the triangles. The perimeter of $\triangle ABC$ is $8 + 6 + 12 = 26$, so set up the proportion and solve:

$$\frac{\text{perimeter } \triangle ABC}{\text{perimeter } \triangle DEF} = \frac{BC}{EF}$$

$$\frac{26}{13} = \frac{12}{EF}$$

$$26(EF) = 156$$

$$EF = 6$$

265.   78°

If two triangles are similar, their angles are congruent. If $\triangle ABC \sim \triangle DEF$, then $\angle A \cong \angle D$; therefore, $m\angle D = 78°$.

266. **61°**

The sum of the three angles of a triangle is 180°:

$$31° + 88° + m\angle F = 180°$$
$$119° + m\angle F = 180°$$
$$m\angle F = 61°$$

If two triangles are similar, their angles are congruent. If $\triangle ABC \sim \triangle DEF$, then $\angle F \cong \angle C$. Therefore, $m\angle C = 61°$.

267. **33**

If two triangles are similar, their sides are in proportion:

$$\frac{AB}{DE} = \frac{AC}{DF}$$
$$\frac{AB}{22} = \frac{24}{16}$$
$$16(AB) = 528$$
$$AB = 33$$

268. **6**

When you connect the midpoints of two sides of a triangle, the new segment is parallel to the third side of the triangle and equals half the length of the third side of the triangle. Set RY equal to half of SE and solve for b:

$$b + 4 = \frac{1}{2}(20)$$
$$b + 4 = 10$$
$$b = 6$$

269. **54**

When you connect the midpoints of two sides of a triangle, the new segment is parallel to the third side of the triangle and equals half the length of the third side of the triangle. Set RY equal to half of SE and solve for b:

$$50 = \frac{1}{2}(2b - 8)$$
$$50 = b - 4$$
$$54 = b$$

270. 7

The midpoint R divides $\overline{SH}$ into two congruent segments, so set HR equal to RS. Now you can solve for x:

$$7x - 28 = 2x - 3$$
$$5x - 28 = -3$$
$$5x = 25$$
$$x = 5$$

Then plug in the value of x to find RS:

$$RS = 2x - 3 = 2(5) - 3 = 7$$

271. 7

When you connect the midpoints of two sides of a triangle, the new segment is parallel to the third side of the triangle and equals half the length of the third side of the triangle. Set RY equal to half of SE and solve for b:

$$8b + 4 = \frac{1}{2}(17b + 1)$$
$$8b + 4 = 8.5b + 0.5$$
$$4 = 0.5b + 0.5$$
$$3.5 = 0.5b$$
$$b = 7$$

272. 4

When you connect the midpoints of two sides of a triangle, the new segment is parallel to the third side of the triangle and equals half the length of the third side of the triangle. Set RY equal to half of SE and solve for b:

$$8 = \frac{1}{2}b^2$$
$$16 = b^2$$
$$4 = b$$

273. 8

When you connect the midpoints of two sides of a triangle, the new segment is parallel to the third side of the triangle and equals half the length of the third side of the triangle. Set RY equal to half of SE and solve for b:

$$b + 24 = \frac{1}{2}b^2$$
$$2b + 48 = b^2$$
$$0 = b^2 - 2b - 48$$
$$0 = (b - 8)(b + 6)$$
$$b = 8 \text{ or } \cancel{b = -6}$$

274.   4

When you connect the midpoints of two sides of a triangle, the new segment is parallel to the third side of the triangle and equals half the length of the third side of the triangle. Set *RY* equal to half of *SE* and solve for *b*:

$$128 = \frac{1}{2}b^4$$
$$256 = b^4$$
$$\sqrt[4]{256} = \sqrt[4]{b^4}$$
$$4 = b$$

275.   12

A line that intersects two sides of a triangle and is parallel to the third side of the triangle creates two similar triangles. If two triangles are similar, their sides are in proportion:

$$\frac{DE}{BC} = \frac{AD}{AB}$$
$$\frac{8}{BC} = \frac{6}{9}$$
$$6(BC) = 72$$
$$BC = 12$$

276.   16

If two triangles are similar, their sides are in proportion:

$$\frac{DE}{BC} = \frac{AD}{AB}$$
$$\frac{8}{BC} = \frac{7}{7+7}$$
$$7(BC) = 8(14)$$
$$7(BC) = 112$$
$$BC = 16$$

277.   15

If two triangles are similar, their sides are in proportion:

$$\frac{AD}{AB} = \frac{AE}{AC}$$
$$\frac{3}{9} = \frac{5}{AC}$$
$$3(AC) = 45$$
$$AC = 15$$

278. 4.8

If two triangles are similar, their sides are in proportion:

$$\frac{AD}{AB} = \frac{AE}{AC}$$

$$\frac{25}{30} = \frac{24}{24 + EC}$$

$$25(24 + EC) = 24(30)$$

$$600 + 25(EC) = 720$$

$$25(EC) = 120$$

$$EC = 4.8$$

279. 10

A line that intersects two sides of a triangle and is parallel to the third side of the triangle creates two similar triangles. If two triangles are similar, their sides are in proportion:

$$\frac{BC}{MT} = \frac{AB}{AM}$$

$$\frac{6}{8} = \frac{30}{30 + BM}$$

$$6(30 + BM) = 30(8)$$

$$180 + 6(BM) = 240$$

$$6(BM) = 60$$

$$BM = 10$$

280. 10

If two triangles are similar, their sides are in proportion:

$$\frac{BC}{MT} = \frac{AB}{AM}$$

$$\frac{2.5}{3} = \frac{AB}{AB + 2}$$

$$2.5(AB + 2) = 3(AB)$$

$$2.5(AB) + 5 = 3(AB)$$

$$5 = 0.5(AB)$$

$$10 = AB$$

281. 6

If two triangles are similar, their sides are in proportion:

$$\frac{BC}{MT} = \frac{AB}{AM}$$

$$\frac{8}{17} = \frac{2x+12}{(2x+12)+(27)}$$

$$\frac{8}{17} = \frac{2x+12}{2x+39}$$

$$8(2x+39) = 17(2x+12)$$

$$16x+312 = 34x+204$$

$$312 = 18x+204$$

$$108 = 18x$$

$$6 = x$$

282. 4

If two triangles are similar, their sides are in proportion:

$$\frac{BC}{MT} = \frac{AB}{AM}$$

$$\frac{3}{6} = \frac{3x}{3x+12}$$

$$18x = 3(3x+12)$$

$$18x = 9x+36$$

$$9x = 36$$

$$x = 4$$

283. 12

If two triangles are similar, their sides are in proportion:

$$\frac{BC}{MT} = \frac{AB}{AM}$$

$$\frac{6}{x+3} = \frac{x}{(x)+(x+6)}$$

$$\frac{6}{x+3} = \frac{x}{2x+6}$$

$$6(2x+6) = x(x+3)$$

$$12x+36 = x^2+3x$$

$$36 = x^2-9x$$

$$0 = x^2-9x-36$$

$$0 = (x-12)(x+3)$$

$$x = 12 \text{ or } \cancel{x=-3}$$

284. 8

If two triangles are similar, their sides are in proportion:

$$\frac{BC}{MT} = \frac{AB}{AM}$$

$$\frac{4}{x+2} = \frac{x}{(x)+(x+4)}$$

$$\frac{4}{x+2} = \frac{x}{2x+4}$$

$$4(2x+4) = x(x+2)$$

$$8x+16 = x^2+2x$$

$$16 = x^2-6x$$

$$0 = x^2-6x-16$$

$$0 = (x-8)(x+2)$$

$$x = 8 \text{ or } \cancel{x=-2}$$

285. 30

If two triangles are similar, their sides are in proportion. Suppose you call the original triangle A and the new triangle B. The proportion would be as follows:

$$\frac{\text{small } A}{\text{small } B} = \frac{\text{big } A}{\text{big } B}$$

$$\frac{3}{x} = \frac{5}{50}$$

$$5x = 150$$

$$x = 30$$

286. 124

If two triangles are similar, their sides are in proportion. Suppose you call the original triangle A and the new triangle B. The proportion would be as follows:

$$\frac{\text{big } A}{\text{big } B} = \frac{\text{perimeter } A}{\text{perimeter } B}$$

$$\frac{31}{x} = \frac{16+23+31}{280}$$

$$\frac{31}{x} = \frac{70}{280}$$

$$70x = 8,680$$

$$x = 124$$

287. 16

If two triangles are similar, their sides are in proportion. Suppose you call the original triangle A and the new triangle B. The proportion would be as follows:

$$\frac{\text{altitude } A}{\text{altitude } B} = \frac{\text{perimeter } A}{\text{perimeter } B}$$

$$\frac{21}{12} = \frac{28}{x}$$

$$21x = 336$$

$$x = 16$$

288. 3

A line that intersects two sides of a triangle and is parallel to the third side of the triangle creates two similar triangles. If two triangles are similar, their sides are in proportion, and the parts of their sides are also in proportion:

$$\frac{SW}{WR} = \frac{SY}{YT}$$

$$\frac{5}{(x+7)} = \frac{x}{(x+3)}$$

$$5(x+3) = x(x+7)$$

$$5x + 15 = x^2 + 7x$$

$$0 = x^2 + 2x - 15$$

$$0 = (x+5)(x-3)$$

$$\cancel{x = -5} \text{ or } x = 3$$

289. 18 ft^2

The ratio of the area of similar triangles is equal to the square of the ratio of the sides:

$$\frac{\text{area } \triangle LIN}{\text{area } \triangle DER} = \left(\frac{LI}{DE}\right)^2$$

$$\frac{32}{x} = \left(\frac{4}{3}\right)^2$$

$$\frac{32}{x} = \frac{16}{9}$$

$$16x = 288$$

$$x = 18$$

290. 100 ft²

The ratio of the area of similar triangles is equal to the square of the ratio of the sides:

$$\frac{\text{area } \triangle LIN}{\text{area } \triangle DER} = \left(\frac{LI}{DE}\right)^2$$

$$\frac{36}{x} = \left(\frac{3}{5}\right)^2$$

$$\frac{36}{x} = \frac{9}{25}$$

$$9x = 900$$

$$x = 100$$

291. ∠EGO

When two parallel lines are cut by a transversal, corresponding angles are formed.

292. **If two angles are corresponding angles, then they're congruent.**

293. ∠E ≅ ∠E

The reflexive property states that an angle is congruent to itself.

294. AA

If two angles of a triangle are congruent to two angles of a different triangle, the two triangles are similar.

295. ∠LMR

When two lines intersect, they form vertical angles across from each other.

296. ∠SMI ≅ ∠LMR

∠SMI and ∠LMR are vertical angles. If two angles are vertical angles, they're congruent, so ∠SMI ≅ ∠LMR.

297. **Perpendicular lines form right angles.**

When two perpendicular lines intersect, they create right angles.

298. AA

If two angles of a triangle are congruent to two angles of a different triangle, the two triangles are similar.

299. If two triangles are similar, their sides are in proportion.

300. $\angle EMA \cong \angle ENA$

You're given that $\overline{ME} \cong \overline{NE}$. When two sides of a triangle are congruent, the angles opposite those sides are also congruent, so $\angle EMA \cong \angle ENA$.

301. Perpendicular lines form right angles.

You're given that $\overline{EA} \perp \overline{MN}$ and $\overline{ST} \perp \overline{EM}$. When two lines that are perpendicular intersect, they form right angles.

302. If two angles are right angles, they're congruent.

$\angle EAN$ and $\angle STM$ are right angles, so they're congruent.

303. $\triangle EAN \sim \triangle STM$

If two angles of one triangle are congruent to two angles of another triangle, the two triangles are similar; therefore, $\triangle EAN \sim \triangle STM$.

304. If two triangles are similar, their sides are in proportion.

305. $AN \times SM = EN \times TM$

In a proportion, the product of the means equals the product of the extremes. This means that once the proportion is set up from the similar triangles, you can cross-multiply and their products are equal.

306. $\angle SRT \cong \angle STR$

You're given that $\triangle RST$ is isosceles with vertex S. This means that $\overline{RS} \cong \overline{TS}$. In an isosceles triangle, the angles opposite these congruent sides are also congruent.

307. $\triangle RPT \sim \triangle TMS$

If two angles of one triangle are congruent to two angles of another triangle, the two triangles are similar.

308. *SM*

If two triangles are similar, their corresponding sides of those triangles are in proportion.

309. If two triangles are similar, the corresponding sides of those triangles are in proportion.

310. TM

In a proportion, the product of the means equals the product of the extremes. This means that after you set up the proportion from the similar triangles, you can cross-multiply and their products are equal.

311. In a proportion, the product of the means equals the product of the extremes.

312. $\angle E \cong \angle E$

You need to use the reflexive property to prove that an angle that's a part of both $\triangle EAB$ and $\triangle EGO$ is congruent to itself. $\angle E$ is a part of both triangles.

313. AA

In order to prove two triangles similar, you need two angles of one triangle to be congruent to two angles of another triangle. AAS is a method of proving two triangles congruent, not similar.

314. $\dfrac{EA}{AB} = \dfrac{EG}{GO}$

When setting up the proportion for the sides of the similar triangles, make sure you're setting up the proportion with the correct corresponding sides. EA and AB are the two sides of the smaller triangle. EG and GO are the two sides of the bigger triangle.

315.

Statements	Reasons
1. In Circle O, chords $\overline{AE}$ and $\overline{BD}$ intersect at C. $\overline{AB}$ and $\overline{DE}$ are chords.	1. Given
2. $\angle ACB$ and $\angle ECD$ are vertical angles.	2. Intersecting lines form vertical angles.
3. $\angle ACB \cong \angle ECD$	3. If two angles are vertical angles, they're congruent.
4. $\angle B = \frac{1}{2}\,\widehat{AD}$ $\angle E = \frac{1}{2}\,\widehat{AD}$	4. An inscribed angle is equal to half the arc it intercepts.
5. $\angle B \cong \angle E$	5. Substitution postulate
6. $\triangle ABC \sim \triangle DEC$	6. AA
7. $\dfrac{AB}{BC} = \dfrac{ED}{DC}$	7. If two triangles are similar, their sides are in proportion.
8. $AB \times DC = ED \times BC$	8. In a proportion, the product of the means equals the product of the extremes.

316. 10

The Pythagorean theorem states that $a^2 + b^2 = c^2$. The a and b represent the legs of the right triangle, and the c represents the hypotenuse, the side opposite the right angle. Substitute the values into the theorem and solve for c:

$$6^2 + 8^2 = c^2$$
$$36 + 64 = c^2$$
$$100 = c^2$$
$$\sqrt{100} = \sqrt{c^2}$$
$$10 = c$$

317. 12

The Pythagorean theorem states that $a^2 + b^2 = c^2$. Substitute the values into the theorem and solve for b:

$$5^2 + b^2 = 13^2$$
$$25 + b^2 = 169$$
$$b^2 = 144$$
$$\sqrt{b^2} = \sqrt{144}$$
$$b = 12$$

318. $2\sqrt{5}$

The Pythagorean theorem states that $a^2 + b^2 = c^2$. Substitute the values into the theorem and solve for c:

$$4^2 + 2^2 = c^2$$
$$16 + 4 = c^2$$
$$20 = c^2$$
$$\sqrt{20} = \sqrt{c^2}$$
$$\sqrt{4 \times 5} = c$$
$$2\sqrt{5} = c$$

319. 7

The Pythagorean theorem states that $a^2 + b^2 = c^2$. Substitute the values into the theorem and solve for b:

$$24^2 + b^2 = 25^2$$
$$576 + b^2 = 625$$
$$b^2 = 49$$
$$\sqrt{b^2} = \sqrt{49}$$
$$b = 7$$

320. $\sqrt{34}$

The Pythagorean theorem states that $a^2 + b^2 = c^2$. Substitute the values into the theorem and solve for c:

$$3^2 + 5^2 = c^2$$
$$9 + 25 = c^2$$
$$34 = c^2$$
$$\sqrt{34} = \sqrt{c^2}$$
$$\sqrt{34} = c$$

321. 15

The Pythagorean theorem states that $a^2 + b^2 = c^2$. Substitute the values into the theorem and solve for x:

$$(x-7)^2 + x^2 = 17^2$$
$$(x-7)(x-7) + x^2 = 289$$
$$x^2 - 14x + 49 + x^2 = 289$$
$$2x^2 - 14x + 49 = 289$$
$$2x^2 - 14x - 240 = 0$$
$$2(x^2 - 7x - 120) = 0$$
$$2(x-15)(x+8) = 0$$
$$x = 15 \text{ or } \cancel{x = -8}$$

322. 28 in.

A diagonal of a rectangle divides the rectangle into two congruent right triangles. That diagonal is the hypotenuse of the right triangle, and the sides of the rectangle are the two legs of the right triangle.

The Pythagorean theorem states that $a^2 + b^2 = c^2$. The a and b represent the legs of the right triangle, and the c represents the hypotenuse, the side opposite the right angle. Substitute the values into the theorem and solve for b:

$$8^2 + b^2 = 10^2$$
$$64 + b^2 = 100$$
$$b^2 = 36$$
$$\sqrt{b^2} = \sqrt{36}$$
$$b = 6$$

If the legs of the right triangle measure 8 and 6, then the rectangle has sides measuring 8, 8, 6, and 6. The perimeter of a rectangle is the sum of the sides of the rectangle. Therefore, the perimeter of this rectangle is 28.

323. $\sqrt{74}$

The diagonals of a rhombus are perpendicular bisectors. This means that they divide each other into two congruent parts, forming four right triangles. The legs of the right triangles are 5 and 7, and the hypotenuse of the right triangle is the side of the rhombus. To find the length of a side of the rhombus, use the Pythagorean theorem.

The Pythagorean theorem states that $a^2 + b^2 = c^2$. Substitute the values into the theorem and solve for c:

$$7^2 + 5^2 = c^2$$
$$49 + 25 = c^2$$
$$74 = c^2$$
$$\sqrt{74} = \sqrt{c^2}$$
$$\sqrt{74} = c$$

324. 12

The altitude of a triangle forms right angles when it intersects the opposite side. The altitude of this isosceles triangle will also bisect the opposite side, $\overline{AC}$. It therefore forms two congruent right triangles, with the altitude being one of its legs, the other leg being 5, and the hypotenuse measuring 13. Use the Pythagorean theorem to solve for the altitude.

The Pythagorean theorem states that $a^2 + b^2 = c^2$. Substitute the values into the theorem and solve for b:

$$5^2 + b^2 = 13^2$$
$$25 + b^2 = 169$$
$$b^2 = 144$$
$$\sqrt{b^2} = \sqrt{144}$$
$$b = 12$$

325. False

To determine whether a triangle is a right triangle, see whether it satisfies Pythagorean theorem. The Pythagorean theorem states that $a^2 + b^2 = c^2$. The a and b represent the legs of the right triangle, and the c represents the hypotenuse, the side opposite the right angle. Substituting the values into the theorem results in the following:

$$10^2 + 20^2 \stackrel{?}{=} 25^2$$
$$100 + 400 \stackrel{?}{=} 625$$
$$500 \neq 625$$

Because the Pythagorean theorem doesn't hold true, this triangle is not a right triangle.

326. **Always**

Because two angles of $\triangle MAT$ are congruent to two angles of $\triangle AHT$, the two triangles are similar:

$$\angle MAT \cong \angle AHT$$
$$\angle T \cong \angle T$$
$$\triangle MAT \sim \triangle AHT$$

Similarly, because two angles of $\triangle MAT$ are congruent to two angles of $\triangle MHA$, the two triangles are similar:

$$\angle MAT \cong \angle MHA$$
$$\angle M \cong \angle M$$
$$\triangle MAT \sim \triangle MHA$$

This means that all three triangles are similar to each other.

327. *c*

When you draw the altitude to the hypotenuse of a right triangle, similar triangles are formed. If all three triangles are similar, then all three triangles are right triangles.

The Pythagorean theorem states that $a^2 + b^2 = c^2$. Substituting in for the little right triangle on the left hand side, you get $a^2 + c^2 = d^2$.

328. *a*

When you draw an altitude to the hypotenuse of a right triangle, you create two right triangles that are similar to each other and to the original right triangle. Because these triangles are similar, you can set up proportions relating the corresponding sides.

The leg of a right triangle is the mean proportional between the hypotenuse and the projection of the leg on the hypotenuse:

$$\frac{\text{Leg}}{\text{Hypotenuse}} = \frac{\text{Projection}}{\text{Leg}}$$

In the figure, this would mean that

$$\frac{d}{a+b} = \frac{a}{d}$$

329. *e*

When you draw an altitude to the hypotenuse of a right triangle, you create two right triangles that are similar to each other and to the original right triangle. Because these triangles are similar, you can set up proportions relating the corresponding sides.

The leg of a right triangle is the mean proportional between the hypotenuse and the projection of the leg on the hypotenuse:

$$\frac{\text{Leg}}{\text{Hypotenuse}} = \frac{\text{Projection}}{\text{Leg}}$$

In the figure, this would mean that

$$\frac{e}{a+b} = \frac{b}{e}$$

---

330. e^2

When you draw the altitude to the hypotenuse of a right triangle, similar triangles are formed. If all three triangles are similar, then all three triangles are right triangles.

The Pythagorean theorem states that $a^2 + b^2 = c^2$. The a and b represent the legs of the right triangle, and the c represents the hypotenuse, the side opposite the right angle. Substituting in for the little right triangle on the right hand side, you get $b^2 + c^2 = e^2$.

---

331.   6

When you draw an altitude to the hypotenuse of a right triangle, you create two right triangles that are similar to each other and to the original right triangle. Because these triangles are similar, you can set up proportions relating the corresponding sides.

The altitude to the hypotenuse of a right triangle is the mean proportional between the two segments that the hypotenuse is divided into:

$$\frac{\text{Altitude}}{\text{Part of Hypotenuse}} = \frac{\text{Other Part of Hypotenuse}}{\text{Altitude}}$$

In the figure, this would mean that

$$\frac{9}{x} = \frac{x}{4}$$

Cross-multiplying gives you the following:

$$x^2 = 36$$
$$\sqrt{x^2} = \sqrt{36}$$
$$x = 6$$

---

332.   4

Because these triangles are similar, you can set up proportions relating the corresponding sides. The leg of a right triangle is the mean proportional between the hypotenuse and the projection of the leg on the hypotenuse:

$$\frac{\text{Leg}}{\text{Hypotenuse}} = \frac{\text{Projection}}{\text{Leg}}$$

In the figure, this would mean that

$$\frac{x}{8} = \frac{2}{x}$$

Cross-multiplying gives you the following:

$$x^2 = 16$$
$$\sqrt{x^2} = \sqrt{16}$$
$$x = 4$$

333.   4

The leg of a right triangle is the mean proportional between the hypotenuse and the projection of the leg on the hypotenuse:

$$\frac{\text{Leg}}{\text{Hypotenuse}} = \frac{\text{Projection}}{\text{Leg}}$$

In the figure, this would mean that

$$\frac{8}{x+12} = \frac{x}{8}$$

Cross-multiplying gives you the following:

$$x^2 + 12x = 64$$
$$x^2 + 12x - 64 = 0$$
$$(x + 16)(x - 4) = 0$$
$$x + 16 = 0 \text{ or } x - 4 = 0$$
$$\cancel{x = -16} \text{ or } x = 4$$

334. $2\sqrt{35}$

The altitude to the hypotenuse of a right triangle is the mean proportional between the two segments that the hypotenuse is divided into:

$$\frac{\text{Altitude}}{\text{Part of Hypotenuse}} = \frac{\text{Other Part of Hypotenuse}}{\text{Altitude}}$$

In the figure, this would mean that

$$\frac{x}{10} = \frac{14}{x}$$

Cross-multiplying gives you the following:

$$x^2 = 140$$
$$\sqrt{x^2} = \sqrt{140}$$
$$x = \sqrt{4}\sqrt{35}$$
$$x = 2\sqrt{35}$$

335. 1

The altitude to the hypotenuse of a right triangle is the mean proportional between the two segments that the hypotenuse is divided into:

$$\frac{\text{Altitude}}{\text{Part of Hypotenuse}} = \frac{\text{Other Part of Hypotenuse}}{\text{Altitude}}$$

In the figure, this would mean that

$$\frac{3}{x} = \frac{9}{3}$$

Cross-multiplying gives you the following:

$$9x = 9$$
$$x = 1$$

336. 16

The leg of a right triangle is the mean proportional between the hypotenuse and the projection of the leg on the hypotenuse:

$$\frac{\text{Leg}}{\text{Hypotenuse}} = \frac{\text{Projection}}{\text{Leg}}$$

In the figure, this would mean that

$$\frac{20}{x+9} = \frac{x}{20}$$

Cross-multiplying gives you the following:

$$x^2 + 9x = 400$$
$$x^2 + 9x - 400 = 0$$
$$(x+25)(x-16) = 0$$
$$x+25 = 0 \text{ or } x-16 = 0$$
$$\cancel{x = -25} \text{ or } x = 16$$

337. $4\sqrt{3}$

The leg of a right triangle is the mean proportional between the hypotenuse and the projection of the leg on the hypotenuse:

$$\frac{\text{Leg}}{\text{Hypotenuse}} = \frac{\text{Projection}}{\text{Leg}}$$

In the figure, this would mean that

$$\frac{x}{12} = \frac{4}{x}$$

Cross-multiplying gives you the following:

$$\begin{aligned} x^2 &= 48 \\ \sqrt{x^2} &= \sqrt{48} \\ x &= \sqrt{16}\sqrt{3} \\ x &= 4\sqrt{3} \end{aligned}$$

338. **24**

Let

$$\begin{aligned} RI &= x \\ GI &= x + 21 \end{aligned}$$

When you draw an altitude to the hypotenuse of a right triangle, you create two right triangles that are similar to each other and to the original right triangle. Because these triangles are similar, you can set up proportions relating the corresponding sides.

The leg of a right triangle is the mean proportional between the hypotenuse and the projection of the leg on the hypotenuse:

$$\frac{\text{Leg}}{\text{Hypotenuse}} = \frac{\text{Projection}}{\text{Leg}}$$

In the figure, this would mean that

$$\begin{aligned} \frac{9}{x + x + 21} &= \frac{x}{9} \\ \frac{9}{2x + 21} &= \frac{x}{9} \end{aligned}$$

Cross-multiplying gives you the following:

$$\begin{aligned} 2x^2 + 21x &= 81 \\ 2x^2 + 21x - 81 &= 0 \end{aligned}$$

You can solve for x using factoring by grouping. Multiplying the coefficient of the x^2 term by the constant gives you 2(–81) = –162. Two numbers that multiply to –162 and that add up to the middle coefficient, 21, are 27 and –6, so split the x term as follows. Then solve for x:

$$\begin{aligned} 2x^2 + 27x - 6x - 81 \\ x(2x + 27) - 3(2x + 27) &= 0 \\ (2x + 27)(x - 3) &= 0 \\ 2x + 27 = 0 \text{ or } x - 3 &= 0 \\ 2x = -27 \text{ or } x &= 3 \\ \cancel{x = \frac{-27}{2}} \text{ or } x &= 3 \end{aligned}$$

Therefore, $GI = x + 21 = 3 + 21 = 24$.

339.   25

Let

$$RI = x$$
$$GI = x + 21$$

When you draw an altitude to the hypotenuse of a right triangle, you create two right triangles that are similar to each other and to the original right triangle. Because these triangles are similar, you can set up proportions relating the corresponding sides.

The altitude to the hypotenuse of a right triangle is the mean proportional between the two segments that the hypotenuse is divided into:

$$\frac{\text{Altitude}}{\text{Part of Hypotenuse}} = \frac{\text{Other Part of Hypotenuse}}{\text{Altitude}}$$

In the figure, this would mean that

$$\frac{10}{x} = \frac{x+21}{10}$$

Cross-multiply and solve for x:

$$x^2 + 21x = 100$$
$$x^2 + 21x - 100 = 0$$
$$(x+25)(x-4) = 0$$
$$x + 25 = 0 \text{ or } x - 4 = 0$$
$$\cancel{x = -25} \text{ or } x = 4$$

Therefore, $GI = x + 21 = 4 + 21 = 25$.

340.   6

Let

$$IT = x$$
$$GI = x + 2$$

$\triangle GIT$ is a right triangle; therefore, you can use the Pythagorean theorem.

The Pythagorean theorem states that $a^2 + b^2 = c^2$. Substitute the values into the theorem and solve for x:

$$x^2 + (x+2)^2 = 10^2$$
$$x^2 + (x+2)(x+2) = 100$$
$$x^2 + x^2 + 4x + 4 = 100$$
$$2x^2 + 4x + 4 = 100$$
$$2x^2 + 4x - 96 = 0$$
$$2(x^2 + 2x - 48) = 0$$
$$2(x-6)(x+8) = 0$$
$$x - 6 = 0 \ \text{ or } \ x + 8 = 0$$
$$x = 6 \ \text{ or } \ \cancel{x = -8}$$

$IT = x$, so the answer is 6.

341. 12

When you draw an altitude to the hypotenuse of a right triangle, you create two right triangles that are similar to each other and to the original right triangle. Because these triangles are similar, you can set up proportions relating the corresponding sides.

The altitude to the hypotenuse of a right triangle is the mean proportional between the two segments that the hypotenuse is divided into:

$$\frac{\text{Altitude}}{\text{Part of Hypotenuse}} = \frac{\text{Other Part of Hypotenuse}}{\text{Altitude}}$$

In the figure, this would mean that

$$\frac{x}{9} = \frac{16}{x}$$

Cross-multiplying gives you the following:

$$x^2 = 144$$
$$\sqrt{x^2} = \sqrt{144}$$
$$x = 12$$

342. 9

Because these triangles are similar, you can set up proportions relating the corresponding sides. The leg of a right triangle is the mean proportional between the hypotenuse and the projection of the leg on the hypotenuse:

$$\frac{\text{Leg}}{\text{Hypotenuse}} = \frac{\text{Projection}}{\text{Leg}}$$

In the figure, this would mean that

$$\frac{6}{x} = \frac{4}{6}$$

Cross-multiplying gives you the following:

$$4x = 36$$
$$x = 9$$

---

343.   50

The altitude to the hypotenuse of a right triangle is the mean proportional between the two segments that the hypotenuse is divided into:

$$\frac{\text{Altitude}}{\text{Part of Hypotenuse}} = \frac{\text{Other Part of Hypotenuse}}{\text{Altitude}}$$

In the figure, this would mean that

$$\frac{25}{x} = \frac{x}{25}$$

Cross-multiplying gives you the following:

$$x^2 = 625$$
$$\sqrt{x^2} = \sqrt{625}$$
$$x = 25$$

Therefore, the hypotenuse is $x + x = 25 + 25 = 50$.

---

344.   9

Let

$$DC = x$$
$$WC = x + 8$$

When you draw an altitude to the hypotenuse of a right triangle, you create two right triangles that are similar to each other and to the original right triangle. Because these triangles are similar, you can set up proportions relating the corresponding sides.

The altitude to the hypotenuse of a right triangle is the mean proportional between the two segments that the hypotenuse is divided into:

$$\frac{\text{Altitude}}{\text{Part of Hypotenuse}} = \frac{\text{Other Part of Hypotenuse}}{\text{Altitude}}$$

In the figure, this would mean that

$$\frac{3}{x} = \frac{x+8}{3}$$

Cross-multiply to solve for x:

$$x^2 + 8x = 9$$
$$x^2 + 8x - 9 = 0$$
$$(x+9)(x-1) = 0$$
$$x + 9 = 0 \text{ or } x - 1 = 0$$
$$\cancel{x = -9} \text{ or } x = 1$$

Therefore, $WC = x + 8 = 1 + 8 = 9$.

345.   18

Let

$$WC = x$$
$$DC = x + 10$$

The altitude to the hypotenuse of a right triangle is the mean proportional between the two segments that the hypotenuse is divided into:

$$\frac{\text{Altitude}}{\text{Part of Hypotenuse}} = \frac{\text{Other Part of Hypotenuse}}{\text{Altitude}}$$

In the figure, this would mean that

$$\frac{12}{x} = \frac{x+10}{12}$$

Cross-multiply and solve for x:

$$x^2 + 10x = 144$$
$$x^2 + 10x - 144 = 0$$
$$(x+18)(x-8) = 0$$
$$x + 18 = 0 \text{ or } x - 8 = 0$$
$$\cancel{x = -18} \text{ or } x = 8$$

Therefore, $DC = x + 10 = 8 + 10 = 18$.

346.   3

Let

$$AD = x$$
$$BD = x + 3$$

The leg of a right triangle is the mean proportional between the hypotenuse and the projection of the leg on the hypotenuse:

$$\frac{\text{Leg}}{\text{Hypotenuse}} = \frac{\text{Projection}}{\text{Leg}}$$

In the figure, this would mean that

$$\frac{\sqrt{27}}{x+x+3} = \frac{x}{\sqrt{27}}$$

$$\frac{\sqrt{27}}{2x+3} = \frac{x}{\sqrt{27}}$$

Cross-multiplying gives you

$$2x^2 + 3x = 27$$

$$2x^2 + 3x - 27 = 0$$

Use factoring by grouping to solve for x. Multiplying the coefficient of the x^2 term by the constant gives you $2(-27) = -54$. Two numbers that multiply to -54 and that add up to the middle coefficient, 3, are 9 and -6, so split the x term as follows. Then solve for x:

$$2x^2 + 9x - 6x - 27 = 0$$

$$x(2x+9) - 3(2x+9) = 0$$

$$(2x+9)(x-3) = 0$$

$$2x + 9 = 0 \quad \text{or} \quad x - 3 = 0$$

$$2x = -9 \quad \text{or} \quad x = 3$$

$$x \neq \frac{9}{2} \quad \text{or} \quad x = 3$$

Therefore, $AD = 3$.

---

347.   6

Let

$$AD = x$$

$$BD = x + 4$$

The leg of a right triangle is the mean proportional between the hypotenuse and the projection of the leg on the hypotenuse:

$$\frac{\text{Leg}}{\text{Hypotenuse}} = \frac{\text{Projection}}{\text{Leg}}$$

In the figure, this would mean that

$$\frac{4}{x+x+4} = \frac{x}{4}$$

$$\frac{4}{2x+4} = \frac{x}{4}$$

Cross-multiply and solve for x:

$$2x^2 + 4x = 16$$
$$2x^2 + 4x - 16 = 0$$
$$2(x^2 + 2x - 8) = 0$$
$$2(x+4)(x-2) = 0$$
$$x+4 = 0 \text{ or } x-2 = 0$$
$$\cancel{x = -4} \text{ or } x = 2$$

Therefore, $BD = x + 4 = 2 + 4 = 6$.

---

348. **32 mi**

When you draw an altitude to the hypotenuse of a right triangle, you create two right triangles that are similar to each other and to the original right triangle. Because these triangles are similar, you can set up proportions relating the corresponding sides.

The altitude to the hypotenuse of a right triangle is the mean proportional between the two segments that the hypotenuse is divided into:

$$\frac{\text{Altitude}}{\text{Part of Hypotenuse}} = \frac{\text{Other Part of Hypotenuse}}{\text{Altitude}}$$

In the figure, this would mean that

$$\frac{24}{18} = \frac{x}{24}$$

Cross-multiplying gives you

$$18x = 576$$
$$x = 32$$

The coffee shop and the pizza place are 32 miles apart.

---

349. **30 mi**

The Pythagorean theorem states that $a^2 + b^2 = c^2$. Substitute the values into the theorem and solve for c:

$$18^2 + 24^2 = c^2$$
$$900 = c^2$$
$$\sqrt{900} = \sqrt{c^2}$$
$$30 = c$$

The deli is 30 miles away from the office.

350. **40 mi**

First find how far away the pizza place is from the coffee shop. When you draw an altitude to the hypotenuse of a right triangle, you create two right triangles that are similar to each other and to the original right triangle. Because these triangles are similar, you can set up proportions relating the corresponding sides.

The altitude to the hypotenuse of a right triangle is the mean proportional between the two segments that the hypotenuse is divided into:

$$\frac{\text{Altitude}}{\text{Part of Hypotenuse}} = \frac{\text{Other Part of Hypotenuse}}{\text{Altitude}}$$

$$\frac{24}{18} = \frac{x}{24}$$

Cross-multiplying gives you

$$18x = 576$$
$$x = 32$$

You can now use the Pythagorean theorem to find the distance from the pizza place to the office. The Pythagorean theorem states that $a^2 + b^2 = c^2$. Substitute the values into the theorem and solve for x:

$$32^2 + 24^2 = x^2$$
$$1,600 = x^2$$
$$\sqrt{1,600} = \sqrt{x^2}$$
$$x = 40$$

The pizza place is 40 miles from Gavin's office. (Hope his boss doesn't expect him back anytime soon.)

351. **$10\sqrt{2}$**

The pattern of the sides of an isosceles right triangle is as follows:

- Let $x =$ the measure of the sides opposite each 45° angle.
- Let $x\sqrt{2} =$ the measure of the hypotenuse.

In this problem, $x = 10$, which means hypotenuse $OR = 10\sqrt{2}$.

352. **18**

The pattern of the sides of an isosceles right triangle is as follows:

- Let $x =$ the measure of the sides opposite each 45° angle.
- Let $x\sqrt{2} =$ the measure of the hypotenuse.

In this problem, hypotenuse $OR = 18\sqrt{2}$. This means that $x = 18$, which is the measure of the length of each leg of the right triangle.

353. 24

The pattern of the sides of an isosceles right triangle is as follows:

- Let x = the measure of the sides opposite each 45° angle.
- Let $x\sqrt{2}$ = the measure of the hypotenuse.

In this problem, $x = 12\sqrt{2}$, which means that hypotenuse $RO = 12\sqrt{2} \cdot \sqrt{2} = 12(2) = 24$.

354. $10\sqrt{2}$

The pattern of the sides of an isosceles right triangle is as follows:

- Let x = the measure of the sides opposite each 45° angle.
- Let $x\sqrt{2}$ = the measure of the hypotenuse.

In this problem, the following is true:

$$x\sqrt{2} = 20$$
$$x = \frac{20}{\sqrt{2}} \cdot \frac{\sqrt{2}}{\sqrt{2}}$$
$$x = \frac{20\sqrt{2}}{2} = 10\sqrt{2}$$

355. $7\sqrt{3}$

The pattern of the sides of a 30°-60°-90° triangle is as follows:

- Let x = the measure of the side opposite the 30° angle.
- Let $x\sqrt{3}$ = the measure of the side opposite the 60° angle.
- Let $2x$ = the measure of the side opposite the 90° angle.

$\overline{IX}$ is the side opposite the 30° angle; therefore, $x = 7$.

$\overline{SI}$ is the side opposite the 60° angle, so it's equal to $x\sqrt{3}$; therefore, $SI = 7\sqrt{3}$.

356. 10

The pattern of the sides of a 30°-60°-90° triangle is as follows:

- Let x = the measure of the side opposite the 30° angle.
- Let $x\sqrt{3}$ = the measure of the side opposite the 60° angle.
- Let $2x$ = the measure of the side opposite the 90° angle.

$\overline{IX}$ is the side opposite the 30° angle; therefore, $x = 5$.

$\overline{SX}$ is the side opposite the 90° angle, so it's equal to $2x$; therefore, $SX = 5(2) = 10$.

357. **30**

The pattern of the sides of a 30°-60°-90° triangle is as follows:

- Let x = the measure of the side opposite the 30° angle.
- Let $x\sqrt{3}$ = the measure of the side opposite the 60° angle.
- Let $2x$ = the measure of the side opposite the 90° angle.

$\overline{SI}$ is the side opposite the 60° angle; therefore,

$$x\sqrt{3} = 15\sqrt{3}$$
$$x = 15$$

$\overline{SX}$ is the side opposite the 90° angle, so it's equal to $2x$; therefore, $2(15) = 30$.

358. **$13\sqrt{3}$**

The pattern of the sides of a 30°-60°-90° triangle is as follows:

- Let x = the measure of the side opposite the 30° angle.
- Let $x\sqrt{3}$ = the measure of the side opposite the 60° angle.
- Let $2x$ = the measure of the side opposite the 90° angle.

$\overline{SX}$ is the side opposite the 90° angle; therefore,

$$2x = 26$$
$$x - 13$$

$\overline{SI}$ is the side opposite the 60° angle, so it's equal to $x\sqrt{3}$; therefore, $SI - 13\sqrt{3}$

359. **$10\sqrt{3}$**

The pattern of the sides of a 30°-60°-90° triangle is as follows:

- Let x = the measure of the side opposite the 30° angle.
- Let $x\sqrt{3}$ = the measure of the side opposite the 60° angle.
- Let $2x$ = the measure of the side opposite the 90° angle.

$\overline{SI}$ is the side opposite the 60° angle; therefore,

$$x\sqrt{3} = 15$$
$$x = \frac{15}{\sqrt{3}} \cdot \frac{\sqrt{3}}{\sqrt{3}}$$
$$x = \frac{15\sqrt{3}}{3} = 5\sqrt{3}$$

$\overline{SX}$ is the side opposite the 90° angle, so it's equal to $2x$; therefore, $SX = 2 \cdot 5\sqrt{3} = 10\sqrt{3}$.

360. 10

The diagonals of a rhombus are perpendicular bisectors, creating four right triangles. The diagonals also bisect the angles of the rhombus. This creates 30°-60°-90° triangles.

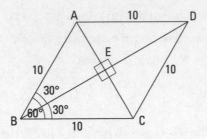

The pattern of the sides of a 30°-60°-90° triangle is as follows:

- Let x = the measure of the side opposite the 30° angle.
- Let $x\sqrt{3}$ = the measure of the side opposite the 60° angle.
- Let $2x$ = the measure of the side opposite the 90° angle.

$\overline{AB}$ is the side opposite the 90° angle; therefore,

$$2x = 10$$
$$x = 5$$

$\overline{AE}$ is the side opposite the 30° angle, so it's equal to x; therefore, $AE = 5$.

AC is double the length of AE; therefore, $AC = 10$.

361. $9\sqrt{2}$

You find the perimeter of a square by adding all four sides of the square together. A square has four equal sides, each measuring x units in this case. If the perimeter is 36, then

$$x + x + x + x = 36$$
$$4x = 36$$
$$x = 9$$

A square has four right angles. The diagonal of a square divides the square into two isosceles right triangles, as in the following figure.

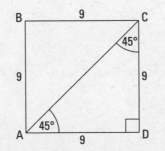

In an isosceles right triangle, if x is the side opposite the 45° angle, then the hypotenuse equals $x\sqrt{2}$. The side of the square is 9, so the diagonal equals $9\sqrt{2}$.

---

362.   8

An *equilateral triangle* is a triangle with three congruent sides and three congruent angles. Each angle therefore measures 60°. When you draw the altitude, it creates two 30°-60°-90° triangles, as in the following figure.

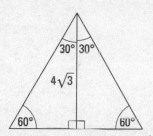

The pattern of the sides of a 30°-60°-90° triangle is as follows:

- Let x = the measure of the side opposite the 30° angle.
- Let $x\sqrt{3}$ = the measure of the side opposite the 60° angle.
- Let $2x$ = the measure of the side opposite the 90° angle.

The altitude is the side opposite the 60° angle; therefore,

$$x\sqrt{3} = 4\sqrt{3}$$
$$x = 4$$

A side of the equilateral triangle is opposite the 90° angle, so it's equal to $2x$; therefore, a side of triangle equals $2 \cdot 4 = 8$.

---

363. $42\sqrt{3}$

An *equilateral triangle* is a triangle with three congruent sides and three congruent angles. Each angle therefore measures 60°. When you draw the altitude, it creates two 30°-60°-90° triangles, as in the following figure.

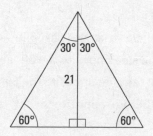

The pattern of the sides of a 30°-60°-90° triangle is as follows:

- Let x = the measure of the side opposite the 30° angle.
- Let $x\sqrt{3}$ = the measure of the side opposite the 60° angle.
- Let $2x$ = the measure of the side opposite the 90° angle.

The altitude is the side opposite the 60° angle; therefore,

$$x\sqrt{3} = 21$$
$$x = \frac{21}{\sqrt{3}}$$
$$x = \frac{21}{\sqrt{3}} \cdot \frac{\sqrt{3}}{\sqrt{3}}$$
$$x = \frac{21\sqrt{3}}{3} = 7\sqrt{3}$$

A side of the equilateral triangle is opposite the 90° angle, so it's equal to $2x$; therefore, a side of triangle equals $2 \cdot 7\sqrt{3} = 14\sqrt{3}$.

If each side of the triangle measures $14\sqrt{3}$, then the perimeter of the triangle, the sum of all three sides, is equal to $3 \cdot 14\sqrt{3} = 42\sqrt{3}$.

364. $14\sqrt{3}$

Consecutive angles of a rhombus are supplementary (add up to 180°). If one pair of opposite angles measure 120°, then the other pair of opposite angles is 60°.

The diagonals of a rhombus are perpendicular bisectors, creating four right triangles. The diagonals also bisect the angles of the rhombus. This creates 30°-60°-90° triangles.

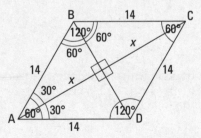

The pattern of the sides of a 30°-60°-90° triangle is as follows:

- Let x = the measure of the side opposite the 30° angle.
- Let $x\sqrt{3}$ = the measure of the side opposite the 60° angle.
- Let $2x$ = the measure of the side opposite the 90° angle.

$\overline{AB}$ (a side of the rhombus) is the side opposite the 90° angle in the 30°-60°-90° triangle; therefore,

$$2x = 14$$
$$x = 7$$

The longer diagonal, $\overline{AC}$, is the diagonal opposite the 120° angle. $\overline{AE}$ is the side opposite the 60° angle, so it's equal to $x\sqrt{3} = 7\sqrt{3}$. AC is double the length of AE, so $AC = 2 \cdot 7\sqrt{3} = 14\sqrt{3}$.

---

365. $3\sqrt{2}$ **feet**

An isosceles trapezoid has a pair of parallel bases and a pair of congruent but not parallel legs. Drawing altitudes in this trapezoid creates two 45°-45°-90° triangles, as in the following figure:

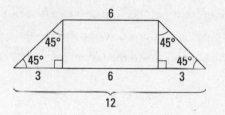

The pattern of the sides of an isosceles right triangle 45°-45°-90° is as follows:

- Let $x =$ the measure of the sides opposite each 45° angle.
- Let $x\sqrt{2} =$ the measure of the hypotenuse.

In this problem, the side opposite the 45° angle is 3; therefore, $x = 3$. The hypotenuse, $x\sqrt{2}$, then equals $3\sqrt{2}$.

---

366. $8\sqrt{3}$

The pattern of the sides of an isosceles right triangle 45°-45°-90° is as follows:

- Let $b =$ the measure of the sides opposite each 45° angle.
- Let $b\sqrt{2} =$ the measure of the hypotenuse.

In this problem, the side opposite the 45° angle is $12\sqrt{2}$; therefore, $b = 12\sqrt{2}$.

The hypotenuse, $\overline{PF}$, then equals $b\sqrt{2} = \left(12\sqrt{2}\right)\sqrt{2} = 12(2) = 24$.

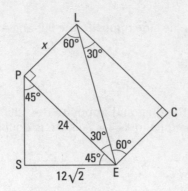

Diagonal $\overline{LE}$ divides this rectangle into two 30°-60°-90° triangles. The pattern of the sides of a 30°-60°-90° triangle is as follows:

- Let x = the measure of the side opposite the 30° angle.
- Let $x\sqrt{3}$ = the measure of the side opposite the 60° angle.
- Let $2x$ = the measure of the side opposite the 90° angle.

$\overline{PE}$, measuring 24, is the side opposite the 60° angle, so it's equal to $x\sqrt{3}$; therefore,

$$x\sqrt{3} = 24$$

$$x = \frac{24}{\sqrt{3}}$$

$$x = \frac{24}{\sqrt{3}} \cdot \frac{\sqrt{3}}{\sqrt{3}}$$

$$x = \frac{24\sqrt{3}}{3} = 8\sqrt{3}$$

$\overline{PL}$ is the side opposite the 60° angle, so it equals x. Therefore, $PL = 8\sqrt{3}$.

367. $16\sqrt{3}$

The pattern of the sides of an isosceles right triangle is as follows:

- Let b = the measure of the sides opposite each 45° angle.
- Let $b\sqrt{2}$ = the measure of the hypotenuse.

In this problem, the side opposite the 45° angle is $12\sqrt{2}$; therefore, $b = 12\sqrt{2}$.

The hypotenuse, $\overline{PE}$, then equals $b\sqrt{2} = \left(12\sqrt{2}\right)\sqrt{2} = 12(2) = 24$.

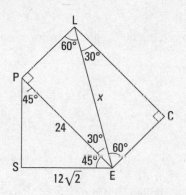

Diagonal *LE* divides this rectangle into two 30°-60°-90° triangles. The pattern of the sides of a 30°-60°-90° triangle is as follows:

- Let $x =$ the measure of the side opposite the 30° angle.
- Let $x\sqrt{3} =$ the measure of the side opposite the 60° angle.
- Let $2x =$ the measure of the side opposite the 90° angle.

$\overline{PE}$, which measures 24, is the side opposite the 60° angle, so it's equal to $x\sqrt{3}$; therefore,

$$x\sqrt{3} = 24$$
$$x = \frac{24}{\sqrt{3}}$$
$$x = \frac{24}{\sqrt{3}} \cdot \frac{\sqrt{3}}{\sqrt{3}}$$
$$x = \frac{24\sqrt{3}}{3} = 8\sqrt{3}$$

$\overline{LE}$ is the side opposite the 90° angle, which equals $2x$. Therefore, $LE = 2 \cdot 8\sqrt{3} = 16\sqrt{3}$.

368. $\frac{32}{3}$

Each similar triangle is a 30°-60°-90° triangle. The pattern of the sides of a 30°-60°-90° triangle is as follows:

- Let $x =$ the measure of the side opposite the 30° angle.
- Let $x\sqrt{3} =$ the measure of the side opposite the 60° angle.
- Let $2x =$ the measure of the side opposite the 90° angle.

In $\triangle ABC$, $\overline{BC}$ is the side opposite the 30° angle; therefore, $x = 3$.

$\overline{AC}$ is the side opposite the 90° angle. Therefore, $AC = 2x = 2(3) = 6$.

In $\triangle ACD$, $\overline{AC}$ is the side opposite the 60° angle; therefore,

$$x\sqrt{3} = 6$$

$$x = \frac{6}{\sqrt{3}}$$

$$x = \frac{6}{\sqrt{3}} \cdot \frac{\sqrt{3}}{\sqrt{3}}$$

$$x = \frac{6\sqrt{3}}{3} = 2\sqrt{3}$$

$\overline{AD}$ is the side opposite the 90° angle; therefore, $AD = 2 \cdot 2\sqrt{3} = 4\sqrt{3}$.

In similar triangles, sides are in proportion. The ratio of $AC : AD$ is

$$\frac{4\sqrt{3}}{6} = \frac{2\sqrt{3}}{3}$$

This means that if you multiply each hypotenuse by $\frac{2\sqrt{3}}{3}$, you'll find the length of the next hypotenuse:

$$AE = 4\sqrt{3} \cdot \frac{2\sqrt{3}}{3} = \frac{8(3)}{3} = 8$$

$$AF = 8 \cdot \frac{2\sqrt{3}}{3} = \frac{16\sqrt{3}}{3}$$

$$AG = \frac{16\sqrt{3}}{3} \cdot \frac{2\sqrt{3}}{3} = \frac{32(3)}{9} = \frac{32}{3}$$

369. $2\sqrt{2}$

The pattern of the sides of an isosceles right triangle (45°-45°-90°) is as follows:

- Let x = the measure of the sides opposite each 45° angle.
- Let $x\sqrt{2}$ = the measure of the hypotenuse.

$\overline{AB}$ is the hypotenuse; therefore,

$$x\sqrt{2} = 8$$

$$x = \frac{8}{\sqrt{2}}$$

$$x = \frac{8}{\sqrt{2}} \cdot \frac{\sqrt{2}}{\sqrt{2}}$$

$$x = \frac{8\sqrt{2}}{2} = 4\sqrt{2}$$

$\overline{AC}$ is the side opposite the 45° angle. Therefore, $AC = x = 4\sqrt{2}$.

In similar triangles, sides are in proportion. The ratio of $AC : AB$ is

$$\frac{4\sqrt{2}}{8} = \frac{\sqrt{2}}{2}$$

This means that if you multiply each hypotenuse by $\frac{\sqrt{2}}{2}$, you'll find the length of the next hypotenuse:

$$AD - 4\sqrt{2} \cdot \frac{\sqrt{2}}{2} - \frac{4(2)}{2} = 4$$

$$AE = 4 \cdot \frac{\sqrt{2}}{2} = 2\sqrt{2}$$

370. $\frac{7}{25}$

In a right triangle, the following trigonometric proportion is true:

$$\sin\theta = \frac{\text{opposite}}{\text{hypotenuse}}$$

The side opposite $\angle T$ is 7, and the hypotenuse is 25. Using the trigonometric ratio, the sine of $\angle T$ is

$$\sin T = \frac{7}{25}$$

371. $\frac{24}{7}$

In a right triangle, the following trigonometric proportion is true:

$$\tan\theta = \frac{\text{opposite}}{\text{adjacent}}$$

The side opposite $\angle R$ is 24, and the side adjacent to $\angle R$ is 7. Using the trigonometric ratio, the tangent of $\angle R$ is

$$\tan R = \frac{24}{7}$$

372. $\frac{24}{25}$

In a right triangle, the following trigonometric proportion is true:

$$\cos\theta = \frac{\text{adjacent}}{\text{hypotenuse}}$$

The side adjacent to $\angle T$ is 24, and the hypotenuse is 25. Using the trigonometric ratio, the cosine of $\angle T$ is

$$\cos T = \frac{24}{25}$$

373. 6

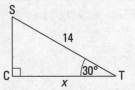

In a right triangle, the following trigonometric proportions are true:

$$\sin\theta = \frac{\text{opposite}}{\text{hypotenuse}}$$

$$\cos\theta = \frac{\text{adjacent}}{\text{hypotenuse}}$$

$$\tan\theta = \frac{\text{opposite}}{\text{adjacent}}$$

In this triangle, you're looking for information about the side opposite $\angle T$, and you're given information about the hypotenuse. Therefore, use the sine function:

$$\sin 30° = \frac{x}{12}$$

Multiplying both sides by 12 gives you

$$x = 12\sin 30° = 6$$

374. 12.1

In a right triangle, the following trigonometric proportions are true:

$$\sin\theta = \frac{\text{opposite}}{\text{hypotenuse}}$$

$$\cos\theta = \frac{\text{adjacent}}{\text{hypotenuse}}$$

$$\tan\theta = \frac{\text{opposite}}{\text{adjacent}}$$

In this triangle, you're looking for information about the side adjacent to $\angle T$, and you're given information about the hypotenuse. Therefore, use the cosine function:

$$\cos 30° = \frac{x}{14}$$

Multiplying both sides by 14 gives you

$$x = 14\cos 30° \approx 12.124 \approx 12.1$$

375.   **54°**

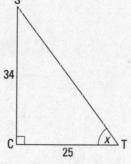

In a right triangle, the following trigonometric proportions are true:

$$\sin\theta = \frac{\text{opposite}}{\text{hypotenuse}}$$

$$\cos\theta = \frac{\text{adjacent}}{\text{hypotenuse}}$$

$$\tan\theta = \frac{\text{opposite}}{\text{adjacent}}$$

In this triangle, you're looking for information about $\angle T$, and you're given information about the side opposite $\angle T$ and the side adjacent to $\angle T$. Therefore, use the tangent function:

$$\tan x = \frac{34}{25}$$

$$\tan^{-1}\frac{34}{25} = x$$

$$54° \approx x$$

376.   **65°**

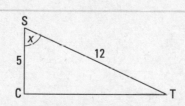

In a right triangle, the following trigonometric proportions are true:

$$\sin\theta = \frac{\text{opposite}}{\text{hypotenuse}}$$

$$\cos\theta = \frac{\text{adjacent}}{\text{hypotenuse}}$$

$$\tan\theta = \frac{\text{opposite}}{\text{adjacent}}$$

Answers
301–400

In this triangle, you're looking for information about $\angle S$, and you're given information about the side adjacent to $\angle S$ and the hypotenuse. Therefore, use the cosine function:

$$\cos x = \frac{5}{12}$$

$$\cos^{-1}\frac{5}{12} = x$$

$$65° \approx x$$

377.   **36.6**

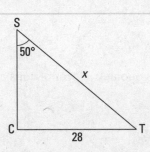

In a right triangle, the following trigonometric proportions are true:

$$\sin\theta = \frac{\text{opposite}}{\text{hypotenuse}}$$

$$\cos\theta = \frac{\text{adjacent}}{\text{hypotenuse}}$$

$$\tan\theta = \frac{\text{opposite}}{\text{adjacent}}$$

In this triangle, you're looking for information about the hypotenuse, and you're given information about the side opposite $\angle S$. Therefore, use the sine function:

$$\sin 50° = \frac{28}{x}$$

Solving for x gives you

$$x \sin 50° = 28$$

$$x = \frac{28}{\sin 50°}$$

$$x \approx 36.6$$

378. **26 ft**

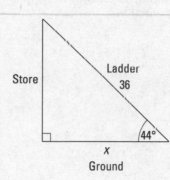

In a right triangle, the following trigonometric proportions are true:

$$\sin\theta = \frac{\text{opposite}}{\text{hypotenuse}}$$

$$\cos\theta = \frac{\text{adjacent}}{\text{hypotenuse}}$$

$$\tan\theta = \frac{\text{opposite}}{\text{adjacent}}$$

In this triangle, you're looking for information about the side adjacent to the 44° angle and are given information about the hypotenuse. Therefore, use the cosine function:

$$\cos 44° = \frac{x}{36}$$

Multiplying both sides by 36 gives you

$$36\cos 44° = x$$
$$25.89623281 \approx x$$
$$26 \approx x$$

379. **27°**

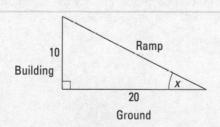

In a right triangle, the following trigonometric proportions are true:

$$\sin\theta = \frac{\text{opposite}}{\text{hypotenuse}}$$

$$\cos\theta = \frac{\text{adjacent}}{\text{hypotenuse}}$$

$$\tan\theta = \frac{\text{opposite}}{\text{adjacent}}$$

In this triangle, you're looking for information about the angle the ramp makes with the ground, and you're given information about the side opposite it and the side adjacent to it. Therefore, use the tangent function:

$$\tan x = \frac{10}{20}$$

$$x = \tan^{-1}\frac{10}{20}$$

$$x \approx 26.56505118° \approx 27°$$

380. **304.7 ft**

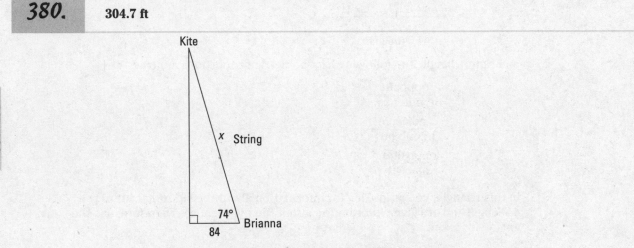

In a right triangle, the following trigonometric proportions are true:

$$\sin\theta = \frac{\text{opposite}}{\text{hypotenuse}}$$

$$\cos\theta = \frac{\text{adjacent}}{\text{hypotenuse}}$$

$$\tan\theta = \frac{\text{opposite}}{\text{adjacent}}$$

In this triangle, you're looking for information about the side adjacent to the hypotenuse, and you're given information about the side adjacent to the 74° angle. Therefore, use the cosine function

$$\cos 74° = \frac{84}{x}$$

Solving for x gives you

$$x\cos 74° = 84$$

$$x = \frac{84}{\cos 74°}$$

$$x \approx 304.7482434 \approx 304.7$$

381. $\overline{AC}$

The sum of the angles of a triangle is 180°, so set up an equation and solve for $m\angle B$:

$$40° + 30° + m\angle B = 180°$$
$$70° + m\angle B = 180°$$
$$m\angle B = 110°$$

$\angle B$ is the largest angle of the triangle; therefore, AC, the side opposite $\angle B$, is the longest side of the triangle.

382. $\overline{MN}$

The sum of the angles of a triangle is 180°, so set up an equation and solve for $m\angle M$:

$$45° + 80° + m\angle M = 180°$$
$$125° + m\angle M = 180°$$
$$m\angle M = 55°$$

$\angle O$ is the largest angle of the triangle; therefore, $\overline{MN}$, the side opposite $\angle O$, is the longest side of the triangle.

383. $\overline{BR}$

The sum of the angles of a triangle is 180°, so set up an equation and solve for $m\angle I$:

$$105° + 50° + m\angle I = 180°$$
$$155° + m\angle I = 180°$$
$$m\angle I = 25°$$

$\angle I$ is the smallest angle of the triangle; therefore, $\overline{BR}$, the side opposite $\angle I$, is the shortest side of the triangle.

384. $\angle P$

The smallest angle of a triangle is opposite the smallest side. You know *PI* and *PN*, so use the Pythagorean theorem to find the length of $\overline{NI}$:

$$a^2 + b^2 = c^2$$
$$15^2 + b^2 = 17^2$$
$$225 + b^2 = 289$$
$$b^2 = 64$$
$$\sqrt{b^2} = \sqrt{64}$$
$$b = 8$$

Because $\overline{NI}$ is the shortest side of the triangle, $\angle P$ is the smallest angle.

385. $\overline{AB}$

The longest side of a triangle is opposite the largest angle. The sum of the angles of a triangle is $180°$, so set up an equation and solve for x:

$$\left(x^2 + 68\right) + \left(x + 5\right) + 87 = 180°$$
$$x^2 + 68 + x + 5 + 87 = 180°$$
$$x^2 + x + 160 = 180°$$
$$x^2 + x - 20 = 0$$
$$\left(x + 5\right)\left(x - 4\right) = 0$$
$$\cancel{x = -5} \text{ or } x = 4$$

Plug in $x = 4$ to find the angle measures:

$$m\angle A = \left(4\right)^2 + 68 = 84°$$
$$m\angle B = \left(4\right) + 5 = 9°$$
$$m\angle C = 87°$$

The largest angle is $\angle C$, so the longest side is $\overline{AB}$.

386. $\angle E$

Plug the given value of x into each expression representing the length of each side of a triangle:

$$EF = \left(-2\right)^3 + 50 = -8 + 50 = 42$$
$$FG = 20\left(-2\right)^2 = 20\left(4\right) = 80$$
$$GE = \left(-2\right) + 62 = 60$$

The largest angle of a triangle is opposite the longest side of the triangle. $\overline{FG}$ is the longest side of the triangle, so $\angle E$ is the largest angle of the triangle.

387. $m\angle L < m\angle K < m\angle M$

The smallest angle of a triangle is found opposite the shortest side of the triangle. $\overline{KM}$ is the shortest side of the triangle; therefore, $\angle L$ is the smallest angle.

---

388. $m\angle Z < m\angle X < m\angle Y$

The smallest angle of a triangle is found opposite the shortest side of the triangle, so compare the lengths of the sides:

$$XY = \sqrt{27} = \sqrt{9}\sqrt{3} = 3\sqrt{3}$$

$$XZ = \frac{5}{2}\sqrt{12} = \frac{5}{2}\sqrt{4}\sqrt{3} = \frac{5}{2}\left(2\sqrt{3}\right) = 5\sqrt{3}$$

$$YZ = 4\sqrt{3}$$

$\overline{XY}$ is the shortest side of the triangle; therefore, $\angle Z$ is the smallest angle. Because $\overline{XZ}$ is the longest side of the triangle, $\angle Y$ is the largest angle of the triangle.

---

389. $\overline{NF} < \overline{NL} < \overline{FL}$

The sum of the angles of a triangle is $180°$, so find the measure of $\angle L$:

$$74 + 57.9 + m\angle L = 180°$$

$$131.9 + m\angle L = 180°$$

$$m\angle L = 48.1°$$

$\angle L$ is the smallest angle of the triangle; therefore, $\overline{NF}$, the side opposite $\angle L$, is the shortest side of the triangle. $\angle N$ is the largest angle of the triangle; therefore, $\overline{FL}$, the side opposite $\angle N$, is the longest side of the triangle.

---

390. $\overline{MB} < \overline{ML} < \overline{LB}$

The sum of the angles of a triangle is $180°$, so find the angle measures:

$$(x) + (x + 5) + 65 = 180°$$

$$x + x + 5 + 65 = 180°$$

$$2x + 70 = 180°$$

$$2x = 110°$$

$$x = 55°$$

$$m\angle L = x = 55°$$

$$m\angle B = x + 5 = 55 + 5 = 60°$$

$$m\angle M = 65°$$

$\angle L$ is the smallest angle of the triangle; therefore, $\overline{MB}$, the side opposite $\angle L$, is the shortest side of the triangle. $\angle M$ is the largest angle of the triangle; therefore, $\overline{LB}$, the side opposite $\angle M$, is the longest side of the triangle.

391.   **True**

The sum of the two shorter sides of a triangle must be greater than the longest side of the triangle:

$$2+3\overset{?}{>}4.995$$
$$5>4.995$$

This statement is true, so a triangle can have these side lengths.

392.   **True**

The sum of the two shorter sides of a triangle must be greater than the longest side of the triangle:

$$5\sqrt{2}+5\sqrt{2}\overset{?}{>}5\sqrt{2}$$
$$10\sqrt{2}>5\sqrt{2}$$

This statement is true, so a triangle can have these side lengths.

393.   **False**

The sum of the two shorter sides of a triangle must be greater than the longest side of the triangle:

$$\sqrt{5}+\sqrt{7}\overset{?}{>}\sqrt{26}$$
$$4.9\not>5.1$$

You can't form a triangle with these side lengths.

394.   **False**

The sum of the two shorter sides of a triangle must be greater than the longest side of the triangle:

$$a+4+a+3\overset{?}{>}2a+7$$
$$2a+7\not>2a+7$$

You can't form a triangle with these side lengths.

395.   **True**

The sum of the two shorter sides of a triangle must be greater than the longest side of the triangle. $\sqrt{221}\approx14.9$, so it's the longest side:

$$7+8\overset{?}{>}14.9$$
$$15>14.9$$

A triangle can have these side lengths.

396. 12

Isosceles triangles have two sides equal in length. The sides of this triangle are either 6, 6, and 12 or 6, 12, and 12.

The sum of the two smaller sides of a triangle must be greater than the longest side of the triangle. The triangle can't have sides measuring 6, 6, and 12 because

$$6+6 \overset{?}{>} 12$$
$$12 \not> 12$$

The triangle can measure 6, 12, and 12 because

$$6+12 \overset{?}{>} 12$$
$$18 > 12$$

397. 14

Isosceles triangles have two sides equal in length. The sides of this triangle are either 5, 5, and 14 or 5, 14, and 14.

The sum of the two shorter sides of a triangle must be greater than the longest side of the triangle. The triangle can't have sides measuring 5, 5, and 14 because

$$5+5 \overset{?}{>} 14$$
$$10 \not> 14$$

The triangle can measure 5, 14, and 14 because

$$5+14 \overset{?}{>} 14$$
$$19 > 14$$

398. {7, 8, 9}

Let x be the third side of the triangle. If 2 and 8 are the two smaller sides of the triangle, then

$$2+8 > x$$
$$10 > x$$

If 8 is the longest side, then

$$2+x > 8$$
$$x > 6$$

Because x has to be greater than 6 but less than 10, x can be 7, 8, or 9.

399. 25

You're looking for the smallest possible perimeter, which means you want the sides to be as short as possible. You also have to remember that the sum of the two shorter sides of a triangle has to be greater than the longest side. Suppose that 12 is the longest side. You'd need the sum of the two shorter sides to be an integer that's larger than 12. If their sum is 13 (say, with side lengths 6 and 7), the perimeter would be $13 + 12 = 25$.

400. $13 < x < 37$

Let x be the third side of the triangle. If 12 and 25 are the two shorter sides of the triangle, then

$$12 + 25 > x$$
$$37 > x$$

If 25 is the longest side, then

$$12 + x > 25$$
$$x > 13$$

Because x has to be greater than 13 but less than 37, the inequality is $13 < x < 37$.

401. $6 < x < 36$

Let x be the third side of the triangle. If 15 and 21 are the two shorter sides of the triangle, then

$$15 + 21 > x$$
$$36 > x$$

If 21 is the longest side, then

$$15 + x > 21$$
$$x > 6$$

Because x has to be greater than 6 but less than 36, the inequality is $6 < x < 36$.

402. 15

Isosceles triangles have two sides equal in length. The sides of this triangle are either 7, 7, and 15 or 7, 15, and 15.

The sum of the two shorter sides of a triangle must be greater than the longest side of the triangle. The triangle can't have sides measuring 7, 7, and 15 because

$$7 + 7 \overset{?}{>} 15$$
$$14 \not> 15$$

The triangle can measure 7, 15, and 15 because

$$7+15\overset{?}{>}15$$
$$22>15$$

---

403. **78.75°**

Isosceles triangles have two sides equal in length. The sides of this triangle are either 3.9, 3.9, and 10 or 3.9, 10, and 10.

The sum of the two shorter sides of a triangle must be greater than the longest side of the triangle. The triangle can't have sides measuring 3.9, 3.9, and 10 because

$$3.9+3.9\overset{?}{>}10$$
$$7.8\not>10$$

The triangle can measure 3.9, 10, and 10 because

$$3.9+10\overset{?}{>}10$$
$$13.9>10$$

This means that $AB=10$. The triangle is isosceles and $AB=AC$, so then $m\angle B=m\angle C$. The three angles of a triangle add to 180°, so

$$22.5°+x+x=180°$$
$$22.5°+2x=180°$$
$$2x=157.5°$$
$$x=78.75°$$

---

404. **False**

Isosceles triangles have two sides equal in length. If $\overline{AC}\cong\overline{BC}$ and $BC=10$, then the length of $\overline{AC}$ would also equal 10. The three sides of the triangle would be 10, 10, and 20.

The sum of the two shorter sides of the triangle must be greater than the longest side. $10+10\not>20$, so $BC\neq10$.

---

405. **False**

Even though the sum of the two shorter sides of the triangle is greater than the longest side, this triangle still can't be created. $\overline{AC}\cong\overline{BC}$, which means $\angle A\cong\angle B$. The angles of a triangle add up to 180°, so

$$x+x+70°=180°$$
$$2x+70°=180°$$
$$2x=110°$$
$$x=55°$$

Using the knowledge that the longest side of a triangle is opposite the largest angle of the triangle, $\overline{BC}$ must be shorter than $\overline{AB}$ and therefore cannot have a length of 25.

406. 50°

The exterior angle of a triangle is equal to the sum of the two nonadjacent interior angles of the triangle. Therefore,

$$120° = 70° + x$$
$$50° = x$$

407. 55°

$\overline{WY} \cong \overline{XY}$, which means that $\angle W \cong \angle X$. The exterior angle of a triangle is equal to the sum of the two nonadjacent interior angles of the triangle. Therefore,

$$110° = x + x$$
$$110° = 2x$$
$$55° = x$$

408. 105°

Three angles of a triangle add up to 180°. First solve for the value of b:

$$(2b) + (b + 30) + (3b) = 180°$$
$$2b + b + 30 + 3b = 180°$$
$$6b + 30 = 180°$$
$$6b = 150°$$
$$b = 25°$$

The exterior angle of a triangle is equal to the sum of the two nonadjacent interior angles of the triangle, so

$$m\angle XYZ = m\angle W + m\angle X$$
$$= b + 30 + 2b$$
$$= 25 + 30 + 2(25)$$
$$= 105°$$

409. 70°

The exterior angle of a triangle is equal to the sum of the two nonadjacent interior angles of the triangle, so set up an equation and solve for x:

$$4x + 30 = x + 50 + 2x$$
$$4x + 30 = 3x + 50$$
$$x + 30 = 50$$
$$x = 20$$

Plug in the value for x to find $m\angle XYZ$:

$$m\angle XYZ = 4x + 30 = 4(20) + 30 = 110°$$

$\angle WYX$ and $\angle XYZ$ are supplementary, so

$$m\angle WYX = 180° - m\angle XYZ = 180° - 110° = 70°$$

410. **60°**

The exterior angle of a triangle is equal to the sum of the two nonadjacent interior angles of the triangle:

$$3a = (4a - 50) + (a + 10)$$
$$3a = 4a - 50 + a + 10$$
$$0 = 2a - 40$$
$$40 = 2a$$
$$20 = a$$

Plug in the value of a:

$$m\angle XYZ = 3a = 3(20) = 60°$$

411. **120°**

The exterior angle of a triangle is equal to the sum of the two nonadjacent interior angles of the triangle. Use this relationship to solve for a:

$$b = 2a + b - 50$$
$$0 = 2a - 50$$
$$50 = 2a$$
$$25 = a$$

Plug in the value of a to find $m\angle W$ and $m\angle WYX$:

$$m\angle W = 2a = 2(25) = 50°$$
$$m\angle WYX = 3a - 15 = 3(25) - 15 = 60°$$

The angles of a triangle add up to 180°. Therefore,

$$50 + 60 + (b - 50) = 180°$$
$$50 + 60 + b - 50 = 180°$$
$$60 + b = 180°$$
$$b = 120°$$

412.   50°

The exterior angle of a triangle is equal to the sum of the two nonadjacent interior angles of the triangle. Therefore,

$$3c - 35 = c + c + 15$$
$$3c - 35 = 2c + 15$$
$$c - 35 = 15$$
$$c = 50$$

413. $\angle XRT$

$\angle E$ and $\angle EXR$ are interior angles in $\triangle EXR$. The exterior angle of a triangle is equal to the sum of the two nonadjacent interior angles of the triangle. $\angle XRT$ is the exterior angle of $\triangle EXR$. Therefore, $m\angle E + m\angle EXR = m\angle XRT$.

414. $\angle XRE$

$\angle T$ and $\angle RXT$ are interior angles in $\triangle RXT$. The exterior angle of a triangle is equal to the sum of the two nonadjacent interior angles of the triangle. $\angle XRE$ is the exterior angle of $\triangle RXT$. Therefore, $m\angle T + m\angle RXT = m\angle XRE$.

415.   **Sometimes**

$\angle XRE$ and $\angle XRT$ form a linear pair. If $\angle XRE$ is acute, then $\angle XRT$ would be obtuse. But the opposite is also true. Therefore, it's only sometimes true that $m\angle XRE > m\angle XRT$.

416.   **Always**

The exterior angle of a triangle is equal to the sum of the two nonadjacent interior angles of the triangle, so $m\angle XRT = m\angle E + m\angle RXE$. A whole is greater than any of its parts. Therefore, $m\angle XRT$ is greater than both $m\angle E$ and $m\angle RXE$.

417.   **Sometimes**

The only relationship between $\angle XRT$ and $\angle RXT$ is that they are two of the three angles in $\triangle RXT$. Although it's possible that $m\angle XRT > m\angle RXT$, it isn't always the case.

418.   **Never**

The exterior angle of a triangle is equal to the sum of the two nonadjacent interior angles of the triangle. Therefore, $m\angle ERX = m\angle RXT + m\angle T$. A whole is greater than any of its parts. Therefore, $m\angle ERX$ is greater than both $m\angle T$ and $m\angle RXT$. So if $m\angle ERX > m\angle T$, it would be impossible for $m\angle T$ to be greater than $m\angle ERX$.

419. **Given**

The very first step of a geometric proof is to state all the given information.

420. ∠6

∠2 is the exterior angle to △*VTE*. Because the exterior angle of a triangle is equal to the sum of the two nonadjacent interior angles, $m\angle 2 = m\angle 5 + m\angle 6$.

421. ∠2

Because $m\angle 2 = m\angle 5 + m\angle 6$, you know that ∠2 is larger than both ∠5 and ∠6.

422. **Transitive property**

You're given that $m\angle 4 > m\angle 2$, and in Statement 3, you found that $m\angle 2 > m\angle 5$. The transitive property then connects those ideas, telling you that $m\angle 4 > m\angle 5$.

423. **In a triangle, the longest side is opposite the largest angle.**

$\overline{TE}$ is opposite ∠4, and $\overline{VE}$ is opposite ∠5. You discovered that $m\angle 4 > m\angle 5$. In a triangle, the longest side is opposite the largest angle, so $TE > VE$.

424. ∠3 ≅ ∠6

A bisector divides an angle into two congruent angles.

425. **The exterior angle of a triangle is equal to the sum of the two nonadjacent interior angles.**

∠4 is the exterior angle to △*VSE*. ∠1 and ∠3 are the two nonadjacent interior angles of △*VSE*. The exterior angle of a triangle is equal to the sum of the two nonadjacent interior angles of the triangle; therefore, $m\angle 4 = m\angle 1 + m\angle 3$.

426. ∠4

You found that $m\angle 4 = m\angle 1 + m\angle 3$. The whole is great than its parts, which means that $m\angle 4 > m\angle 3$.

427. **Transitive property**

Because ∠3 ≅ ∠6 and $m\angle 4 > m\angle 3$, the transitive property finds that $m\angle 4 > m\angle 6$.

428. **In a triangle, the longest side is opposite the largest angle.**

$\overline{ET}$ is opposite ∠4, and $\overline{VT}$ is opposite ∠6. You discovered that $m\angle 4 > m\angle 6$. In a triangle, the longest side is opposite the largest angle, so $ET > TV$.

429.

Statements	Reasons
1. △*SET*, $m\angle 2 > m\angle 4$	1. Given
2. $m\angle 4 = m\angle 1 + m\angle 3$	2. The exterior angle of a triangle is equal to the sum of the two nonadjacent interior angles of the triangle.
3. $m\angle 4 > m\angle 1$	3. A whole is greater than its parts.
4. $m\angle 2 > m\angle 1$	4. Transitive property

430.

Statements	Reasons
1. △*ABC* with exterior $\angle BCD$; $\overline{AECD}$	1. Given
2. $m\angle 3 = m\angle 1 + m\angle A$	2. The exterior angle of a triangle is equal to the sum of the two nonadjacent interior angles of the triangle.
3. $m\angle 3 > m\angle A$	3. A whole is greater than its parts.
4. $m\angle BCD = m\angle 3 + m\angle 4$	4. The exterior angle of a triangle is equal to the sum of the two nonadjacent interior angles of the triangle.
5. $m\angle BCD > m\angle 3$	5. A whole is greater than its parts.
6. $m\angle BCD > m\angle A$	6. Transitive property

431. **Pentagon**

A *pentagon* is a polygon that has five sides.

432. **Hexagon**

A *hexagon* is a polygon that has six sides.

433. **Heptagon**

A *heptagon* is a polygon that has seven sides.

434. **Dodecagon**

A *dodecagon* is a polygon that has 12 sides.

435. **Nonagon**

A *nonagon* is a polygon that has nine sides.

436. **540°**

The sum of the degree measures of the interior angles of a polygon is $180(n-2)$, where n represents the number of sides. Therefore the sum of the angles of a pentagon, which has five sides, is $180(5-2)=540°$

437. **1,440°**

The sum of the degree measures of the interior angles of a polygon is $180(n-2)$, where n represents the number of sides. Therefore, the sum of the interior angles of an octagon, which has eight sides, is $180(8-2)=1,080°$. The sum of the exterior angles of a polygon is 360°. Therefore, the sum of the interior and exterior angles of a polygon is $1,080°+360°=1,440°$.

438. **360°**

The sum of the exterior angles of any polygon is 360°.

439. **3,600°**

The sum of the degree measures of the interior angles of a polygon is $180(n-2)$, where n represents the number of sides. Therefore, the sum of the angles of a 22-sided polygon is $180(22-2)=3,600°$.

440. **360°**

The sum of the exterior angles of any polygon is 360°.

441. **Nonagon**

The sum of the exterior angles of any polygon is 360°. To get the sum of the interior angles of the polygon, subtract 360 from 1,620, the given total:

$$1,620°-360°=1,260°$$

The sum of the interior angles of a polygon is $180(n-2)$, where n represents the number of sides. Plug in the numbers and solve for n:

$$180(n-2)=1,260$$
$$180n-360=1,260$$
$$180n=1,620$$
$$n=9$$

A polygon with nine sides is called a *nonagon*.

442. 135°

The sum of the degree measures of the interior angles of a polygon is $180(n-2)$, where n represents the number of sides. Therefore, the sum of the angles of a regular octagon, which has eight equal sides, is $180(8-2)=1,080°$. A regular octagon also has eight equal angles, so divide by 8:

$$\frac{1,080°}{8} = 135°$$

Therefore, each angle of the octagon is 135°.

443. 150°

The sum of the degree measures of the interior angles of a polygon is $180(n-2)$, where n represents the number of sides. Therefore, the sum of the angles of a regular dodecagon, which has 12 equal sides, is $180(12-2)=1,800°$. A regular dodecagon also has 12 equal angles, so divide by 12:

$$\frac{1,800°}{12} = 150°$$

Therefore, each angle of the dodecagon is 150°.

444. 72°

The sum of the exterior angles of a polygon is 360°. A regular pentagon has five equal sides and five equal angles, so divide by 5:

$$\frac{360°}{5} = 72°$$

Therefore, each exterior angle of the pentagon is 72°.

445. 36°

The sum of the exterior angles of a polygon is 360°. A regular decagon has 10 equal sides and 10 equal angles, so divide by 10:

$$\frac{360°}{10} = 36°$$

Therefore, each exterior angle of the decagon is 36°.

446. 20

The sum of the exterior angles of a polygon is 360°. The regular polygon you're dealing with has n equal sides and angles, so set up the equation and solve for n:

$$\frac{360°}{n} = 18°$$
$$360 = 18n$$
$$20 = n$$

Therefore, the polygon has 20 sides.

447. 30

The sum of the exterior angles of a polygon is 360°. The regular polygon you're dealing with has n equal sides and angles, so set up the equation and solve for n:

$$\frac{360°}{n} = 12°$$
$$360 = 12n$$
$$30 = n$$

Therefore, the polygon has 30 sides.

448. 18

If each interior angle of the polygon is 160°, each exterior angle would have to be 20°, because the interior and exterior angles of a polygon are supplementary.

The sum of the exterior angles of a polygon is 360°. The regular polygon you're dealing with has n equal sides and angles, so set up the equation and solve for n:

$$\frac{360°}{n} = 20°$$
$$360 = 20n$$
$$18 = n$$

Therefore, the polygon has 18 sides.

449. 30

If each interior angle of the polygon is 168°, each exterior angle would have to be 12°, because the interior and exterior angles of a polygon are supplementary.

The sum of the exterior angles of a polygon is 360°. The regular polygon you're dealing with has n equal sides and angles, so set up the equation and solve for n:

$$\frac{360°}{n} = 12°$$
$$360 = 12n$$
$$30 = n$$

Therefore, the polygon has 30 sides.

450. 24°

The interior and exterior angles of a polygon are supplementary:

$$180° - 156° = 24°$$

451. 58°

The sum of the exterior angles of a polygon is $180(n-2)$, where n represents the number of sides. The sum of the angles of a pentagon (five sides) is equal to $180(5-2) = 540°$. The pentagon is missing one interior angle, which you can call y:

$$100° + 120° + 90° + 108° + y = 540°$$
$$418° + y = 540°$$
$$y = 122°$$

The interior and exterior angles of a polygon are supplementary. Therefore,

$$122° + x = 180°$$
$$x = 58°$$

452. 20°

The sum of the interior angles of a polygon is $180(n-2)$, where n represents the number of sides. The sum of the angles of a hexagon (six sides) is equal to $180(6-2) = 720°$. Add the interior angles, set the sum equal to 720, and solve for x:

$$120 + (8x-8) + (4x+14) + 7x + (5x-6) + 6x = 720°$$
$$120 + 8x - 8 + 4x + 14 + 7x + 5x - 6 + 6x = 720$$
$$30x + 120 = 720$$
$$30x = 600$$
$$x = 20$$

453. 24°

The sum of the exterior angles of a polygon is 360°. Add the exterior angles, set the sum equal to 360, and solve for x:

$$4x + 5x + x + 2x + 3x = 360°$$
$$15x = 360$$
$$x = 24$$

454. 30°

The sum of the interior angles of a polygon is $180(n-2)$, where n represents the number of sides. The sum of the angles of a heptagon (seven sides) is equal to $180(7-2) = 900°$. Add the interior angles, set the sum equal to 900, and solve for x:

$$(5x+5) + (2x+22) + (6x-6) + (2x+12) + (4x+20) + (6x-3) + (3x+10) = 900°$$
$$5x + 5 + 2x + 22 + 6x - 6 + x + 12 + 4x + 20 + 7x - 3 + 3x + 10 = 900$$
$$28x + 60 = 900$$
$$28x = 840$$
$$x = 30$$

455.   30°

Notice that two triangles are next to each other in this diagram. △*BAC* has three congruent sides, so it also has three congruent angles. The sum of the three angles of a triangle is 180°. If you call each of the angles *y*, then

$$y + y + y = 180°$$
$$3y - 180°$$
$$y = 60°$$

Therefore, $m\angle BAC = 60°$. $\angle BAC$ and $\angle BAD$ form a linear pair, which means their sum is 180°. Therefore, $m\angle BAD = 120°$. △*BAD* has two congruent sides, which means the angles opposite those sides are congruent. If the sum of the three angles of a triangle is 180°, then you can set up the following equation and solve for *x*:

$$x + x + 120° = 180°$$
$$2x + 120° = 180°$$
$$2x = 60°$$
$$x = 30°$$

456.   50°

The sum of the interior angles of a polygon is $180(n-2)$, where *n* represents the number of sides. The sum of the angles of a heptagon (seven sides) is equal to $180(7-2) - 900°$. This pentagon is missing one interior angle, which is called *b*:

$$114 + 130 + 164 + 132 + 120 + 110 + b = 900°$$
$$770 + b = 900°$$
$$b = 130°$$

The interior and exterior angles of a polygon are supplementary. Therefore,

$$130° + x = 180°$$
$$x = 50°$$

457. $\frac{1}{2}(\text{apothem})(\text{perimeter})$

The formula for the area of a regular polygon is $\frac{1}{2}(\text{apothem})(\text{perimeter})$. The *apothem*, which is the line segment from the center of the polygon to the midpoint of one of the sides, is perpendicular to one of the sides. The *perimeter* is the total distance around the polygon.

458.   100 units²

The formula for the area of a regular polygon is $\frac{1}{2}(\text{apothem})(\text{perimeter})$. The apothem is 5 and the perimeter is 40, so the area is

$$\frac{1}{2}(5)(40) = 100 \text{ units}^2$$

459. 696 ft²

The formula for the area of a regular polygon is $\frac{1}{2}$(apothem)(perimeter). A *regular octagon* is a polygon with eight equal sides. If each side of the polygon is 12 feet, then the perimeter of the triangle is $8(12 \text{ ft}) = 96$ ft. You're given that the apothem is 14.5 feet. When you plug everything into the formula, you get the following:

$$\frac{1}{2}(14.5 \text{ ft})(96 \text{ ft}) = 696 \text{ ft}^2$$

460. $150\sqrt{3}$ units²

The formula for the area of a regular polygon is $\frac{1}{2}$(apothem)(perimeter). A *regular hexagon* is a polygon with six equal sides. You're given that the perimeter of the hexagon is 60 units, which means each side is 10. The apothem is joined to the midpoint of one of the sides and is also perpendicular to the side, forming a 30°-60°-90° triangle. The side opposite the 30° angle is x, the side opposite the 60° angle is $x\sqrt{3}$, and the side opposite the 90° angle is $2x$. The apothem is opposite the 60° angle, so the apothem equals $5\sqrt{3}$ units. When you plug everything into the formula, you get

$$\frac{1}{2}\left(5\sqrt{3} \text{ units}\right)(60 \text{ units}) = 150\sqrt{3} \text{ units}^2$$

461. 36 units

A *regular hexagon* is a polygon with six congruent sides. The area of a polygon is $\frac{1}{2}$(apothem)(perimeter).

The apothem is joined to the midpoint of one of the sides and is also perpendicular to the side, forming a 30°-60°-90° triangle. The side opposite the 30° angle is x, the side opposite the 60° angle is $x\sqrt{3}$, and the side opposite the 90° angle is $2x$. Each side of the hexagon therefore measures $2x$. Therefore, the perimeter is $2x(6) = 12x$ units. The apothem measures $x\sqrt{3}$ units. Plugging into the formula, you get the following:

$$\frac{1}{2}\left(x\sqrt{3}\right)(12x) = 54\sqrt{3}$$
$$6x^2\sqrt{3} = 54\sqrt{3}$$
$$6x^2 = 54$$
$$x^2 = 9$$
$$x = 3$$

If $x = 3$, then a side of the hexagon is $2(3) = 6$ units, and the perimeter of the hexagon would be $6(6) = 36$ units.

462. $\angle 6$

When parallel lines are cut by a transversal, alternate interior angles are formed. If you look for a Z shape, you can find the alternate interior angles at the two corners inside the Z. Alternate interior angles, such as $\angle 3$ and $\angle 6$, are congruent.

463. ∠7

When parallel lines are cut by a transversal, alternate exterior angles are formed. Alternate exterior angles, such as ∠2 and ∠7, are congruent.

464. 125°

∠4 and ∠5 are alternate interior angles. Alternate interior angles are congruent, so ∠4 and ∠5 are congruent.

465. 111.5°

∠1 and ∠8 are alternate exterior angles. Alternate exterior angles are congruent, so ∠1 and ∠8 are congruent.

466. 25

∠4 and ∠5 are alternate interior angles. Alternate interior angles are congruent, so set their measures equal to each other and solve for x:

$$5x - 80 = 2x - 5$$
$$3x - 80 = -5$$
$$3x = 75$$
$$x = 25$$

467. 20

∠3 and ∠6 are alternate interior angles. Alternate interior angles are congruent, so set their measures equal to each other and solve for x:

$$\frac{1}{2}x + 25 = 4x - 45$$
$$25 = 3.5x - 45$$
$$70 = 3.5x$$
$$20 = x$$

468. 50°

∠2 and ∠7 are alternate exterior angles. Alternate exterior angles are congruent, so set their measures equal to each other and solve for x:

$$3x - 40 = x + 20$$
$$2x - 40 = 20$$
$$2x = 60$$
$$x = 30$$

Therefore, $m\angle 7 = x + 20 = 30 + 20 = 50°$.

Answers
401–500

469. 140°

$\angle 2$ and $\angle 7$ are alternate exterior angles. Alternate exterior angles are congruent, so set their measures equal to each other and solve for x:

$$2x - 40 = \frac{1}{2}x + 20$$
$$1.5x - 40 = 20$$
$$1.5x = 60$$
$$x = 40$$

Plug in x to find $m\angle 7$:

$$m\angle 7 = \frac{1}{2}x + 20 = \frac{1}{2}(40) + 20 = 40°$$

$\angle 7$ and $\angle 5$ are supplementary angles, which means they add up to $180°$. Use this info to solve for $m\angle 5$:

$$m\angle 7 + m\angle 5 = 180°$$
$$40° + m\angle 5 = 180°$$
$$m\angle 5 = 140°$$

470. 10

$\angle 4$ and $\angle 5$ are alternate interior angles. Alternate interior angles are congruent, so set their measures equal to each other and solve for x:

$$x^2 = 8x + 20$$
$$x^2 - 8x - 20 = 0$$
$$(x - 10)(x + 2) = 0$$
$$x - 10 = 0 \text{ or } x + 2 = 0$$
$$x = 10 \text{ or } \cancel{x = -2}$$

471. 9

$\angle 3$ and $\angle 6$ are alternate interior angles. Alternate interior angles are congruent, so set their measures equal to each other and solve for x:

$$x^2 - 2 = 3x + 52$$
$$x^2 - 3x - 54 = 0$$
$$(x - 9)(x + 6) = 0$$
$$x - 9 = 0 \text{ or } x + 6 = 0$$
$$x = 9 \text{ or } \cancel{x = -6}$$

472. **Acute**

When parallel lines are cut by a transversal, alternate interior angles are formed. $\angle R$ and $\angle N$ are alternate interior angles, so they're congruent:

$$m\angle R = m\angle N = 65°$$

The angles of a triangle add up to $180°$, so plug in the known values and solve for $m\angle M$:

$$m\angle N + m\angle MON + m\angle M = 180°$$
$$65° + 72° + m\angle M = 180°$$
$$137° + m\angle M = 180°$$
$$m\angle M = 43°$$

Because $\triangle MNO$ contains angles measuring $65°$, $72°$, and $43°$, it's an acute triangle.

473. **Acute isosceles**

$\angle P$ and $\angle M$ are alternate interior angles, as are $\angle R$ and $\angle N$. Alternate interior angles are congruent, so

$$m\angle P = m\angle M = 80°$$
$$m\angle R = m\angle N = 20°$$

The angles of a triangle add up to $180°$, so plug in the known values and solve for $m\angle M$:

$$m\angle N + m\angle MON + m\angle M = 180°$$
$$20° + m\angle MON + 80° = 180°$$
$$m\angle MON + 100° = 180°$$
$$m\angle MON = 80°$$

Because $\triangle MNO$ contains angles measuring $20°$, $80°$, and $80°$, it's an acute isosceles triangle.

474. **Right**

$\angle R$ and $\angle N$ are alternate interior angles. Alternate interior angles are congruent, so

$$m\angle R = m\angle N = 20°$$

The angles of a triangle add up to $180°$, so plug in the known values and solve for $m\angle M$:

$$m\angle N + m\angle NOM + m\angle M = 180°$$
$$55° + 35° + m\angle M = 180°$$
$$m\angle M + 90° = 180°$$
$$m\angle M = 90°$$

Because $\triangle MNO$ contains a right angle, it's a right triangle.

475. **Equiangular**

$\angle P$ and $\angle M$ are alternate interior angles. Alternate interior angles are congruent, so

$$m\angle P = m\angle M = 60°$$

The angles of a triangle add up to 180°, so plug in the known values and solve for $m\angle NOM$:

$$m\angle N + m\angle NOM + m\angle M = 180°$$
$$60° + m\angle NOM + 60° = 180°$$
$$m\angle NOM + 120° = 180°$$
$$m\angle NOM = 60°$$

Because $\triangle MNO$ contains three 60° angles, it's an equiangular triangle.

476. **110°**

$\angle 1$ and $\angle 2$ are supplementary angles, which means they add up to 180°:

$$m\angle 1 + m\angle 2 = 180°$$
$$70° + m\angle 2 = 180°$$
$$m\angle 2 = 110°$$

477. **25°**

$\angle 1$ and $\angle 2$ are supplementary angles, which means they add up to 180°:

$$m\angle 1 + \angle 2 = 180°$$
$$70° + m\angle 2 = 180°$$
$$m\angle 2 = 110°$$

The angles of a triangle add up to 180°, so

$$m\angle 2 + m\angle 3 + m\angle 5 = 180°$$
$$110° + m\angle 3 + 45° = 180°$$
$$m\angle 3 + 155° = 180°$$
$$m\angle 3 = 25°$$

478. **110°**

$\angle 1$ and $\angle EFD$ are alternate interior angles. Alternate interior angles are equal:

$$m\angle 1 = m\angle EFD = 70°$$

$\angle 4$ is supplementary to $\angle EFD$, which means they add up to 180°:

$$m\angle 4 + m\angle EFD = 180°$$
$$m\angle 4 + 70° = 180°$$
$$m\angle 4 = 110°$$

479. 25°

$\overline{AB} \parallel \overline{CD}$ and $\overline{BF}$ is a transversal. This means that $\angle 3$ and $\angle 6$ are alternate interior angles. If two angles are alternate interior angles, then they're congruent. Therefore, $m\angle 3 = m\angle 6$, so $m\angle 6$ also equals 25°.

480. $\angle 5$

When two parallel lines are cut by a transversal, corresponding angles are formed. *Corresponding angles,* such as $\angle 1$ and $\angle 5$, are two angles that are both either above or below the two parallel lines and on the same side of the transversal. Corresponding angles are equal.

481. $\angle 8$

Intersecting lines form vertical angles, and vertical angles are congruent.

482. 60

$\angle 1$ and $\angle 2$ are supplementary angles, which means they add up to 180°:

$$(2x+20)+(x-20)=180°$$
$$2x+20+x-20=180$$
$$3x=180$$
$$x=60$$

483. 112°

$\angle 3$ and $\angle 5$ are same-side interior angles, so they're supplementary angles, which means they add up to 180°. Set up an equation and solve for x:

$$(x+28)+(2x+32)=180°$$
$$x+28+2x+32=180$$
$$3x+60=180$$
$$3x=120$$
$$x=40$$

Therefore, $m\angle 5 = 2x+32 = 2(40)+32 = 112°$.

Answers
401–500

484. 130°

$\angle 4$ and $\angle 8$ are corresponding angles. Corresponding angles are congruent, so set their measures equal to each other and solve for x:

$$3x - 50 = 2x + 10$$
$$x - 50 = 10$$
$$x = 60$$

Therefore, $m\angle 4 = 3x - 50 = 3(60) - 50 = 130°$.

485. 34°

$\angle 6$ and $\angle 7$ are vertical angles. Vertical angles are congruent, so set their measures equal to each other and solve for x:

$$x - 18 = 2x - 70$$
$$-18 = x - 70$$
$$52 = x$$

Therefore, $m\angle 6 = x - 18 = (52) - 18 = 34°$.

486. 154°

$\angle 5$ and $\angle 7$ are supplementary angles. Supplementary angles add up to $180°$. Set up an equation and solve for x:

$$\left(x^2 - 2x - 14\right) + \left(2x - 2\right) = 180$$
$$x^2 - 2x - 14 + 2x - 2 = 180$$
$$x^2 - 16 = 180$$
$$x^2 = 196$$
$$x^2 - 196 = 0$$
$$(x + 14)(x - 14) = 0$$
$$\cancel{x = -14} \text{ or } x = 14$$

Therefore, $m\angle 5 = x^2 - 2x - 14 = (14)^2 - 2(14) - 14 = 154°$.

487. 5

$\angle 4$ and $\angle 6$ are same-side interior angles, so they're supplementary angles. Supplementary angles add up to $180°$, so

$$x^3 + 55 = 180°$$
$$x^3 = 125$$
$$\sqrt[3]{x} = \sqrt[3]{125}$$
$$x = 5$$

488.   4

∠2 and ∠5 are same-side interior angles, so they're supplementary angles. Supplementary angles add up to 180°, so

$$\left(x^2 + 15x + 70\right) + \left(x^2 - 3x + 30\right) = 180°$$
$$x^2 + 15x + 70 + x^2 - 3x + 30 = 180$$
$$2x^2 + 12x + 100 = 180$$
$$2x^2 + 12x - 80 = 0$$
$$2\left(x^2 + 6x - 40\right) = 0$$
$$2\left(x + 10\right)\left(x - 4\right) = 0$$
$$x + 10 = 0 \ \text{ or } \ x - 4 = 0$$
$$\cancel{x = -10} \ \text{ or } \ x = 4$$

489.   80°

When two parallel lines are cut by a transversal, corresponding angles are formed. ∠*P* and ∠*ORC* are corresponding angles, so they're congruent:

$$m\angle P = m\angle ORC = 40°$$

The three angles of △*ORC* add up to 180°, so

$$m\angle O + m\angle ORC + m\angle C = 180°$$
$$60° + 40° + m\angle C = 180°$$
$$100° + m\angle C = 180°$$
$$m\angle C = 80°$$

490.   55°

When two parallel lines are cut by a transversal, corresponding angles are formed. ∠*P* and ∠*R* are corresponding angles, so they're congruent:

$$m\angle P = m\angle ORC = 55°$$

The three angles of △*ENR* add up to 180°, so

$$m\angle PES + m\angle ORC + m\angle ENR = 180°$$
$$70° + 55° + m\angle ENR = 180°$$
$$125° + m\angle ENR = 180°$$
$$m\angle ENR = 55°$$

∠*O* and ∠*ENR* are corresponding angles, which means they're congruent, so $m\angle ENR = m\angle O = 55°$.

491. 160°

When two parallel lines are cut by a transversal, corresponding angles are formed. $\angle RCO$ and $\angle PES$ are corresponding angles, so they're congruent:

$$m\angle RCO = m\angle PES = 85°$$

The three angles of $\triangle SEP$ add up to 180°, so

$$m\angle PES + m\angle S + m\angle P = 180°$$
$$85° + 75° + m\angle P = 180°$$
$$160° + m\angle P = 180°$$
$$m\angle P = 20°$$

$\angle ORC$ and $\angle SPE$ ($\angle P$) are corresponding angles, so they're congruent:

$$m\angle ORC = m\angle P = 20°$$

$\angle ORC$ and $\angle PRO$ are supplementary angles, which means they add up to 180°. Therefore,

$$m\angle ORC + m\angle PRO = 180°$$
$$20° + m\angle PRO = 180°$$
$$m\angle PRO = 160°$$

492. 60°

When two parallel lines are cut by a transversal, corresponding angles are formed. $\angle RCO$ and $\angle PES$ are corresponding angles, so they're congruent:

$$m\angle RCO = m\angle PES = 58°$$

$\angle ORC$ and $\angle SPE$ ($\angle P$) are corresponding angles, so they're congruent as well:

$$m\angle ORC = m\angle P = 62°$$

The three angles of $\triangle SEP$ add up to 180°, so

$$m\angle PES + m\angle ORC + m\angle RNE = 180°$$
$$58° + 62° + m\angle RNE = 180°$$
$$120° + m\angle RNE = 180°$$
$$m\angle RNE = 60°$$

493. 58°

When two parallel lines are cut by a transversal, alternate interior angles are formed. $\angle O$ and $\angle ONS$ are alternate interior angles, so they're congruent:

$$m\angle O = m\angle ONS = 58°$$

Intersecting lines form vertical angles, and vertical angles are equal. $\angle ONS$ and $\angle RNE$ are vertical angles, so

$$m\angle ONS = m\angle RNE = 58°$$

494. 5°

When two parallel lines are cut by a transversal, corresponding angles are formed. $\angle C$ and $\angle PES$ are corresponding angles, so they're congruent. Set their measures equal to each other and solve for x:

$$m\angle C = m\angle PES$$
$$x^2 - 20 = x$$
$$x^2 - x - 20 = 0$$
$$(x-5)(x+4) = 0$$
$$x - 5 = 0 \text{ or } x + 4 = 0$$
$$x = 5 \text{ or } \quad \cancel{x = -4}$$

495. 60°

A bisector divides an angle into two congruent angles, so

$$\angle WRE \cong \angle TRE$$
$$m\angle WRE = 60°$$

When two parallel lines are cut by a transversal, corresponding angles are formed. $\angle WRE$ and $\angle S$ are corresponding angles, so

$$\angle WRE \cong \angle S$$
$$60° = m\angle S$$

496. 144°

When two parallel lines are cut by a transversal, alternate interior angles are formed. $\angle T$ and $\angle TRE$ are alternate interior angles, so they're congruent:

$$\angle T \cong \angle TRE$$
$$72° = m\angle TRE$$

A bisector divides an angle into two congruent angles, so

$$\angle WRE \cong \angle TRE$$
$$m\angle WRE = 72°$$

$$m\angle WRT = m\angle WRE + m\angle TRE$$
$$= 72° + 72°$$
$$= 144°$$

497. 50°

A bisector divides an angle into two congruent angles, so

$$\angle WRE \cong \angle TRE$$

$\angle WRE$, $\angle TRE$, and $\angle TRS$ form a straight line, which means they add up to 180°. Set up the following equation, letting $m\angle WRE = x$:

$$m\angle WRE + m\angle TRE + \angle TRS = 180°$$
$$x + x + 80 = 180°$$
$$2x + 80 = 180°$$
$$2x = 100°$$
$$x = 50°$$

498. 62°

A bisector divides an angle into two congruent angles, so

$$\angle WRE \cong \angle TRE$$

$\angle WRE$, $\angle TRE$, and $\angle TRS$ form a straight line, which means they add up to 180°. Set up the following equation, letting $m\angle WRE = x$:

$$m\angle WRE + m\angle TRE + m\angle SRT = 180°$$
$$x + x + 56° = 180°$$
$$2x + 56° = 180°$$
$$2x = 124°$$
$$x = 62°$$

When two parallel lines are cut by a transversal, alternate interior angles are formed. $\angle T$ and $\angle TRE$ are alternate interior angles, so they're congruent:

$$\angle T \cong \angle TRE$$
$$m\angle T = 62°$$

499. 35°

$\angle BAE$ and $\angle GEA$ are supplementary, which means they add up to 180°:

$$m\angle BAE + m\angle GEA = 180°$$
$$95° + m\angle GEA = 180°$$
$$m\angle GEA = 85°$$

$\angle DCE$ and $\angle FEC$ are also supplementary:

$$m\angle DCE + m\angle FEC = 180°$$
$$120° + m\angle FEC = 180°$$
$$m\angle FEC = 60°$$

$\angle GEA$, $\angle GEF$, and $\angle FEC$ form a straight line, so they also add up to 180°:

$$m\angle GEA + m\angle GEF + m\angle FEC = 180°$$
$$85° + m\angle GEF + 60° = 180°$$
$$145° + m\angle GEF = 180°$$
$$m\angle GEF = 35°$$

500. 100°

$\angle GEA$, $\angle GEF$, and $\angle FEC$ form a straight line, which means they add up to 180°:

$$m\angle GEA + m\angle GEF + m\angle FEC = 180°$$
$$75° + 25° + m\angle FEC = 180°$$
$$100° + m\angle FEC = 180°$$
$$m\angle FEC = 80°$$

$\angle DCE$ and $\angle FEC$ are supplementary, so they also add up to 180°:

$$m\angle DCE + m\angle FEC = 180°$$
$$m\angle DCE + 80° = 180°$$
$$m\angle DCE = 100°$$

501. 105°

$\angle GEA$, $\angle GEF$, and $\angle FEC$ form a straight line, which means they add up to 180°:

$$m\angle GEA + m\angle GEF + m\angle FEC = 180°$$
$$m\angle GEA + 45° + 60° = 180°$$
$$m\angle GEA + 105° = 180°$$
$$m\angle GEA = 75°$$

$\angle BAE$ and $\angle GEA$ are supplementary, so they also add up to 180°:

$$m\angle BAE + m\angle GEA = 180°$$
$$m\angle BAE + 75° = 180°$$
$$m\angle BAE = 105°$$

502. 135°

$\angle AEF$ and $\angle FEC$ form a linear pair, which means they add up to 180°:

$$m\angle AEF + m\angle FEC = 180°$$
$$135° + m\angle FEC = 180°$$
$$m\angle FEC = 45°$$

$\angle FEC$ and $\angle DCE$ are supplementary, so they also add up to 180°:

$$m\angle FEC + m\angle DCE = 180°$$
$$45° + m\angle DCE = 180°$$
$$m\angle DCE = 135°$$

503. **110°**

A bisector divides an angle into two congruent angles, so

$$m\angle AEG = \frac{1}{2}m\angle AEF$$
$$m\angle AEG = \frac{1}{2}(140°)$$
$$m\angle AEG = 70°$$

$\angle BAE$ and $\angle AEG$ are supplementary, which means they add up to 180°:

$$m\angle BAE + m\angle AEG = 180°$$
$$m\angle BAE + 70° = 180°$$
$$m\angle BAE = 110°$$

504. **80°**

Let

$$m\angle AEG = 3x$$
$$m\angle FEC = 2x$$

$m\angle BAE$ and $\angle AEG$ are supplementary, which means they add up to 180°:

$$m\angle BAE + m\angle AEG = 180°$$
$$120° + 3x = 180°$$
$$3x = 60°$$
$$x = 20°$$

$$m\angle AEG = 3x = 3(20) = 60°$$
$$m\angle FEC = 2x = 2(20) = 40°$$

$\angle GEA$, $\angle GEF$, and $\angle FEC$ form a straight line, so they also add up to 180°:

$$m\angle GEA + m\angle GEF + m\angle FEC = 180°$$
$$60° + m\angle GEF + 40° = 180°$$
$$m\angle GEF + 100° = 180°$$
$$m\angle GEF = 80°$$

505.

Statements	Reasons
1. $\overline{AR} \parallel \overline{PL}$ with transversal $\overline{PR}$, and $\overline{AR} \cong \overline{PL}$	1. Given
2. $\angle ARP$ and $\angle LPR$ are alternate interior angles.	2. If two parallel lines are cut by a transversal, then alternate interior angles are formed.
3. $\angle ARP \cong \angle LPR$	3. Alternate interior angles are congruent.
4. $\overline{PR} \cong \overline{PR}$	4. Reflexive property
5. $\triangle PAR \cong \triangle RLP$	5. SAS

506. $\angle DAC$

$\angle BCA$ and $\angle DAC$ are alternate interior angles because they're both inside the two parallel lines and are on opposite sides of the transversal.

507. **Alternate interior angles are congruent.**

If two parallel lines are cut by a transversal, then alternate interior angles are formed. Alternate interior angles are congruent to each other.

508. **Reflexive property**

A line segment is congruent to itself.

509. **Addition postulate**

The *addition postulate* states that if two segments are congruent to two other segments, then the sums of the segments are also congruent to each other.

510. $\triangle BCE \cong \triangle DAF$

Two sides and the included angle of one triangle are congruent to two sides and the included angle of the other triangle.

511. **CPCTC**

Corresponding parts of congruent triangles are congruent to each other.

512. **18**

Opposite sides of a parallelogram are congruent, so $AT = MH$:

$$8x + 2 = 5x + 8$$
$$3x + 2 = 8$$
$$3x = 6$$
$$x = 2$$

Therefore, $MH = 5x + 8 = 5(2) + 8 = 18$.

513. **25**

Opposite sides of a parallelogram are congruent, so $AM = TH$:

$$40 = 2x - 10$$
$$50 = 2x$$
$$25 = x$$

514. 30

Opposite angles of a parallelogram are congruent, so $m\angle AMH = m\angle HTA$:

$$80 = x + 50$$
$$30 = x$$

515. 70°

Consecutive angles of a parallelogram are supplementary, which means they add up to 180°:

$$x + 40° + 110° = 180°$$
$$x + 150° = 180°$$
$$x = 30°$$

Therefore, $m\angle AMH = x + 40° = (30°) + 40° = 70°$.

516. 95°

Consecutive angles of a parallelogram are supplementary, which means they add up to 180°:

$$2x + 25 + 3x + 5 = 180°$$
$$5x + 30 = 180°$$
$$5x = 150°$$
$$x = 30°$$

Therefore, $m\angle MHT = 3x + 5 = 3(30°) + 5 = 95°$.

517. 8

The diagonals of a parallelogram bisect each other. This means they divide each other into two congruent segments. Therefore, $ME = TE = 8$.

518. 30

The diagonals of a parallelogram bisect each other. This means they divide each other into two congruent segments. Therefore, $TE = ME$:

$$5x - 20 = x + 20$$
$$4x - 20 = 20$$
$$4x = 40$$
$$x = 10$$

That means $TE = 5x - 20 = 5(10) - 20 = 30$.

519. 40

The diagonals of a parallelogram bisect each other. This means they divide each other into two congruent segments. Therefore, $MT = 2(TE)$:

$$2(x+8) = 4x-8$$
$$2x+16 = 4x-8$$
$$16 = 2x-8$$
$$24 = 2x$$
$$12 = x$$

That means $MT = 4x-8 = 4(12)-8 = 40$.

520. 14

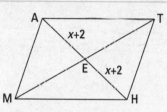

The diagonals of a parallelogram bisect each other. This means they divide each other into two congruent segments. Therefore, $AH = 2(AE)$:

$$3x-8 = 2(x+2)$$
$$3x-8 = 2x+4$$
$$x-8 = 4$$
$$x = 12$$

This means $AE = x+2 = 12+2 = 14$.

521. 7

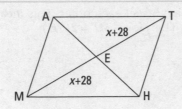

The diagonals of a parallelogram bisect each other. This means they divide each other into two congruent segments. Therefore, $MT = 2(ME)$:

$$2(x+28) = x^2 + 3x$$
$$2x + 56 = x^2 + 3x$$
$$56 = x^2 + x$$
$$0 = x^2 + x - 56$$
$$0 = (x+8)(x-7)$$
$$x + 8 = 0 \ \text{ or } \ x - 7 = 0$$
$$\cancel{x = -8} \ \text{ or } \ x = 7$$

522. 11

Opposite sides of a parallelogram are congruent. Therefore, $AM = TH$ and $AT = MH$. Solve $4a + b = 26$ for b:

$$4a + b = 26$$
$$b = 26 - 4a$$

Now plug the value of b into $2a + 3b = 28$ and solve for a:

$$2a + 3b = 28$$
$$2a + 3(26 - 4a) = 28$$
$$2a + 78 - 12a = 28$$
$$-10a + 78 = 28$$
$$-10a = -50$$
$$a = 5$$

Plug in the value of a to find b:

$$b = 26 - 4a = 26 - 4(5) = 6$$

Therefore, $a + b = 5 + 6 = 11$.

523. 79°

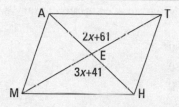

∠*TEA* and ∠*MEH* are vertical angles. Vertical angles are congruent, so

$$m\angle TEA = m\angle MEH$$
$$2x + 61 = 3x + 41$$
$$61 = x + 41$$
$$20 = x$$

Therefore, $m\angle MEH = 3x + 41 = 3(20) + 41 = 101°$. ∠*AEM* and ∠*MEH* are supplementary angles, which means they add up to 180°:

$$m\angle AEM + m\angle MEH = 180°$$
$$m\angle AEM + 101° = 180°$$
$$m\angle AEM = 79°$$

524. 80°

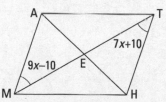

In a parallelogram, opposite sides are parallel and the diagonal acts as a transversal. When parallel lines are cut by a transversal, alternate interior angles are formed, and they're congruent:

$$m\angle AMT = m\angle HTM$$
$$9x - 10 = 7x + 10$$
$$2x - 10 = 10$$
$$2x = 20$$
$$x = 10$$

Therefore, $m\angle HTM = 7x + 10 = 7(10) + 10 = 80°$.

525. 145°

Consecutive angles of a parallelogram are supplementary, which means they add up to 180°:

$$m\angle MAT + m\angle HTA = 180°$$
$$7x + 5 + x + 15 = 180°$$
$$8x + 20 = 180°$$
$$8x = 160°$$
$$x = 20°$$

Therefore, $m\angle MAT = 7x + 5 = 7(20°) + 5 = 145°$.

In a parallelogram, opposite angles are congruent, so

$$m\angle MAT = m\angle MHT$$
$$145° = m\angle MHT$$

526. 120°

Let

$$m\angle B = x$$
$$m\angle A = 2x$$

Consecutive angles of a parallelogram are supplementary, which means they add up to 180°:

$$m\angle A + m\angle B = 180°$$
$$2x + x = 180°$$
$$3x = 180°$$
$$x = 60°$$

Therefore, $m\angle A = 2x = 2(60°) = 120°$.

527. 2

Opposite sides of a parallelogram are congruent, so

$$DR = WE$$
$$9x - 5 = 3x + 7$$
$$6x - 5 = 7$$
$$6x = 12$$
$$x = 2$$

528.   105°

Opposite angles of a parallelogram are congruent, so

$$m\angle O = m\angle A$$
$$6x + 30 = 8x + 15$$
$$30 = 2x + 15$$
$$15 = 2x$$
$$7.5 = x$$

Therefore, $m\angle O = 6x + 30 = 6(7.5) + 30 = 75°$.

Consecutive angles of a parallelogram are supplementary, which means they add up to 180°:

$$m\angle O + m\angle J = 180°$$
$$75 + m\angle J = 180°$$
$$m\angle J = 105°$$

529.   107.5°

Consecutive angles of a parallelogram are supplementary, which means they add up to 180°:

$$m\angle C + m\angle D = 180°$$
$$3a - 10 + a + 80 = 180°$$
$$4a + 70 = 180°$$
$$4a = 110°$$
$$a = 27.5°$$

Therefore, $m\angle D = a + 80 = 27.5° + 80 = 107.5°$.

530.   12

Opposite sides of a parallelogram are congruent, so

$$SH = ER$$
$$x^2 + 4x = x^3 + x^2$$
$$4x = x^3$$
$$0 = x^3 - 4x$$
$$0 = x(x^2 - 4)$$
$$0 = x(x-2)(x+2)$$
$$x = 0 \text{ or } x - 2 = 0 \text{ or } x + 2 = 0$$
$$\cancel{x = 0} \text{ or } x = 2 \text{ or } \cancel{x = 2}$$

Therefore, $ER = x^3 + x^2 = 2^3 + 2^2 = 8 + 4 = 12$.

531.   20

Opposite angles of a parallelogram are congruent, so

$$m\angle L = m\angle N$$
$$x + y = 50°$$

Consecutive angles of a parallelogram are supplementary, which means they add up to 180°:

$$m\angle I + m\angle N = 180°$$
$$(3x + 2y) + 50° = 180°$$
$$3x + 2y = 130°$$

You now have a system of equations, $x + y = 50$ and $3x + 2y = 130$. Solve using the elimination method. Multiply the first equation by –3 to get $-3(x + y = 50) = -3x - 3y = -150$. Then add the equations:

$$-3x - 3y = -150$$
$$\underline{3x + 2y = 130}$$
$$-y = -20$$
$$y = 20$$

532.   15

Opposite sides of a rectangle are congruent, so

$$RS = TW$$
$$y + 25 = 40$$
$$y = 15$$

533.   4

All four angles of a rectangle are right angles. Right angles measure 90°:

$$m\angle R = 90°$$
$$10y + 50 = 90$$
$$10y = 40$$
$$y = 4$$

534. 1

Opposite sides of a rectangle are congruent, so

$$ST = RW$$
$$8y - 6 = 5y - 3$$
$$3y - 6 = -3$$
$$3y = 3$$
$$y - 1$$

535. 15

All four angles of a rectangle are right angles. Right angles measure 90°:

$$m\angle T = 90°$$
$$3y + 45 = 90$$
$$3y = 45$$
$$y = 15$$

536. 3

Opposite sides of a rectangle are congruent, so

$$ST = RW$$
$$7y + 10 = -y + 34$$
$$8y + 10 = 34$$
$$8y = 24$$
$$y = 3$$

537. 13

A rectangle contains four right angles, and the diagonal of a rectangle creates a right triangle. This particular right triangle has legs measuring 5 and 12, with the diagonal being the hypotenuse. Use the Pythagorean theorem to solve for the hypotenuse:

$$a^2 + b^2 = c^2$$
$$5^2 + 12^2 = c^2$$
$$169 = c^2$$
$$\sqrt{169} = \sqrt{c^2}$$
$$13 = c$$

538. 25

A rectangle contains four right angles, and the diagonal of a rectangle creates a right triangle. This particular right triangle has legs measuring 15 and 20, with the diagonal being the hypotenuse. Use the Pythagorean theorem to solve for the hypotenuse:

$$a^2 + b^2 = c^2$$
$$15^2 + 20^2 = c^2$$
$$625 = c^2$$
$$\sqrt{625} = \sqrt{c^2}$$
$$25 = c$$

539. 25

A rectangle contains four right angles, and the diagonal of a rectangle creates a right triangle. This particular right triangle has legs measuring 7 and 24, with the diagonal being the hypotenuse. Use the Pythagorean theorem to solve for the hypotenuse:

$$a^2 + b^2 = c^2$$
$$7^2 + 24^2 = c^2$$
$$625 = c^2$$
$$\sqrt{625} = \sqrt{c^2}$$
$$25 = c$$

540. $\sqrt{41}$

Opposite sides of a rectangle are congruent, so

$$ST = RW = 5$$
$$SR = TW = 4$$

A rectangle contains four right angles, and the diagonal of a rectangle creates a right triangle. This right triangle has legs measuring 5 and 4, with the diagonal being the hypotenuse. Use the Pythagorean theorem to solve for the hypotenuse:

$$a^2 + b^2 = c^2$$
$$4^2 + 5^2 = x^2$$
$$41 = x^2$$
$$\sqrt{41} = \sqrt{x^2}$$
$$\sqrt{41} = x$$

541. 13

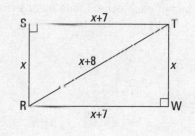

Let

$$x = SR$$
$$x + 7 = RW$$
$$x + 8 = RT$$

Opposite sides of a rectangle are congruent, which means that $ST = x + 7$ and that $TW = x$.

A rectangle contains four right angles, and the diagonal of a rectangle creates a right triangle. This right triangle has legs measuring x and $x + 7$, with the hypotenuse measuring $x + 8$. Use the Pythagorean theorem to solve for x:

$$a^2 + b^2 = c^2$$
$$(x)^2 + (x+7)^2 = (x+8)^2$$
$$x^2 + (x+7)(x+7) = (x+8)(x+8)$$
$$x^2 + x^2 + 14x + 49 = x^2 + 16x + 64$$
$$2x^2 + 14x + 49 = x^2 + 16x + 64$$
$$x^2 + 14x + 49 = 16x + 64$$
$$x^2 - 2x + 49 = 64$$
$$x^2 - 2x - 15 = 0$$
$$(x-5)(x+3) = 0$$
$$x - 5 = 0 \text{ or } x + 3 = 0$$
$$x = 5 \text{ or } \quad \cancel{x = -3}$$

If $x = 5$, then $RT = x + 8 = 5 + 8 = 13$. $\overline{RT}$ and $\overline{SW}$ are diagonals of the rectangle. Because the diagonals of a rectangle are congruent, $SW = 13$ as well.

542. 10

The diagonals of a rectangle bisect each other. This means that $\overline{RA} \cong \overline{AC}$. You're given that $RA = 4y + 16$, so $AC = 4y + 16$ also. Therefore,

$$RA + AC = RC$$
$$4y + 16 + 4y + 16 = 12y - 8$$
$$8y + 32 = 12y - 8$$
$$32 = 4y - 8$$
$$40 = 4y$$
$$10 = y$$

543. 16

The diagonals of a rectangle are congruent, so

$$RC = TE$$
$$y + 20 = 2y + 4$$
$$20 = y + 4$$
$$16 = y$$

544. 1

The diagonals of a rectangle are congruent, and they also bisect each other. This means that the pieces of the diagonal are congruent to each other:

$$RA = AE$$
$$7y - 3 = 2y + 2$$
$$5y - 3 = 2$$
$$5y = 5$$
$$y = 1$$

545. 12

The diagonals of a rectangle bisect each other. This means that $\overline{TA} \cong \overline{AE}$. Therefore, $AE = y + 8$. Diagonal $\overline{TE}$ has a length of $y + 8 + y + 8 = 2y + 16$. The diagonals of a rectangle are congruent, so

$$RC = TE$$
$$5y - 20 = 2y + 16$$
$$3y - 20 = 16$$
$$3y = 36$$
$$y = 12$$

546. 7

The diagonals of a rectangle are congruent, so

$$RC = ET$$
$$y^2 = y + 42$$
$$y^2 - y - 42 = 0$$
$$(y-7)(y+6) = 0$$
$$y - 7 = 0 \text{ or } y + 6 = 0$$
$$y = 7 \text{ or } \cancel{y = -6}$$

547. 8

The diagonals of a rectangle bisect each other. $AC = RA = 3y + 8$, so

$$AC + RA = RC$$
$$3y + 8 + 3y + 8 = y^2$$
$$6y + 16 = y^2$$
$$0 = y^2 - 6y - 16$$
$$0 = (y-8)(y+2)$$
$$y - 8 = 0 \text{ or } y + 2 = 0$$
$$y = 8 \text{ or } \cancel{y = -2}$$

548. 10

The diagonals of a rhombus are perpendicular bisectors, which means they form right angles at their point of intersection. Therefore,

$$m\angle NPO = 90°$$
$$2x + 70 = 90$$
$$2x = 20$$
$$x = 10$$

549. 25

The diagonals of a rhombus bisect the angles of the rhombus, and a bisector divides an angle into two congruent angles. Therefore,

$$m\angle LOP = m\angle NOP$$
$$3x = 2x + 25$$
$$x = 25$$

550. 10

The diagonals of a rhombus bisect the angles of the rhombus, and a bisector divides an angle into two congruent angles. Therefore,

$$m\angle MNP = m\angle ONP$$
$$4x - 12 = x + 18$$
$$3x - 12 = 18$$
$$3x = 30$$
$$x = 10$$

551. 8

The diagonals of a rhombus bisect the angles of the rhombus, and a bisector divides an angle into two congruent angles. $m\angle NMO = m\angle LMO = 3x$, so

$$m\angle LMO + m\angle NMO = \angle LMN$$
$$3x + 3x = x^2 - 16$$
$$6x = x^2 - 16$$
$$0 = x^2 - 6x - 16$$
$$0 = (x - 8)(x + 2)$$
$$x - 8 = 0 \text{ or } x + 2 = 0$$
$$x = 8 \text{ or } \cancel{x = -2}$$

552. 10

The diagonals of a rhombus bisect the angles of the rhombus, and a bisector divides an angle into two congruent angles. $m\angle MNP = m\angle ONP = x + 10$, so

$$m\angle MNP + m\angle ONP = m\angle MNO$$
$$x + 10 + x + 10 = 4x$$
$$2x + 20 = 4x$$
$$20 = 2x$$
$$10 = x$$

553. 5

The diagonals of a rhombus are perpendicular bisectors, which means they form right angles at their point of intersection. Therefore,

$$m\angle MPL = 90°$$
$$x^2 + 3x + 50 = 90$$
$$x^2 + 3x - 40 = 0$$
$$(x + 8)(x - 5) = 0$$
$$x + 8 = 0 \text{ or } x - 5 = 0$$
$$\cancel{x = -8} \text{ or } x = 5$$

554. 20

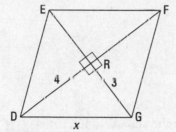

The diagonals of a rhombus are perpendicular bisectors, which means they form right angles at their point of intersection. This creates four right triangles within the rhombus. Using the Pythagorean theorem to find the hypotenuse of one of the right triangles will give you the length of one of the sides of the rhombus:

$$a^2 + b^2 = c^2$$
$$3^2 + 4^2 = x^2$$
$$25 = x^2$$
$$\sqrt{25} = \sqrt{x^2}$$
$$5 = x$$

All four sides of a rhombus are congruent, so in this case, each of the sides of the rhombus is equal to 5. The perimeter of a rhombus is equal to the sum of the four sides of the rhombus:

$$5 + 5 + 5 + 5 = 20$$

555. 68

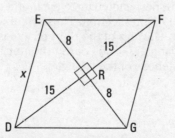

The diagonals of a rhombus bisect each other. This means they divide each other into two congruent segments:

$$DR = RF = 15$$
$$ER = RG = 8$$

The diagonals of a rhombus are perpendicular to each other, which means they form right angles at their point of intersection. These right angles create four right triangles in the rhombus.

The legs of the right triangles are 8 and 15, and the hypotenuse of the right triangle is a side of the rhombus. Use the Pythagorean theorem to solve for the hypotenuse of the triangle:

$$a^2 + b^2 = c^2$$
$$8^2 + 15^2 = c^2$$
$$289 = c^2$$
$$\sqrt{289} = \sqrt{c^2}$$
$$17 = c$$

The perimeter of a rhombus is the sum of all the sides of a rhombus. The four sides of a rhombus are equal in length; therefore, all 4 sides measure 17. The perimeter of the rhombus is

$$17 + 17 + 17 + 17 = 68$$

556.   156

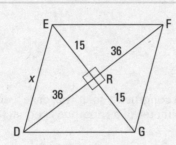

The diagonals of a rhombus bisect each other. This means they divide each other into two congruent segments. Therefore,

$$DR = RF = 36$$

The diagonals of a rhombus are perpendicular to each other, which means they form right angles at their point of intersection. These right angles create four right triangles in the rhombus. The legs of the right triangles are 36 and 15, and the hypotenuse of the right triangle is a side of the rhombus. Use the Pythagorean theorem to solve for the hypotenuse of the triangle:

$$a^2 + b^2 = c^2$$
$$36^2 + 15^2 = c^2$$
$$1{,}521 = c^2$$
$$\sqrt{1{,}521} = \sqrt{c^2}$$
$$39 = c$$

The perimeter of a rhombus is the sum of all the sides of a rhombus. The four sides of a rhombus are equal in length; therefore, each side measures 39. The perimeter of the rhombus is

$$39 + 39 + 39 + 39 = 156$$

557. 52

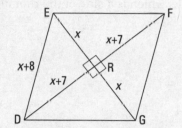

The diagonals of a rhombus are perpendicular to each other, which means they form right angles at their point of intersection. These right angles create four right triangles in the rhombus. The legs of the right triangles are x and $x+7$. The hypotenuse of the right triangle is $x+8$. Use the Pythagorean theorem to solve for x:

$$a^2 + b^2 = c^2$$
$$(x)^2 + (x+7)^2 = (x+8)^2$$
$$x^2 + (x+7)(x+7) = (x+8)(x+8)$$
$$x^2 + x^2 + 7x + 7x + 49 = x^2 + 8x + 8x + 64$$
$$2x^2 + 14x + 49 = x^2 + 16x + 64$$
$$x^2 - 2x - 15 = 0$$
$$(x-5)(x+3) = 0$$
$$x - 5 = 0 \text{ or } x + 3 = 0$$
$$x = 5 \text{ or } \quad \cancel{x = -3}$$

The perimeter of a rhombus is the sum of all the sides of a rhombus. The four sides of a rhombus are equal in length. Each side measures $x+8$, which equals $5+8 = 13$. The perimeter of the rhombus is

$$13 + 13 + 13 + 13 = 52$$

558. 68

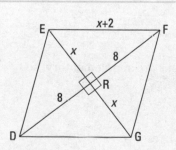

The diagonals of a rhombus bisect each other. This means they divide each other into two congruent segments:

$$DR = RF = 8$$

The diagonals of a rhombus are perpendicular to each other, which means they form right angles at their point of intersection. These right angles create four right triangles in the rhombus. The legs of the right triangles are 8 and x, and the hypotenuse of the right triangle, which is a side of the rhombus, is $x + 2$. Use the Pythagorean theorem to solve for x:

$$a^2 + b^2 = c^2$$
$$8^2 + x^2 = (x+2)^2$$
$$64 + x^2 = (x+2)(x+2)$$
$$64 + x^2 = x^2 + 2x + 2x + 4$$
$$64 + x^2 = x^2 + 4x + 4$$
$$64 = 4x + 4$$
$$60 = 4x$$
$$15 = x$$

The perimeter of a rhombus is the sum of all the sides of a rhombus. The four sides of a rhombus are equal in length; therefore, each side measures $x + 2 = 15 + 2 = 17$. The perimeter of the rhombus is

$$17 + 17 + 17 + 17 = 68$$

559. **14**

All four sides of a rhombus are congruent, so

$$DR = DW$$
$$x + 8 = 3x - 4$$
$$8 = 2x - 4$$
$$12 = 2x$$
$$6 = x$$

Therefore, $DW = 3x - 4 = 3(6) - 4 = 14$.

560. **136°**

Let

$$m\angle D = x$$
$$m\angle R = 3x + 4$$

Consecutive angles of a rhombus are supplementary, so

$$m\angle D + m\angle R = 180°$$
$$x + 3x + 4 = 180°$$
$$4x + 4 = 180°$$
$$4x = 176°$$
$$x = 44°$$

Therefore, $m\angle R = 3x + 4 = 3(44°) + 4 = 136°$.

561.   40°

Consecutive angles of a rhombus are supplementary, so

$$m\angle D + m\angle W = 180°$$
$$m\angle D + 140° = 180°$$
$$m\angle D = 40°$$

562.   26

The diagonals of a rhombus bisect each other, which means they divide each other into two congruent segments. If you call this point of intersection A, then

$$DA = EA = 24$$
$$RA = AW = 10$$

The diagonals of a rhombus are perpendicular to each other, which means they form right angles at their point of intersection. These right angles create four right triangles in the rhombus. The legs of the right triangles are 10 and 24, and the hypotenuse of the right triangle is a side of the rhombus. Use the Pythagorean theorem to solve for the hypotenuse of the triangle:

$$a^2 + b^2 = c^2$$
$$10^2 + 24^2 = c^2$$
$$676 = c^2$$
$$\sqrt{676} = \sqrt{c^2}$$
$$26 = c$$

563.   80

The diagonals of a rhombus bisect the angles of the rhombus, which means that $\overline{DE}$ divides $\angle REW$ into two congruent angles:

$$m\angle RED = m\angle WED = 60°$$

The diagonals of a rhombus bisect each other. This means they divide each other into two congruent segments. If you call this point of intersection A, then

$$DA = EA = 10$$

The diagonals of a rhombus are perpendicular to each other, which means they form right angles at their point of intersection. $\triangle RAE$ is a 30°-60°-90° triangle. The side opposite the 30° ($\angle ERW$) is equivalent to x. This means that $x = 10$. The side of the rhombus is the hypotenuse (opposite the 90° angle), which is equal to $2x$. Because $x = 10$, the hypotenuse is $2(10) = 20$.

You can find the perimeter of a rhombus by adding the four congruent sides:

$$20 + 20 + 20 + 20 = 80$$

564.   **10**

The diagonals of a rhombus bisect the angles of the rhombus, which means that $\overline{DE}$ divides $\angle REW$ into two congruent angles:

$$m\angle RED = m\angle WED = 60°$$

The diagonals of a rhombus are perpendicular to each other, which means they form right angles at their point of intersection. Call this point of intersection A. A rhombus has four equal sides, each equaling 10 in this case. $\triangle RAE$ is a 30°-60°-90° triangle. $\overline{RE}$ is the hypotenuse (opposite the 90° angle), which is equal to $2x$. Because $RE = 10$, x must equal 5. $\overline{AE}$, the side opposite the 30° angle, has a length equivalent to x. This means $AE = 5$. Because the diagonals of a rhombus bisect each other, you know that

$$DA = AE = 5$$
$$DE = 5 + 5 = 10$$

565.   **46**

A square has four congruent sides, so

$$KG = IN$$
$$2x + 34 = 7x + 4$$
$$34 = 5x + 4$$
$$30 = 5x$$
$$6 = x$$

Therefore, $KG = 2x + 34 = 2(6) + 34 = 46$.

566.   **–6**

A square has four congruent sides, so

$$NG = NI$$
$$x^2 + 4x = 12$$
$$x^2 + 4x - 12 = 0$$
$$(x + 6)(x - 2) = 0$$
$$x + 6 = 0 \text{ or } x - 2 = 0$$
$$x = -6 \text{ or } \cancel{x = 2}$$

567.   5

The diagonals of a square are congruent, so

$$KN = IG$$
$$x^2 - 3x = x + 5$$
$$x^2 - 4x = 5$$
$$x^2 - 4x - 5 = 0$$
$$(x-5)(x+1) = 0$$
$$x - 5 = 0 \text{ or } x + 1 = 0$$
$$x = 5 \text{ or } \quad \cancel{x = -1}$$

568.   100 cm

A square contains four congruent sides. If you call the length of one of the sides s, you can write the perimeter as follows:

$$4s = 400$$
$$s = 100$$

569. $25\sqrt{2}$ ft

A square contains four congruent sides. If you call the length of one of the sides s, you can write the perimeter as follows:

$$4s = 100$$
$$s = 25$$

The diagonal of a square divides the square into two right triangles. The diagonal is the hypotenuse of the right triangle. Use the Pythagorean theorem to solve for the missing piece of the triangle:

$$a^2 + b^2 = c^2$$
$$25^2 + 25^2 = c^2$$
$$1,250 = c^2$$
$$\sqrt{1,250} = \sqrt{c^2}$$
$$\sqrt{625}\sqrt{2} = c$$
$$25\sqrt{2} = c$$

570.   20 units

A square contains four congruent sides. The formula for the area of a square is $A = s^2$, where s represents the length of a side of the square:

$$s^2 = 400$$
$$\sqrt{s^2} = \sqrt{400}$$
$$s = 20$$

571. 38

A square contains four right angles, so

$$m\angle SRW = 90°$$
$$2a + 14 = 90$$
$$2a = 76$$
$$a = 38$$

572. 12

The diagonals of a square are perpendicular to each other. This means that they form right angles at their points of intersection, so

$$m\angle REW = 90°$$
$$8x - 6 = 90$$
$$8x = 96$$
$$x = 12$$

573. 17

The diagonals of a square bisect each other. This means that they divide each other into two congruent segments:

$$SE = WE$$
$$2x - 1 = x + 8$$
$$x - 1 = 8$$
$$x = 9$$

Therefore, $WE = x + 8 = 9 + 8 = 17$.

574. 45°

The diagonals of a square bisect the angles of the square, so they divide the angles into two congruent angles. A square contains four right angles. When the diagonal bisects the 90° angle, it cuts it in half, so each angle formed by the diagonal measures 45°.

575. 32

The diagonals of a square bisect each other. This means they divide each other into two congruent segments. $RE = ET = 3x + 5$, so

$$RT = RE + ET$$
$$7x - 22 = 3x + 5 + 3x + 5$$
$$7x - 22 = 6x + 10$$
$$x - 22 = 10$$
$$x = 32$$

576. 50.5

A square contains four right angles, so

$$m\angle STW = 90^\circ$$
$$2x - 11 = 90$$
$$2x = 101$$
$$x = 50.5$$

577. 36

A square contains four congruent sides. Set the two known side lengths equal to each other and solve for x:

$$JO = ON$$
$$7x + 2 = 8x + 1$$
$$2 = x + 1$$
$$1 = x$$

Therefore, $JO = 7x + 2 = 7(1) + 2 = 9$. The perimeter is then

$$4s = 4(9) = 36$$

578. 10

A square contains four congruent sides. The diagonal of a square divides the square into two right triangles. Use the Pythagorean theorem to solve for the missing legs of the square:

$$a^2 + b^2 = c^2$$
$$x^2 + x^2 = \left(10\sqrt{2}\right)^2$$
$$2x^2 = 100(2)$$
$$2x^2 = 200$$
$$x^2 = 100$$
$$\sqrt{x^2} = \sqrt{100}$$
$$x = 10$$

579. 6

The diagonals of a square bisect each other, which means they divide each other into two congruent segments:

$$AT = OT = x + 2$$

The diagonals of a square are congruent, so

$$JN = OA$$
$$4x - 8 = x + 2 + x + 2$$
$$4x - 8 = 2x + 4$$
$$2x - 8 = 4$$
$$2x = 12$$
$$x = 6$$

580. 9

A square contains four right angles. The diagonals of a square bisect the angles of the square, so

$$m\angle JOA = 45°$$
$$x^2 - 4x = 45$$
$$x^2 - 4x - 45 = 0$$
$$(x - 9)(x + 5) = 0$$
$$x - 9 = 0 \text{ or } x + 5 = 0$$
$$x = 9 \text{ or } \cancel{x = -5}$$

581. 26

The median of a trapezoid is equal to the average of the bases of the trapezoid:

$$\frac{20 + 32}{2} = 26$$

582. 19.5

The median of a trapezoid is equal to the average of the bases of the trapezoid:

$$\frac{17 + 22}{2} = 19.5$$

583. 26

The median of a trapezoid is equal to the average of the bases of the trapezoid:

$$\frac{x + 50}{2} = 38$$
$$x + 50 = 76$$
$$x = 26$$

584. 12

The median of a trapezoid is equal to the average of the bases of the trapezoid:

$$\frac{2x+8+6x-2}{2} = 51$$

$$\frac{8x+6}{2} = 51$$

$$8x+6 = 102$$

$$8x = 96$$

$$x = 12$$

585. 5

The median of a trapezoid is equal to the average of the bases of the trapezoid:

$$\frac{4x-1+5x}{2} = 2x+2$$

$$\frac{9x-1}{2} = 2x+2$$

$$9x-1 = 2(2x+2)$$

$$9x-1 = 4x+4$$

$$5x-1 = 4$$

$$5x = 5$$

$$x = 1$$

Therefore, $PA = 5x = 5(1) = 5$.

586. 65°

Consecutive angles between the two parallel bases of a trapezoid are supplementary, which means they add up to 180°:

$$m\angle T + m\angle P = 180°$$

$$115° + m\angle P = 180°$$

$$m\angle P = 65°$$

587. 55

Consecutive angles between the two parallel bases of a trapezoid are supplementary, which means they add up to 180°:

$$m\angle R + m\angle A = 180°$$

$$2x+4+x+11 = 180$$

$$3x+15 = 180$$

$$3x = 165$$

$$x = 55$$

588. 53°

Consecutive angles between the two parallel bases of a trapezoid are supplementary, which means they add up to 180°:

$$m\angle T + m\angle P = 180°$$
$$4x + 3 + 2x - 9 = 180$$
$$6x - 6 = 180$$
$$6x = 186$$
$$x = 31$$

Therefore, $m\angle P = 2x - 9 = 2(31) - 9 = 53°$.

589. 19

The diagonals of an isosceles trapezoid are congruent, so

$$TA = RP$$
$$5x = 95$$
$$x = 19$$

590. 3.6

The diagonals of an isosceles trapezoid are congruent, so

$$TA = RP$$
$$7x - 21 = 2x - 3$$
$$5x - 21 = -3$$
$$5x = 18$$
$$x = 3.6$$

591. 6

The diagonals of an isosceles trapezoid are congruent, so

$$TA = RP$$
$$x^2 + 3x = 9x$$
$$x^2 - 6x = 0$$
$$x(x - 6) = 0$$
$$\cancel{x = 0} \text{ or } x - 6 = 0$$
$$x = 6$$

592. 7

Use the distance formula to solve this problem:

$$d = \sqrt{(x_2 - x_1)^2 + (y_2 - y_1)^2}$$
$$= \sqrt{(-3-(-3))^2 + (2-(-5))^2}$$
$$= \sqrt{0^2 + 7^2}$$
$$= \sqrt{49}$$
$$= 7$$

593. $2\sqrt{13}$

The origin is the point $(0,0)$. Use the distance formula to solve this problem:

$$d = \sqrt{(x_2 - x_1)^2 + (y_2 - y_1)^2}$$
$$= \sqrt{(0-6)^2 + (0-(-4))^2}$$
$$= \sqrt{(-6)^2 + 4^2}$$
$$= \sqrt{52}$$
$$= \sqrt{4}\sqrt{13}$$
$$= 2\sqrt{13}$$

594. 10

Use the distance formula:

$$d = \sqrt{(x_2 - x_1)^2 + (y_2 - y_1)^2}$$
$$= \sqrt{(-4-2)^2 + (3-11)^2}$$
$$= \sqrt{(-6)^2 + (-8)^2}$$
$$= \sqrt{100}$$
$$= 10$$

595. *b*

Use the distance formula:

$$d = \sqrt{(x_2 - x_1)^2 + (y_2 - y_1)^2}$$
$$= \sqrt{(a-(a-b))^2 + (c-c)^2}$$
$$= \sqrt{(a-a+b)^2 + 0^2}$$
$$= \sqrt{b^2 + 0^2}$$
$$= \sqrt{b^2}$$
$$= b$$

596. $\sqrt{a^2 + b^2}$

Use the distance formula:

$$d = \sqrt{(x_2 - x_1)^2 + (y_2 - y_1)^2}$$
$$= \sqrt{(4a-3a)^2 + (b-0)^2}$$
$$= \sqrt{(a)^2 + (b)^2}$$
$$= \sqrt{a^2 + b^2}$$

597. **Isosceles**

To classify the triangle, you need to know the length of each side. Use the distance formula to find the side lengths:

$$d = \sqrt{(x_2 - x_1)^2 + (y_2 - y_1)^2}$$
$$JK = \sqrt{(5-4)^2 + (0-4)^2}$$
$$= \sqrt{1^2 + (-4)^2}$$
$$= \sqrt{17}$$
$$JL = \sqrt{(5-6)^2 + (0-4)^2}$$
$$= \sqrt{(-1)^2 + (-4)^2}$$
$$= \sqrt{17}$$
$$KL = \sqrt{(4-6)^2 + (4-4)^2}$$
$$= \sqrt{(-2)^2 + 0^2}$$
$$= \sqrt{4}$$
$$= 2$$

Two of the sides are equal in length; therefore, the triangle is isosceles.

The triangle is not a right isosceles triangle because the lengths of the sides do not satisfy the Pythagorean theorem:

$$a^2 + b^2 = c^2$$

$$2^2 + \left(\sqrt{17}\right)^2 \overset{?}{=} \left(\sqrt{17}\right)^2$$

$$4 + 17 \overset{?}{=} 17$$

$$21 \neq 17$$

---

598. Equilateral

To classify the triangle, you need to know the length of each side. Use the distance formula to find the side lengths:

$$d = \sqrt{\left(x_2 - x_1\right)^2 + \left(y_2 - y_1\right)^2}$$

$$DI = \sqrt{\left(4 - (-6)\right)^2 + \left(0 - 0\right)^2}$$

$$= \sqrt{10^2 + 0^2}$$

$$= \sqrt{100}$$

$$= 10$$

$$IS = \sqrt{\left(-6 - (-1)\right)^2 + \left(0 - \left(-5\sqrt{3}\right)\right)^2}$$

$$= \sqrt{\left(-5\right)^2 + \left(5\sqrt{3}\right)^2}$$

$$= \sqrt{25 + 25(3)}$$

$$= \sqrt{100}$$

$$= 10$$

$$DS = \sqrt{\left(4 - (-1)\right)^2 + \left(0 - \left(5\sqrt{3}\right)\right)^2}$$

$$= \sqrt{5^2 + \left(-5\sqrt{3}\right)^2}$$

$$= \sqrt{25 + 25(3)}$$

$$= \sqrt{100}$$

$$= 10$$

Because all three sides of the triangle are equal in length, the triangle is equilateral.

599. 2 and 10

Use the distance formula:

$$d = \sqrt{\left(x_2 - x_1\right)^2 + \left(y_2 - y_1\right)^2}$$
$$5 = \sqrt{\left(2 - 5\right)^2 + \left(6 - y\right)^2}$$
$$5 = \sqrt{\left(-3\right)^2 + \left(6 - y\right)^2}$$
$$5 = \sqrt{9 + \left(6 - y\right)^2}$$
$$5^2 = \left(\sqrt{9 + \left(6 - y\right)^2}\right)^2$$
$$25 = 9 + \left(6 - y\right)^2$$
$$25 = 9 + \left(6 - y\right)\left(6 - y\right)$$
$$25 = 9 + 36 - 12y + y^2$$
$$25 = 45 - 12y + y^2$$
$$0 = y^2 - 12y + 20$$
$$0 = \left(y - 10\right)\left(y - 2\right)$$
$$y - 10 = 0 \;\; \text{or} \;\; y - 2 = 0$$
$$y = 10 \;\; \text{or} \;\; y = 2$$

600. (11, 6)

Use the midpoint formula:

$$M = \left(\frac{x_1 + x_2}{2}, \; \frac{y_1 + y_2}{2}\right)$$
$$= \left(\frac{14 + 8}{2}, \; \frac{2 + 10}{2}\right)$$
$$= \left(11, \; 6\right)$$

601. (1, 13.5)

Use the midpoint formula:

$$M = \left(\frac{x_1 + x_2}{2}, \frac{y_1 + y_2}{2}\right)$$
$$= \left(\frac{-2 + 4}{2}, \frac{21 + 6}{2}\right)$$
$$= \left(1, 13.5\right)$$

602. (5, 5)

The diagonals of a rectangle bisect each other, which means that they intersect at the midpoint of either diagonal. Here's how to use the midpoint formula to find the midpoint of diagonal $\overline{DA}$:

$$M = \left(\frac{x_1 + x_2}{2}, \ \frac{y_1 + y_2}{2} \right)$$

$$= \left(\frac{1+9}{2}, \ \frac{2+8}{2} \right)$$

$$= (5, \ 5)$$

603. (2, 2)

Use the midpoint formula to find the middle of the diameter:

$$M = \left(\frac{x_1 + x_2}{2}, \ \frac{y_1 + y_2}{2} \right)$$

$$= \left(\frac{10 + (-6)}{2}, \ \frac{-4 + 8}{2} \right)$$

$$= (2, \ 2)$$

604. $\left(3a + 2b, \ 2a \right)$

Use the midpoint formula:

$$M = \left(\frac{x_1 + x_2}{2}, \ \frac{y_1 + y_2}{2} \right)$$

$$= \left(\frac{4a + b + 2a + 3b}{2}, \ \frac{3a + b + a - b}{2} \right)$$

$$= \left(\frac{6a + 4b}{2}, \ \frac{4a}{2} \right)$$

$$= \left(3a + 2b, \ 2a \right)$$

605. (16, −7)

Use the midpoint formula:

$$M = \left(\frac{x_1 + x_2}{2}, \ \frac{y_1 + y_2}{2} \right)$$

$$(10, -3) = \left(\frac{4 + x}{2}, \ \frac{1 + y}{2} \right)$$

Set the x coordinates equal to each other and solve for x:

$$\frac{4+x}{2} = 10$$
$$4 + x = 20$$
$$x = 16$$

Set the y coordinates equal to each other and solve for y:

$$\frac{1+y}{2} = -3$$
$$1 + y = -6$$
$$y = -7$$

Therefore, K is at $(16, -7)$.

606. $(-8, -14)$

Use the midpoint formula:

$$M = \left(\frac{x_1 + x_2}{2}, \frac{y_1 + y_2}{2} \right)$$
$$(-2, -8) = \left(\frac{4+x}{2}, \frac{-2+y}{2} \right)$$

Set the x coordinates equal to each other and solve for x:

$$\frac{4+x}{2} = -2$$
$$4 + x = -4$$
$$x = -8$$

Set the y coordinates equal to each other and solve for y:

$$\frac{-2+y}{2} = -8$$
$$-2 + y = -16$$
$$y = -14$$

Therefore, P is at $(-8, -14)$.

607. $(-6, 10)$

The center of a circle is the midpoint of the diameter. This means that the midpoint of $\overline{AC}$ is $(-4, 7)$. Use the midpoint formula:

$$M = \left(\frac{x_1 + x_2}{2}, \frac{y_1 + y_2}{2} \right)$$
$$(-4, 7) = \left(\frac{-2+x}{2}, \frac{4+y}{2} \right)$$

Set the x coordinates equal to each other and solve for x:

$$\frac{-2+x}{2} = -4$$

$$-2+x = -8$$

$$x = -6$$

Set the y coordinates equal to each other and solve for y:

$$\frac{4+y}{2} = 7$$

$$4 + y = 14$$

$$y = 10$$

Therefore, C is at $(-6, 10)$.

---

608.   **2**

Use the slope formula:

$$m = \frac{y_2 - y_1}{x_2 - x_1} = \frac{0 - 8}{5 - 9} = \frac{-8}{-4} = 2$$

---

609.   **9**

Use the slope formula:

$$m = \frac{y_2 - y_1}{x_2 - x_1} = \frac{-10 - 8}{-4 - (-2)} = \frac{-18}{-2} = 9$$

---

610. $\frac{3}{4}$

Use the slope formula:

$$m = \frac{y_2 - y_1}{x_2 - x_1} = \frac{8 - 2}{9 - 1} = \frac{6}{8} = \frac{3}{4}$$

---

611.   **True**

Points that are *collinear* lie on the same line. If the points lie along the same line, the slope between any two of the points will be the same. Use the slope formula to find the slope of each line segment:

$$m = \frac{y_2 - y_1}{x_2 - x_1}$$

$$m_{\overline{AB}} = \frac{5 - 3}{3 - (-2)} = \frac{2}{5}$$

$$m_{\overline{BC}} = \frac{7 - 5}{8 - 3} = \frac{2}{5}$$

$$m_{\overline{AC}} = \frac{7 - 3}{8 - (-2)} = \frac{4}{10} = \frac{2}{5}$$

The slopes are all the same, so the points are collinear.

612. 6

Use the slope formula:

$$m = \frac{y_2 - y_1}{x_2 - x_1}$$

$$-2 = \frac{10 - y}{2 - 4}$$

$$-2 = \frac{10 - y}{-2}$$

$$4 = 10 - y$$

$$-6 = -y$$

$$6 = y$$

613. 4

Slope that is undefined (or *no slope*) means that the slope formula has a denominator of 0. The slope formula is

$$m = \frac{y_2 - y_1}{x_2 - x_1}$$

Set the denominator equal to 0 and solve for x:

$$x_2 - x_1 = 0$$

$$x - 4 = 0$$

$$x = 4$$

Another approach is to recognize that a vertical line is the only line that has an undefined slope. If a line is vertical, the x coordinates of the points must be the same, so they must both equal 4.

614. Parallel

If two lines are parallel, their slopes are equal. If two lines are perpendicular, their slopes are negative reciprocals of each other. Use the slope formula to find the slopes of $\overline{RS}$ and $\overline{TV}$:

$$m = \frac{y_2 - y_1}{x_2 - x_1}$$

$$m_{\overline{RS}} = \frac{4 - 3}{5 - 0} = \frac{1}{5}$$

$$m_{\overline{TV}} = \frac{9 - 8}{3 - (-2)} = \frac{1}{5}$$

The slopes are equal, which means the lines are parallel.

615. **Perpendicular**

If two lines are parallel, their slopes are equal. If two lines are perpendicular, their slopes are negative reciprocals of each other. Use the slope formula to find the slopes of $\overline{MA}$ and $\overline{TH}$:

$$m = \frac{y_2 - y_1}{x_2 - x_1}$$
$$m_{\overline{MA}} = \frac{0-1}{10-3} = \frac{-1}{7}$$
$$m_{\overline{TH}} = \frac{-3-4}{6-7} = \frac{-7}{-1} = 7$$

Because the slopes are negative reciprocals of each other, the lines are perpendicular.

616. $\frac{1}{3}$

First find the coordinates of M and N using the midpoint formula:

$$M = \left(\frac{x_1 + x_2}{2}, \frac{y_1 + y_2}{2} \right) = \left(\frac{2+4}{2}, \frac{0+8}{2} \right) = (3, 4)$$
$$N = \left(\frac{4+8}{2}, \frac{8+2}{2} \right) = (6, 5)$$

Now that you know the coordinates of Points M and N, you can use the slope formula to find the slope of $\overline{MN}$:

$$m = \frac{y_2 - y_1}{x_2 - x_1}$$
$$m_{\overline{MN}} = \frac{5-4}{6-3} = \frac{1}{3}$$

617. $-\frac{3}{2}$

The slope-intercept form of a line is $y = mx + b$, where m represents the slope and b represents the y-intercept. The equation shows that the slope of the given line is $-\frac{3}{2}$. Parallel lines have equal slopes, so the slope of the second line must also be $-\frac{3}{2}$.

618. $-\frac{3}{5}$

Perpendicular lines have slopes that are negative reciprocals of each other. If the slope of one line is $\frac{5}{3}$, then the slope of the other line has to be $-\frac{3}{5}$.

619. $\frac{1}{2}$

The slope-intercept form of a line is $y = mx + b$, where m represents the slope and b represents the y-intercept. The equation shows that the slope of the given line is $\frac{1}{2}$. Parallel lines have equal slopes, so the slope of the second line must also be $\frac{1}{2}$.

620. −2

The slope-intercept form of a line is $y = mx + b$, where m represents the slope and b represents the y-intercept. The equation shows that the slope of the given line is $\frac{1}{2}$. Perpendicular lines have slopes that are negative reciprocals of each other, so the slope of the second line must be −2.

621. Perpendicular

Lines that are parallel have equal slopes. Perpendicular lines have slopes that are negative reciprocals of each other. To determine the slope of each line, first put the equations in slope-intercept form:

$$2y + 3 = 4x$$
$$2y = 4x - 3$$
$$\frac{2y}{2} = \frac{4x}{2} - \frac{3}{2}$$
$$y = 2x - \frac{3}{2}$$

$$4y + 2x = 12$$
$$4y = -2x + 12$$
$$\frac{4y}{4} = \frac{-2x}{4} + \frac{12}{4}$$
$$y = \frac{-1}{2}x + 3$$

The slope-intercept form of a line is $y = mx + b$, where m represents the slope and b represents the y-intercept. The first equation shows that the slope of the line is 2. The second equation shows that the slope of the line is $-\frac{1}{2}$. Because the two slopes are negative reciprocals of each other, the lines must be perpendicular.

622. Parallel

Lines that are parallel have equal slopes. Lines that are perpendicular have slopes that are negative reciprocals of each other. To determine the slope of the second line, put its equation in slope-intercept form:

$$6y = 4x + 3$$
$$\frac{6y}{6} = \frac{4x}{6} + \frac{3}{6}$$
$$y = \frac{2}{3}x + \frac{1}{2}$$

The slope-intercept form of a line is $y = mx + b$, where m represents the slope and b represents the y-intercept. The first equation shows that the slope of the line is $\frac{2}{3}$, and the second equation shows that the slope of the other line is also $\frac{2}{3}$. Because the two slopes are equal, the lines must be parallel.

623. **Neither**

Lines that are parallel have equal slopes. Lines that are perpendicular have slopes that are negative reciprocals of each other. To determine the slope of the second line, put its equation in slope-intercept form:

$$10y - 2x = 3$$
$$10y = 2x + 3$$
$$\frac{10y}{10} = \frac{2x}{10} + \frac{3}{10}$$
$$y = \frac{1}{5}x + \frac{3}{10}$$

The slope-intercept form of a line is $y = mx + b$, where m represents the slope and b represents the y-intercept. The first equation shows that the slope of the line is 5. The second equation shows that the slope of the other line is $\frac{1}{5}$.

Because the two slopes are neither equal nor negative reciprocals, the lines are neither perpendicular nor parallel. If the slope of the second line were negative instead of positive, the lines would have been perpendicular.

624. **Perpendicular**

Lines that are parallel have equal slopes. Lines that are perpendicular have slopes that are negative reciprocals of each other. The line $x = 3$ is a vertical line. Pick any two points on this line. Because every coordinate along this line has an x value of 3, you can match the x value of 3 with any y value you would like. Use $(3, 1)$ and $(3, 4)$. Now you can find the slope of this line using the slope formula:

$$m = \frac{y_2 - y_1}{x_2 - x_1} = \frac{4 - 1}{3 - 3} = \frac{3}{0}$$

The slope is undefined.

The line $y = -4$ is a horizontal line. Pick any two points that land on this line. Because every coordinate along this line has a y value of -4, you can match the y value of -4 with any x value you would like. Use $(1, -4)$ and $(3, -4)$. Now you can find the slope of this line using the slope formula:

$$m = \frac{y_2 - y_1}{x_2 - x_1} = \frac{-4 - (-4)}{3 - 1} = \frac{0}{2}$$

The slope is 0.

The slopes of the lines are reciprocals of each other, so the lines are perpendicular to each other. (Because you're dealing with a value of 0, you'd never be able to see a *negative* reciprocal.)

625. Perpendicular

If two lines are parallel, their slopes are equal. If two lines are perpendicular, their slopes are negative reciprocals of each other. Use the slope formula to find the slope of the line containing the points $(0, 1)$ and $(5, 6)$:

$$m = \frac{y_2 - y_1}{x_2 - x_1}$$

$$m = \frac{6 - 1}{5 - 0} = \frac{5}{5} = 1$$

Use the slope formula to find the slope of the line containing the points $(-1, 5)$ and $(3, 1)$:

$$m = \frac{y_2 - y_1}{x_2 - x_1}$$

$$m = \frac{1 - 5}{3 - (-1)} = \frac{-4}{4} = -1$$

The slopes are negative reciprocals of each other, which means that the two lines are perpendicular.

626. Neither

If two lines are parallel, their slopes are equal. If two lines are perpendicular, their slopes are negative reciprocals of each other. Use the slope formula to find the slope of the line containing the points $(-1, 0)$ and $(1, 1)$:

$$m = \frac{y_2 - y_1}{x_2 - x_1}$$

$$m = \frac{1 - 0}{1 - (-1)} = \frac{1}{2}$$

Use the slope formula to find the slope of the line containing the points $(0, -1)$ and $(1, 1)$:

$$m = \frac{y_2 - y_1}{x_2 - x_1}$$

$$m = \frac{1 - (-1)}{1 - 0} = \frac{2}{1} = 2$$

The slopes aren't negative reciprocals of each other, so the two lines aren't perpendicular. The slopes aren't equal, which means that the two lines aren't parallel.

627. $y = 3x - 4$

The slope-intercept form of a line is $y = mx + b$, where m represents the slope and b represents the y-intercept. This means that a line with a slope of 3 and a y-intercept of -4 would be $y = 3x - 4$.

628. $y = -\dfrac{1}{4}x$

The slope-intercept form of a line is $y = mx + b$, where m represents the slope and b represents the y-intercept. A point that passes through the origin, $(0,0)$, has a y-intercept of 0 because it crosses the y-axis at 0. This means that a line with a slope of $-\dfrac{1}{4}$ and a y-intercept of 0 would have the equation $y = -\dfrac{1}{4}x$.

629. $y = \dfrac{3}{5}x + 4$

The slope-intercept form of a line is $y = mx + b$, where m represents the slope and b represents the y-intercept. This means that a line with a slope of $\dfrac{3}{5}$ would have the equation $y = \dfrac{3}{5}x + b$.

You know that the line passes through the point $(5,7)$. Plug those x and y values into the equation to solve for b:

$$y = \frac{3}{5}x + b$$
$$7 = \frac{3}{5}(5) + b$$
$$7 = 3 + b$$
$$4 = b$$

Now that you know the slope (m) and the y-intercept (b), you can plug both values into the slope-intercept form of a line:

$$y = mx + b$$
$$y = \frac{3}{5}x + 4$$

630. $y = 12$

To determine the equation of a line, you need to know the slope and the y-intercept. Find the slope of the line first:

$$m = \frac{y_2 - y_1}{x_2 - x_1} = \frac{12 - 12}{6 - 4} = \frac{0}{2} = 0$$

The slope-intercept form of a line is $y = mx + b$, where m represents the slope and b represents the y-intercept. This means that a line with a slope of 0 would have the equation $y = 0x + b$. You know that the line passes through the point $(4,12)$. You can plug those x and y values into the equation to solve for b:

$$y = 0x + b$$
$$12 = 0(4) + b$$
$$12 = b$$

Now that you know the slope *(m)* and the *y*-intercept *(b)*, you can plug both values into the slope-intercept form of a line:

$$y = mx + b$$
$$y = 0x + 12$$
$$y = 12$$

631. $x = -1$

To determine the equation of a line, you need to know the slope and the *y*-intercept. Find the slope of the line first:

$$m = \frac{y_2 - y_1}{x_2 - x_1} = \frac{8 - 5}{-1 - (-1)} = \frac{3}{0} = \text{undefined}$$

Because the slope is undefined, you know that this line is a vertical line. Every point on a vertical line contains the same *x* value. Both points on this line contain the *x* value –1, so the equation of this line is $x = -1$.

632. $y = x + 19$

To determine the equation of a line, you need to know the slope and the *y*-intercept. Find the slope of the line first:

$$m = \frac{y_2 - y_1}{x_2 - x_1} = \frac{17 - 16}{-2 - (-3)} = \frac{1}{1} = 1$$

The slope-intercept form of a line is $y = mx + b$, where *m* represents the slope and *b* represents the *y*-intercept. This means that a line with a slope of 1 would have the equation $y = x + b$. You know that the line passes through the point $(-3, 16)$. You can plug those *x* and *y* values into the equation to solve for *b*:

$$y = x + b$$
$$16 = -3 + b$$
$$19 = b$$

Now that you know the slope *(m)* and the *y*-intercept *(b)*, you can plug both values into the slope-intercept form of a line:

$$y = mx + b$$
$$y = x + 19$$

633. $y = -\frac{3}{2}x + \frac{13}{2}$

To determine the equation of a line, you need to know the slope and the *y*-intercept. Find the slope of the line first:

$$m = \frac{y_2 - y_1}{x_2 - x_1} = \frac{5 - 8}{1 - (-1)} = \frac{-3}{2}$$

The slope-intercept form of a line is $y = mx + b$, where m represents the slope and b represents the y-intercept. This means that a line with a slope of $-\frac{3}{2}$ would have the equation $y = -\frac{3}{2}x + b$. You know that the line passes through the point $(-1, 8)$. You can plug those x and y values into the equation to solve for b:

$$y = \frac{-3}{2}x + b$$
$$8 = \frac{3}{2}(-1) + b$$
$$8 = \frac{3}{2} + b$$
$$\frac{13}{2} = b$$

Now that you know the slope (m) and the y-intercept (b), you can plug both values into the slope-intercept form of a line:

$$y = mx + b$$
$$y = \frac{-3}{2}x + \frac{13}{2}$$

634. $y = 3x + 10$

The slope-intercept form of a line is $y = mx + b$, where m represents the slope of the line and b represents the y-intercept. This means that the slope of the given line is 3. Because parallel lines have equal slopes, you now know that the slope of the new line will also be 3. You were given that the y-intercept, b, of the equation is 10. You now have enough information to plug into the slope-intercept form of a line to get the new equation:

$$y = mx + b$$
$$y = 3x + 10$$

635. $y = 2x + 5$

The slope-intercept form of a line is $y = mx + b$, where m represents the slope of the line and b represents the y-intercept. Before you can determine the slope and y-intercept, put the given equation in $y =$ form:

$$\frac{2y}{2} = \frac{4x}{2} - \frac{2}{2}$$
$$y = 2x - 1$$

This means that the slope of the given line is 2. Because parallel lines have equal slopes, you now know that the slope of the new line will also be 2. If you plug that into the slope-intercept form of a line, you get

$$y = 2x + b$$

The question also tells you that the new line passes through the point $\left(\frac{1}{2}, 6\right)$. You can plug those x and y values into the equation to find the y-intercept:

$$y = 2x + b$$
$$6 = 2\left(\frac{1}{2}\right) + b$$
$$6 = 1 + b$$
$$5 = b$$

You now have enough information to plug into the slope-intercept form of a line. The new equation is

$$y = mx + b$$
$$y = 2x + 5$$

636. $x = 8$

Lines that are perpendicular to each other have negative reciprocal slopes. The x-axis is a horizontal line that has a slope of 0. The line perpendicular to the x-axis is a vertical line, which has an undefined slope.

All points along a vertical line have the same x value. You already know that the line passes through a point whose x value is 8. This means that the equation of the line is $x = 8$.

637. $y = 5$

Lines that are parallel to each other have equal slopes. The x-axis is a horizontal line that has a slope of 0. A line parallel to the x-axis will also be a horizontal line.

All points along a horizontal line have the same y value. You already know that the line passes through a point whose y value is 5. This means that the equation of the line is $y = 5$.

638. $y = -\frac{1}{3}x - 2$

The slope-intercept form of a line is $y = mx + b$, where m represents the slope and b represents the y-intercept. Lines that are perpendicular have slopes that are negative reciprocals of each other. Because the slope of the given line is 3, the slope of the line perpendicular to it is $-\frac{1}{3}$. Plugging this slope into the slope-intercept form of a line gives you

$$y = -\frac{1}{3}x + b$$

You know that the line must pass through the point $(9, -5)$. You can plug the x and y values of that point into the equation of the line and solve for the y-intercept:

$$-5 = -\frac{1}{3}(9) + b$$
$$-5 = -3 + b$$
$$-2 = b$$

You now know that the equation of the perpendicular line has a slope of $-\frac{1}{3}$ and a y-intercept of -2. The slope-intercept form is

$$y = -\frac{1}{3}x - 2$$

---

639. $\quad y = -\frac{4}{3}x + 5$

First isolate the y so you can use the slope-intercept form of the equation to determine the slope and y-intercept of the new line:

$$4y\ 3x = 5$$
$$4y = 3x + 5$$
$$\frac{4y}{4} = \frac{3x}{4} + \frac{5}{4}$$
$$y = \frac{3}{4}x + \frac{5}{4}$$

The slope-intercept form of a line is $y = mx + b$, where m represents the slope and b represents the y-intercept. This means that the slope of the given line is $\frac{3}{4}$. Because perpendicular lines have slopes that are negative reciprocals of each other, you know that the slope of the new line is $-\frac{4}{3}$. If you plug that into the slope-intercept form of a line, you get

$$y = -\frac{4}{3}x + b$$

The question tells you that the new line passes through the point $(12, -11)$. You can plug those x and y values into the equation to find the y-intercept:

$$y = -\frac{4}{3}x + b$$
$$-11 = -\frac{4}{3}(12) + b$$
$$-11 = -16 + b$$
$$5 = b$$

You know have enough information to plug into the slope-intercept form of a line:

$$y = mx + b$$
$$y = -\frac{4}{3}x + 5$$

---

640. $\quad y = -2x + 15$

Perpendicular lines have slopes that are negative reciprocals of each other. Use the slope formula to find the slope of $\overline{AB}$:

$$m = \frac{y_2 - y_1}{x_2 - x_1}$$
$$m_{\overline{AB}} = \frac{5-1}{10-2} = \frac{4}{8} = \frac{1}{2}$$

This means that the slope of the new line must be –2. If you plug that into the slope-intercept form of a line, you get

$$y = -2x + b$$

You're also told that the new line is a perpendicular bisector of $\overline{AB}$. This means that the new line will pass through the midpoint of $\overline{AB}$. Use the midpoint formula:

$$M = \left(\frac{x_1 + x_2}{2}, \frac{y_1 + y_2}{2} \right)$$
$$= \left(\frac{2+10}{2}, \frac{1+5}{2} \right)$$
$$= (6, 3)$$

You can plug those x and y values into the equation to find the y-intercept:

$$y = -2x + b$$
$$3 = -2(6) + b$$
$$3 = -12 + b$$
$$15 = b$$

You now have enough information to plug into the slope-intercept form of a line:

$$y = mx + b$$
$$y = -2x + 15$$

641. $y = -\frac{1}{2}x + 12$

Perpendicular lines have slopes that are negative reciprocals of each other. Use the slope formula to find the slope of $\overline{AB}$:

$$m = \frac{y_2 - y_1}{x_2 - x_1}$$
$$m_{\overline{AB}} = \frac{17-1}{10-2} = \frac{16}{8} = 2$$

This means that the slope of the new line must be $-\frac{1}{2}$. If you plug that into the slope-intercept form of a line, you get

$$y = -\frac{1}{2}x + b$$

You're also told that the new line is a perpendicular bisector of $\overline{AB}$. This means that the new line will pass through the midpoint of $\overline{AB}$. Use the midpoint formula:

$$M = \left(\frac{x_1 + x_2}{2}, \frac{y_1 + y_2}{2} \right)$$
$$= \left(\frac{2+10}{2}, \frac{1+17}{2} \right)$$
$$= (6, 9)$$

You can plug those x and y values into the equation to find the y-intercept:

$$y = \frac{-1}{2}x + b$$
$$9 = \frac{-1}{2}(6) + b$$
$$9 = -3 + b$$
$$12 = b$$

You now have enough information to plug into the slope-intercept form of a line:

$$y = mx + b$$
$$y = \frac{-1}{2}x + 12$$

---

642.   **The length of each diagonal is 10.**

Use the distance formula to prove that the line segments are congruent:

$$d = \sqrt{(x_2 - x_1)^2 + (y_2 - y_1)^2}$$
$$RC = \sqrt{(9-1)^2 + (8-2)^2}$$
$$= \sqrt{(8)^2 + (6)^2}$$
$$= \sqrt{100}$$
$$= 10$$
$$ET = \sqrt{(10-0)^2 + (5-5)^2}$$
$$= \sqrt{(10)^2 + (0)^2}$$
$$= \sqrt{100}$$
$$= 10$$

The length of each diagonal is 10; therefore, the diagonals are congruent.

---

643.   **The midpoint of the diagonals is (5, 5).**

When two line segments bisect each other, they share a common midpoint. Use the midpoint formula to find the midpoint of diagonal $\overline{RC}$:

$$M = \left(\frac{x_1 + x_2}{2}, \ \frac{y_1 + y_2}{2} \right)$$
$$= \left(\frac{1+9}{2}, \ \frac{2+8}{2} \right)$$
$$= (5, 5)$$

Use the midpoint formula again to find the midpoint of diagonal $\overline{ET}$:

$$M = \left(\frac{x_1 + x_2}{2},\ \frac{y_1 + y_2}{2} \right)$$

$$= \left(\frac{10 + 0}{2},\ \frac{5 + 5}{2} \right)$$

$$= (5,\ 5)$$

Because the two segments share the same midpoint, they bisect each other.

644. **The slopes of the diagonals are $\frac{3}{4}$ and 0, which are not negative reciprocals.**

Lines that are perpendicular have slopes that are negative reciprocals of each other. Use the slope formula:

$$m = \frac{y_2 - y_1}{x_2 - x_1}$$

$$m_{\overline{RC}} = \frac{8 - 2}{9 - 1} = \frac{6}{8} = \frac{3}{4}$$

$$m_{\overline{ET}} = \frac{5 - 5}{0 - 10} = \frac{0}{-10} = 0$$

The slopes of the lines aren't negative reciprocals of each other; therefore, the lines are not perpendicular to each other.

645. **The slopes of the diagonals are 1 and –1, which are negative reciprocals.**

Lines that are perpendicular have slopes that are negative reciprocals of each other. Use the slope formula:

$$m = \frac{y_2 - y_1}{x_2 - x_1}$$

$$m_{\overline{RO}} = \frac{7 - 1}{10 - 4} = \frac{6}{6} = 1$$

$$m_{\overline{HM}} = \frac{5 - 3}{6 - 8} = \frac{2}{-2} = -1$$

The slopes of the lines are negative reciprocals of each other; therefore, the lines are perpendicular to each other.

646. **The lengths of the diagonals are $6\sqrt{2}$ and $2\sqrt{2}$.**

You would use the distance formula to prove that line segments are congruent:

$$d = \sqrt{\left(x_2 - x_1\right)^2 + \left(y_2 - y_1\right)^2}$$

$$RO = \sqrt{\left(10 - 4\right)^2 + \left(7 - 1\right)^2}$$

$$= \sqrt{\left(6\right)^2 + \left(6\right)^2}$$

$$= \sqrt{72}$$

$$= \sqrt{36}\sqrt{2}$$

$$= 6\sqrt{2}$$

$$HM = \sqrt{(6-8)^2 + (5-3)^2}$$
$$= \sqrt{(-2)^2 + (2)^2}$$
$$= \sqrt{8}$$
$$= \sqrt{4}\sqrt{2}$$
$$= 2\sqrt{2}$$

The lengths of the diagonals are different; therefore, the diagonals are not congruent.

---

647.   **All four sides are $2\sqrt{5}$.**

Use the distance formula to prove that the line segments are congruent:

$$d = \sqrt{(x_2 - x_1)^2 + (y_2 - y_1)^2}$$
$$RH = \sqrt{(8-4)^2 + (3-1)^2}$$
$$= \sqrt{4^2 + 2^2}$$
$$= \sqrt{20}$$
$$= \sqrt{4}\sqrt{5}$$
$$= 2\sqrt{5}$$
$$HO = \sqrt{(10-8)^2 + (7-3)^2}$$
$$= \sqrt{2^2 + 4^2}$$
$$= \sqrt{20}$$
$$= \sqrt{4}\sqrt{5}$$
$$= 2\sqrt{5}$$
$$OM = \sqrt{(6-10)^2 + (5-7)^2}$$
$$= \sqrt{(-4)^2 + (-2)^2}$$
$$= \sqrt{20}$$
$$= \sqrt{4}\sqrt{5}$$
$$= 2\sqrt{5}$$
$$RM = \sqrt{(6-4)^2 + (5-1)^2}$$
$$= \sqrt{2^2 + 4^2}$$
$$= \sqrt{20}$$
$$= \sqrt{4}\sqrt{5}$$
$$= 2\sqrt{5}$$

The side lengths are equal, so the four sides are congruent.

648. **The midpoints of the diagonals are both (7, 4).**

When two line segments bisect each other, they share a common midpoint. Use the midpoint formula to find the midpoint of diagonal $\overline{RO}$:

$$M = \left(\frac{x_1 + x_2}{2}, \ \frac{y_1 + y_2}{2} \right)$$

$$= \left(\frac{4 + 10}{2}, \ \frac{1 + 7}{2} \right)$$

$$= (7, \ 4)$$

Use the midpoint formula to find the midpoint of diagonal $\overline{HM}$:

$$M = \left(\frac{x_1 + x_2}{2}, \ \frac{y_1 + y_2}{2} \right)$$

$$= \left(\frac{8 + 6}{2}, \ \frac{3 + 5}{2} \right)$$

$$= (7, \ 4)$$

Because the two segments share the same midpoint, they bisect each other.

649. **The midpoints of both diagonals are (4, –2.5).**

When two line segments bisect each other, they share a common midpoint. Use the midpoint formula to find the midpoint of diagonal $\overline{PR}$:

$$M = \left(\frac{x_1 + x_2}{2}, \ \frac{y_1 + y_2}{2} \right)$$

$$= \left(\frac{-2 + 10}{2}, \ \frac{5 + (-10)}{2} \right)$$

$$= (4, -2.5)$$

Use the midpoint formula again to find the midpoint of diagonal $\overline{AL}$:

$$M = \left(\frac{x_1 + x_2}{2}, \ \frac{y_1 + y_2}{2} \right)$$

$$= \left(\frac{7 + 1}{2}, \ \frac{-1 + (-4)}{2} \right)$$

$$= (4, -2.5)$$

Because the two segments share the same midpoint, they bisect each other.

650. The slopes of the diagonals are $-\frac{5}{4}$ and $\frac{1}{2}$, which are not negative reciprocals.

Lines that are perpendicular have slopes that are negative reciprocals of each other. Use the slope formula:

$$m = \frac{y_2 - y_1}{x_2 - x_1}$$

$$m_{\overline{PR}} = \frac{-10 - 5}{10 - (-2)} = \frac{-15}{12} = \frac{-5}{4}$$

$$m_{\overline{AL}} = \frac{-4 - (-1)}{1 - 7} = \frac{-3}{-6} = \frac{1}{2}$$

The slopes of the lines aren't negative reciprocals of each other; therefore, the lines are not perpendicular to each other.

651. The lengths of the diagonals are $3\sqrt{41}$ and $3\sqrt{5}$.

Use the distance formula to determine whether the line segments are congruent:

$$d = \sqrt{(x_2 - x_1)^2 + (y_2 - y_1)^2}$$

$$PR = \sqrt{(10 - (-2))^2 + (-10 - 5)^2}$$

$$= \sqrt{(12)^2 + (-15)^2}$$

$$= \sqrt{369}$$

$$= \sqrt{9}\sqrt{41}$$

$$= 3\sqrt{41}$$

$$AL = \sqrt{(1 - 7)^2 + (-4 - (-1))^2}$$

$$= \sqrt{(-6)^2 + (-3)^2}$$

$$= \sqrt{45}$$

$$= \sqrt{9}\sqrt{5}$$

$$= 3\sqrt{5}$$

The lengths of the diagonals are different; therefore, the diagonals are not congruent.

652. Isosceles

Use the distance formula to determine whether the line segments are congruent:

$$d = \sqrt{(x_2 - x_1)^2 + (y_2 - y_1)^2}$$

$$JK = \sqrt{(7-0)^2 + (-3-(-6))^2}$$

$$= \sqrt{7^2 + 3^2}$$

$$= \sqrt{58}$$

$$KL = \sqrt{(0-7)^2 + (0-(-3))^2}$$

$$= \sqrt{(-7)^2 + 3^2}$$

$$= \sqrt{58}$$

$$JL = \sqrt{(0-0)^2 + (0-(-6))^2}$$

$$= \sqrt{0^2 + 6^2}$$

$$= 6$$

Two of the sides have the same length; therefore, the triangle is isosceles.

The triangle isn't a right triangle because the lengths of the sides do not satisfy the Pythagorean theorem:

$$a^2 + b^2 = c^2$$

$$6^2 + \left(\sqrt{58}\right)^2 \overset{?}{=} \left(\sqrt{58}\right)^2$$

$$36 + 58 \overset{?}{=} 58$$

$$94 \neq 58$$

653. Equilateral

Use the distance formula to determine whether the line segments are congruent:

$$d = \sqrt{(x_2 - x_1)^2 + (y_2 - y_1)^2}$$

$$TR = \sqrt{(-6-4)^2 + (0-0)^2}$$

$$= \sqrt{(-10)^2 + 0^2}$$

$$= \sqrt{100}$$

$$= 10$$

$$RS = \sqrt{\left(-1-(-6)\right)^2 + \left(5\sqrt{3}-0\right)^2}$$

$$= \sqrt{5^2 + \left(5\sqrt{3}\right)^2}$$

$$= \sqrt{25 + 25\sqrt{9}}$$

$$= \sqrt{25 + 25(3)}$$

$$= \sqrt{100}$$

$$= 10$$

$$TS = \sqrt{\left(-1-4\right)^2 + \left(5\sqrt{3}-0\right)^2}$$

$$= \sqrt{\left(-5\right)^2 + \left(5\sqrt{3}\right)^2}$$

$$= \sqrt{25 + 25\sqrt{9}}$$

$$= \sqrt{25 + 25(3)}$$

$$= \sqrt{100}$$

$$= 10$$

All three sides have the same length; therefore, the triangle is equilateral.

654.   **Right and scalene**

Use the distance formula to determine whether the line segments are congruent:

$$d = \sqrt{\left(x_2 - x_1\right)^2 + \left(y_2 - y_1\right)^2}$$

$$QR = \sqrt{\left(6-0\right)^2 + \left(-4-(-2)\right)^2}$$

$$= \sqrt{6^2 + \left(-2\right)^2}$$

$$= \sqrt{40}$$

$$= \sqrt{4}\sqrt{10}$$

$$= 2\sqrt{10}$$

$$RS = \sqrt{\left(1-6\right)^2 + \left(1-(-4)\right)^2}$$

$$= \sqrt{\left(-5\right)^2 + 5^2}$$

$$= \sqrt{50}$$

$$= \sqrt{25}\sqrt{2}$$

$$= 5\sqrt{2}$$

$$QS = \sqrt{\left(1-0\right)^2 + \left(1-(-2)\right)^2}$$

$$= \sqrt{1^2 + 3^2}$$

$$= \sqrt{10}$$

The triangle is scalene because the sides are different lengths.

Now plug the side lengths into the Pythagorean theorem to see whether the theorem holds true:

$$a^2 + b^2 = c^2$$
$$\left(\sqrt{10}\right)^2 + \left(\sqrt{40}\right)^2 \overset{?}{=} \left(\sqrt{50}\right)^2$$
$$10 + 40 \overset{?}{=} 50$$
$$50 = 50$$

The triangle satisfies the Pythagorean theorem; therefore, the triangle is also a right triangle.

655. **The lengths of the diagonals are both $\sqrt{145}$.**

Use the distance formula to determine whether the line segments are congruent:

$$d = \sqrt{\left(x_2 - x_1\right)^2 + \left(y_2 - y_1\right)^2}$$
$$TA = \sqrt{\left(4 - (-5)\right)^2 + (10 - 2)^2}$$
$$= \sqrt{9^2 + 8^2}$$
$$= \sqrt{145}$$
$$RP = \sqrt{(-8 - 4)^2 + (6 - 5)^2}$$
$$= \sqrt{(-12)^2 + 1^2}$$
$$= \sqrt{145}$$

The diagonals have the same lengths; therefore, they're congruent.

656. **The slopes of the legs are both $\frac{1}{3}$.**

Use the slope formula:

$$m = \frac{y_2 - y_1}{x_2 - x_1}$$
$$m_{\overline{TR}} = \frac{5 - 2}{4 - (-5)} = \frac{3}{9} = \frac{1}{3}$$
$$m_{\overline{AP}} = \frac{6 - 10}{-8 - 4} = \frac{-4}{-12} = \frac{1}{3}$$

Bases $\overline{TR}$ and $\overline{AP}$ have the same slope; therefore, they're parallel.

657. $r_{y\text{-axis}}$

The figure shows that the y-axis is the line of symmetry for the two triangles. When you reflect over the y-axis, you negate the sign of the original x value. In this figure, $A\,(-3, 0.5)$ maps to $A'\,(3, 0.5)$. Because the x value is negated, the point is reflected over the y-axis.

658. $r_{x\text{-axis}}$

The figure shows that the x-axis is the line of symmetry for the two triangles. When you reflect over the x-axis, you negate the sign of the original y value. In this figure, A (1, 1) maps to A′ (1, −1). Because the y value is negated, the point is reflected over the x-axis.

659. $r_{y\text{-axis}}$

The figure shows that the y-axis is the line of symmetry for the two triangles.

When you reflect over the y-axis, you negate the sign of the original x value. In this figure, A (−3, −1) maps to A′ (3, −1). Because the x value is negated, the point is reflected over the y-axis.

660. $r_{x\text{-axis}}$

The figure shows that the x-axis is the line of symmetry for the two triangles. When you reflect over the x-axis, you negate the sign of the original y value. In this figure, A (−2, 1) maps to A′ (−2, −1). Because the y value is negated, the point is reflected over the x-axis.

661. $r_{x=3}$

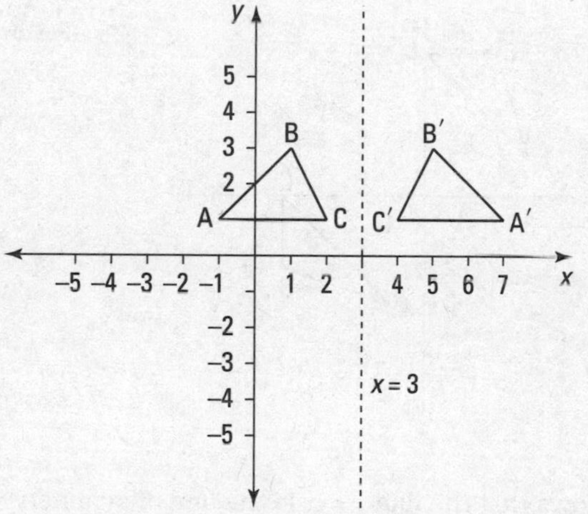

The figure shows that the line $x = 3$ is the line of symmetry for the two triangles.

662. $r_{y=-1}$

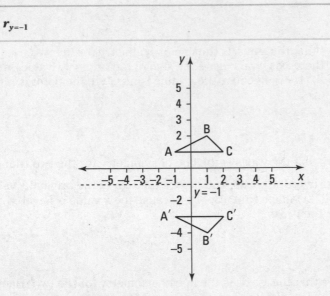

The figure shows that the line $y = -1$ is the line of symmetry for the two triangles.

663. $r_{y=x}$

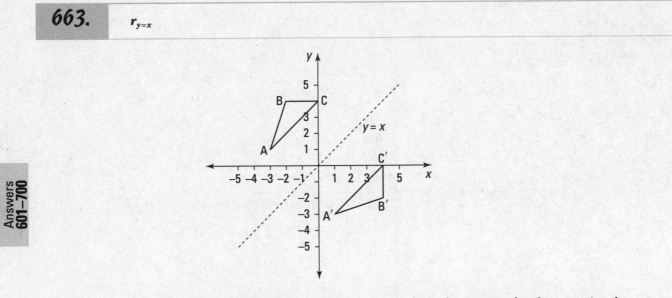

The figure shows that the line $y = x$ is the line of symmetry for the two triangles.

664. $r_{x=-1}$

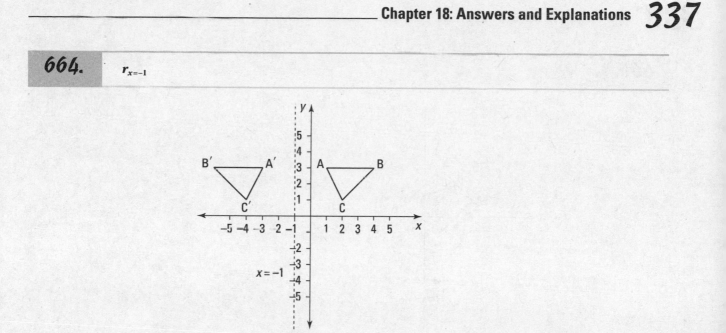

The figure shows that the line $x = -1$ is the line of symmetry for the two triangles.

665. $r_{y=x}$

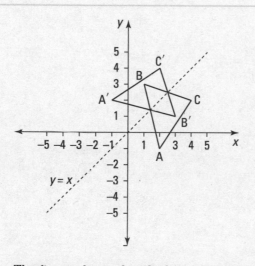

The figure shows that the line $y = x$ is the line of symmetry for the two triangles.

666. $r_{y=-x}$

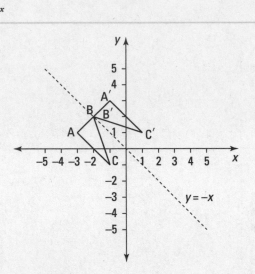

The figure shows that the line $y = -x$ is the line of symmetry for the two triangles.

667. $r_{y=3}$

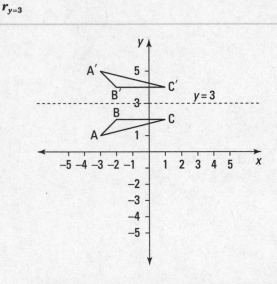

The figure shows that the line $y = 3$ is the line of symmetry for the two triangles.

668. $r_{y=x+2}$

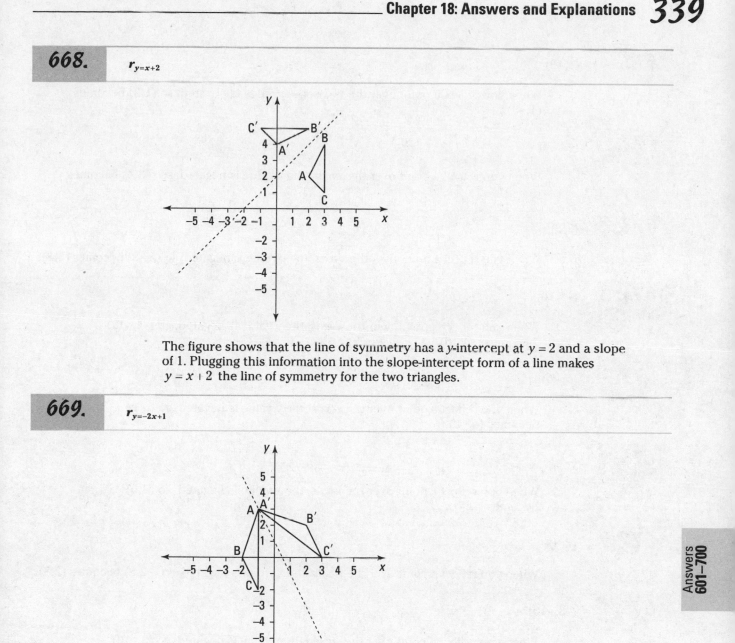

The figure shows that the line of symmetry has a y-intercept at $y = 2$ and a slope of 1. Plugging this information into the slope-intercept form of a line makes $y = x + 2$ the line of symmetry for the two triangles.

669. $r_{y=-2x+1}$

The figure shows that the line of symmetry has a y-intercept at $y = 1$ and a slope of -2. Plugging this information into the slope-intercept form of a line makes $y = -2x + 1$ the line of symmetry for the two triangles.

670. $r_{x\text{-axis}} \circ r_{y\text{-axis}}$ (or $r_{y\text{-axis}} \circ r_{x\text{-axis}}$)

The transformation that maps C $(2, 1)$ onto C' $(-2, -1)$ is a reflection over the origin. When a point is reflected over the origin, both the x and y coordinates are negated. Reflecting over the origin is the same as reflecting over both the x- and y-axes in either order.

671. (1,–3)

When you reflect a point over the x-axis, the y value is negated, so $(1,3)$ becomes $(1,–3)$.

672. (–2,–4)

When you reflect a point over the x-axis, the y value is negated, so $(–2,4)$ becomes $(–2,–4)$.

673. (5,3)

When you reflect a point over the x-axis, the y value is negated, so $(5,–3)$ becomes $(5,3)$.

674. (–4,10)

When you reflect a point over the x-axis, the y value is negated, so $(–4,–10)$ becomes $(–4,10)$.

675. (a,–b)

When you reflect a point over the x-axis, the y value is negated, so (a,b) becomes $(a,–b)$.

676. (–4,5)

When you reflect a point over the y-axis, the x value is negated, so $(4,5)$ becomes $(–4,5)$.

677. (2,8)

When you reflect a point over the y-axis, the x value is negated, so $(–2,8)$ becomes $(2,8)$.

678. (–3,–10)

When you reflect a point over the y-axis, the x value is negated, so $(3,–10)$ becomes $(–3,–10)$.

679. (1,–2)

When you reflect a point over the y-axis, the x value is negated, so $(–1,–2)$ becomes $(1,–2)$.

680. (–a,b)

When you reflect a point over the y-axis, the x value is negated, so (a,b) becomes $(–a,b)$.

681. $r_{y=x}$

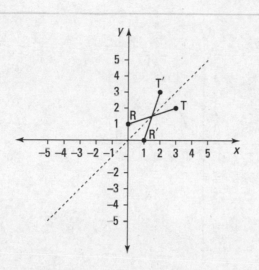

The figure shows that the line of symmetry has a y-intercept at $y = 0$ and a slope of 1. Plugging this information into the slope-intercept form of a line makes $y = x$ the line of symmetry for the two segments.

682. $r_{y=-x}$

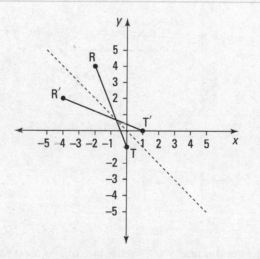

The figure shows that the line of symmetry has a y-intercept at $y = 0$ and a slope of –1. Plugging this information into the slope-intercept form of a line makes $y = -x$ the line of symmetry for the two segments.

683. $r_{x=3}$

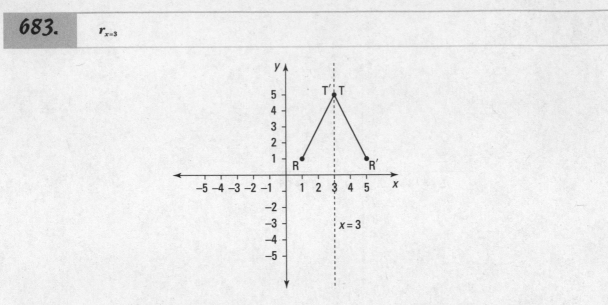

The figure shows that the line $x = 3$ is the line of symmetry for the two segments.

684. $r_{y=\frac{3}{2}}$

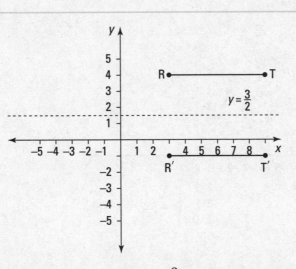

The figure shows that the line $y = \frac{3}{2}$ is the line of symmetry for the two segments.

685. $r_{x=1}$

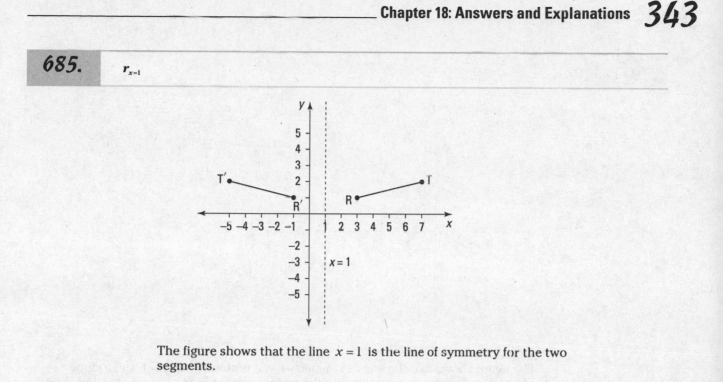

The figure shows that the line $x = 1$ is the line of symmetry for the two segments.

686. $r_{x\text{-axis}}$

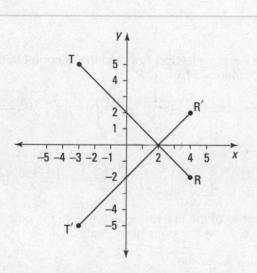

The figure shows that the x-axis is the line of symmetry for the two segments.

687. $r_{y=x+1}$

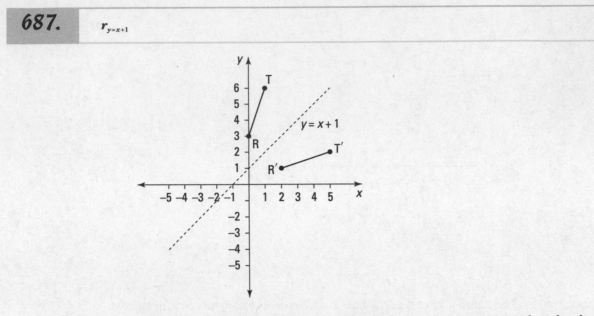

The figure shows that the line of symmetry has a y-intercept at $y = 1$ and a slope of 1. If you plug this information into the slope-intercept form of a line, you find that $y = x + 1$ is the line of symmetry for the two segments.

688. $r_{y=\frac{1}{2}x+2}$

To determine the line of reflection, first find the midpoint between each point and its image point. The midpoint of R and R' is

$$M = \left(\frac{x_1 + x_2}{2}, \frac{y_1 + y_2}{2} \right) = \left(\frac{-1+1}{2}, \frac{4+0}{2} \right) = \left(\frac{0}{2}, \frac{4}{2} \right) = (0, 2)$$

The midpoint of T and T' is

$$M = \left(\frac{x_1 + x_2}{2}, \frac{y_1 + y_2}{2} \right) = \left(\frac{-1+5}{2}, \frac{9+(-3)}{2} \right) = \left(\frac{4}{2}, \frac{6}{2} \right) = (2, 3)$$

Now find the equation of the line that goes through the points $(0, 2)$ and $(2, 3)$. To do this, you need to know the slope of the line:

$$m = \frac{y_2 - y_1}{x_2 - x_1} = \frac{3-2}{2-0} = \frac{1}{2}$$

You can now use the point-slope form of a line to get the equation of the line:

$$y - y_1 = m(x - x_1)$$

$$y - 2 = \frac{1}{2}(x - 0)$$

$$y - 2 = \frac{1}{2}x$$

$$y = \frac{1}{2}x + 2$$

689.   H

A figure that has point symmetry looks exactly the same after it has been rotated 180°. If you turn H upside down without flipping it over (a rotation of 180°), it looks exactly the same.

690.   W

A figure that has point symmetry looks exactly the same after it has been rotated 180°. If you turn W upside down without flipping it over (a rotation of 180°), it doesn't look exactly the same; it looks like an M.

691.   (–2, 5)

When you reflect a point in the origin, the x and y coordinates are both negated.

692.   (6, –3)

When you reflect a point in the origin, the x and y coordinates are both negated.

693.   180°

When you reflect a point in the origin, the x and y coordinates are both negated. Similarly, when you rotate a point 180°, both the x and the y are negated. This means that reflecting over the origin and rotating 180° are equivalent transformations.

694.   **Reflection in the origin**

The x and y coordinates of each point are the opposite sign. When you reflect a point in the origin, the x and y coordinates are both negated.

695.   **Dilation**

A *rigid motion* is a transformation in which distance between points is preserved. A dilation changes the size of the shape, making it either bigger or smaller; therefore, dilation isn't an example of a rigid motion.

696.   (7, –1)

The x coordinate for Point P is 3. It's being reflected over a point whose x coordinate is 5. The x coordinate of P is 2 units away from the point of reflection. This means that the new x coordinate for P will be 2 more units away from the point of reflection; $5 + 2 = 7$, so the x coordinate for the image is 7.

The y coordinate for Point P is 5. It's being reflected over a point whose y coordinate is 2. The y coordinate of P is 3 units away from the point of reflection. This means that the new y coordinate for P will be 3 more units away from the point of reflection; $2 - 3 = -1$, so the y coordinate for the image is –1.

Therefore, the image for P is (7, –1).

697. $T_{0,-1}$

Point A has an x coordinate of 1. A' also has an x coordinate of 1. This means that the x value hasn't changed at all.

Point A has a y coordinate of 1. A' has a y coordinate of 0. This means that the y value has decreased by 1.

$T_{0,-1}$ tells you to leave the original x alone and subtract 1 from the original y to determine where the image will land.

698. $T_{3,0}$

Point S has an x coordinate of –1. S' has an x coordinate of 2. This means that the x value has increased by 3.

Point S has a y coordinate of 3. S' also has a y coordinate of 3. This means that the y value hasn't changed at all.

$T_{3,0}$ tells you to add 3 to the original x and leave the original y alone to determine where the image will land.

699. $T_{4,-1}$

Point N has an x coordinate of –2. N' has an x coordinate of 2. This means that the x value has increased by 4.

Point N has a y coordinate of 3.5. N' has a y coordinate of 2.5. This means that the y value has decreased by 1.

$T_{4,-1}$ tells you to add 4 to the original x and to subtract 1 from the original y to determine where the image will land.

700. $T_{1,4}$

Point R has an x coordinate of –2. R' has an x coordinate of –1. This means that the x value has increased by 1.

Point R has a y coordinate of –1. R' has a y coordinate of 3. This means that the y value has increased by 4.

$T_{1,4}$ tells you to add 1 to the original x and to add 4 to the original y to determine where the image will land.

701. $T_{-3,-2}$

Point B has an x coordinate of 2. B' has an x coordinate of –1. This means that the x value has decreased by 3.

Point B has a y coordinate of 1. B' has a y coordinate of –1. This means that the y value has decreased by 2.

$T_{-3,-2}$ tells you to subtract 3 from the original x and to subtract 2 from the original y to determine where the image will land.

702. **(6, 2)**

The rule says to add 4 to the x coordinate and to subtract 3 from the y coordinate:

$$x = 2 + 4 = 6$$
$$y = 5 - 3 = 2$$

$(6, 2)$ is the image.

703. **(2, –3)**

The rule says to add 4 to the x coordinate and to subtract 3 from the y coordinate:

$$x = -2 + 4 = 2$$
$$y = 0 - 3 = -3$$

$(2, -3)$ is the image.

704. **(11, –6)**

The rule says to add 4 to the x coordinate and to subtract 3 from the y coordinate:

$$x = 7 + 4 = 11$$
$$y = -3 - 3 = -6$$

$(11, -6)$ is the image.

705. **(3, 12)**

The rule says to add 4 to the x coordinate and to subtract 3 from the y coordinate:

$$x = -1 + 4 = 3$$
$$y = 15 - 3 = 12$$

$(3, 12)$ is the image.

706. **$(x + 1, y - 1)$**

Point T has an x coordinate of 4. T' has an x coordinate of 5. This means that the x value has increased by 1.

Point T has a y coordinate of 6. T' has a y coordinate of 5. This means that the y value has decreased by 1.

$(x + 1, y - 1)$ tells you to add 1 to the original x and to subtract 1 from the original y to determine where the image will land.

707. $(x+8, \ y+2)$

Point T has an x coordinate of –2. T' has an x coordinate of 6. This means that the x value has increased by 8.

Point T has a y coordinate of 10. T' has a y coordinate of 12. This means that the y value has increased by 2.

$(x+8, \ y+2)$ tells you to add 8 to the original x and to add 2 to the original y to determine where the image will land.

708. $(x-6, \ y-1)$

Point T has an x coordinate of 5. T' has an x coordinate of –1. This means that the x value has decreased by 6.

Point T has a y coordinate of –1. T' has a y coordinate of –2. This means that the y value has decreased by 1.

$(x-6, \ y-1)$ tells you to subtract 6 from the original x and to subtract 1 from the original y to determine where the image will land.

709. $(x+16, \ y+7)$

Point T has an x coordinate of –8. T' has an x coordinate of 8. This means that the x value has increased by 16.

Point T has a y coordinate of –3. T' has a y coordinate of 4. This means that the y value has increased by 7.

$(x+16, \ y+7)$ tells you to add 16 to the original x and to add 7 to the original y to determine where the image will land.

710. $(0,6)$

$(x+1, \ y+4)$ means to add 1 to the original x value and to add 4 to the original y value to determine where the image will land:

$$x = -1+1 = 0$$
$$y = 2+4 = 6$$

$(0,6)$ is the image of W.

711. $(11,3)$

$T_{10,-2}$ means to add 10 to the original x value and to subtract 2 from the original y value to determine where the image will land:

$$x = 1+10 = 11$$
$$y = 5-2 = 3$$

$(11,3)$ is the image of X

712. (2, 3)

$(x-4, y+3)$ means to subtract 4 from the original x value and to add 3 to the original y value to determine where the image will land:

$$x = 6-4 = 2$$
$$y = 0+3 = 3$$

$(2,3)$ is the image of Y.

713. $T_{4,15}$

When doing composition of translations, combine the x values you're adding together and also combine the y values together:

$$x = -2 + 6 - 4$$
$$y = 3+12 = 15$$

Therefore, $T_{-2,3} \circ T_{6,12} = T_{4,15}$.

714. (9, −2)

Point P has an x coordinate of 6. P' has an x coordinate of 11. This means that the x value has increased by 5.

Point P has a y coordinate of −2. P' has a y coordinate of −5. This means that the y value has decreased by 3.

$(x+5, y-3)$ tells you to add 5 to the original x and to subtract 3 from the original y to determine where the image will land.

715. (−2, −1)

Point B has an x coordinate of 3. B' also has an x coordinate of 3. This means that the x value has not changed at all.

Point B has a y coordinate of 0. B' has a y coordinate of −5. This means that the y value has decreased by 5.

$(x, y-5)$ tells you not to change the x coordinate and to subtract 5 from the y coordinate to determine where the image will land. Because G has an x coordinate of −2 and a y coordinate of 4, its image will be $(-2,-1)$.

716. (24, 50)

A dilation changes the distance between points by multiplying the x and y coordinates by the scale factor. In this problem, the scale factor is 2:

$$x = 12(2) = 24$$
$$y = 25(2) = 50$$

The image is therefore $(24,50)$.

717. (–9, 1)

A dilation changes the distance between points by multiplying the x and y coordinates by the scale factor. In this problem, the scale factor is $\frac{-1}{3}$:

$$x = 27\left(\frac{-1}{3}\right) = -9$$
$$y = -3\left(\frac{-1}{3}\right) = 1$$

The image is therefore (–9, 1).

718. (6, 33)

A dilation changes the distance between points by multiplying the x and y coordinates by the scale factor. In this problem, the scale factor is –3:

$$x = -2(-3) = 6$$
$$y = -11(-3) = 33$$

The image is therefore (6, 33).

719. (–1, 7)

A dilation changes the distance between points by multiplying the x and y coordinates by the scale factor. In this problem, the scale factor is $\frac{1}{2}$:

$$x = -2\left(\frac{1}{2}\right) = -1$$
$$y = 14\left(\frac{1}{2}\right) = 7$$

The image is therefore (–1, 7).

720. 3

A dilation changes the distance between points by multiplying the x and y coordinates by the scale factor. The x coordinate for A is 3, and the x coordinate for A' is 9. If you call the scale factor k, then

$$3k = 9$$
$$k = 3$$

You can also determine the scale factor by looking at the y coordinates. The y coordinate for A is 2, and the y coordinate for A' is 6, so

$$2k = 6$$
$$k = 3$$

721. $\frac{2}{3}$

A dilation changes the distance between points by multiplying the x and y coordinates by the scale factor. The x coordinate for B is -12, and the x coordinate for B' is -8. If you call the scale factor k, then

$$-12k = -8$$
$$k = \frac{2}{3}$$

You can also determine the scale factor by looking at the y coordinates. The y coordinate for B is 6, and the y coordinate for B' is 4, so

$$6k = 4$$
$$k = \frac{2}{3}$$

722. $(-8, 12)$

A dilation changes the distance between points by multiplying the x and y coordinates by the scale factor. The x coordinate for A is 4, and the x coordinate for A' is 16. If you call the scale factor k, then

$$4k = 16$$
$$k = 4$$

Multiply the coordinates of B by the scale factor of 4:

$$x = -2(4) = -8$$
$$y = 3(4) = 12$$

The image is therefore $(-8, 12)$.

723. $(-1, 5)$

A dilation changes the distance between points by multiplying the x and y coordinates by the scale factor. The x coordinate for J is 6, and the x coordinate for J' is 3. If you call the scale factor k, then

$$6k = 3$$
$$k = \frac{1}{2}$$

Multiply the coordinates of K by the scale factor of $\frac{1}{2}$:

$$x = -2\left(\frac{1}{2}\right) = -1$$
$$y = 10\left(\frac{1}{2}\right) = 5$$

The image is therefore $(-1, 5)$.

724. **False**

A dilation changes the distance between points by multiplying the x and y coordinates by the scale factor. Although angle measure and midpoint are preserved (they stay the same), distance is not.

725. *N*

When doing the composition of transformations, you perform the transformation closest to the point first. Therefore, you perform the transformations from right to left.

Rotate Point S counterclockwise 90° first. Then take the new point and reflect it over line p:

$$R_{90°}(S) = G$$

$$r_p(G) = N$$

There are eight vertices in an octagon, and $\frac{360°}{8} = 45°$. Each vertex therefore represents a 45° rotation.

726. *A*

Instead of performing both rotations separately, you can do one rotation with the sum of the angles:

$$135° + (-180°) = -45°$$

$$R_{-45°}(T) = A$$

727. *O*

When doing the composition of transformations, you perform the transformation closest to the point first. Therefore, you perform the transformations from right to left.

Rotate Point A counterclockwise 405° first. Rotating 405° means you're rotating more than a full revolution:

$$405° - 360° = 45°$$

This means you're really just rotating the point 45° counterclockwise:

$$R_{45°}(A) = T$$

Take the new point and reflect it over line l:

$$r_l(T) = N$$

Take the new point and rotate it 135° clockwise:

$$R_{-135°}(N) = O$$

728.

C

When doing the composition of transformations, you perform the transformation closest to the point first. Therefore, you perform the transformations from right to left.

Instead of performing both rotations separately, you can do one rotation with the sum of the angles:

$$45° + (-225°) = -180°$$
$$R_{-180°}(R) = A$$

Take the new point and reflect it over line *m*:

$$r_m(A) = O$$

Take the new point and rotate it counterclockwise 315°:

$$R_{315°}(O) = C$$

729.

$(-1, 2)$

Rotating the point 90° counterclockwise about the origin is the same as reflecting over the line $y = x$ and then reflecting over the *y*-axis. This means that the point (x, y) will become the point $(-y, x)$. In this question, $(2, 1)$ will become $(-1, 2)$.

730.

$(5, -10)$

Rotating 180° about the origin yields the same result as reflecting over the origin. This means that the point (x, y) will become the point $(-x, -y)$. In this question, $(-5, 10)$ will become $(5, -10)$.

731.

$(-10, -5)$

Rotating 90° counterclockwise about the origin is the same as reflecting over the line $y = x$ and then reflecting over the *y*-axis. This means that the point (x, y) will become the point $(-y, x)$. In this question, $(-5, 10)$ will become $(-10, -5)$.

732.

$(-6, -4)$

All the rules for rotations are written so that when you're rotating counterclockwise, a full revolution is 360°. Rotating 90° clockwise is the same as rotating 270° counterclockwise. Rotating 270° counterclockwise about the origin is the same as reflecting over the line $y = x$ and then reflecting over the *x*-axis. This means that the point (x, y) will become the point $(y, -x)$. In this question, $(4, -6)$ will become $(-6, -4)$.

733.

240°

A full revolution is 360°. Rotating 120° clockwise is the same as rotating 240° counterclockwise.

734. $R_{230°}$

Instead of performing all rotations separately, you can do one rotation with the sum of the angles:

$$-120° + 100° + 250° = 230°$$
$$R_{-120°} \circ R_{100°} \circ R_{250°} = R_{230°}$$

735. **II**

A full revolution is 360°. Rotating 270° clockwise is the same as rotating 90° counterclockwise. Rotating 90° counterclockwise about the origin is the same as reflecting over the line $y = x$ and then reflecting over the y-axis. This means that the point (x, y) will become the point $(-y, x)$. A $(4, 3)$ will become A' $(-3, 4)$. B $(5, 7)$ will become B' $(-7, 5)$. The following figure shows the naming of the quadrants:

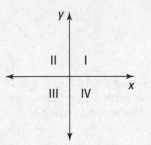

The segment created when you connect A' and B' is in Quadrant II.

736. $r_{x\text{-axis}}$ **and** $T_{-1,1}$

Reflecting over the x-axis negates the y coordinate. $T_{-1,1}$ tells you to follow the rule $(x - 1, y + 1)$.

737. $r_{y\text{-axis}}$ **and** $T_{-2,2}$

Reflecting over the y-axis negates the x coordinate. $T_{-2,2}$ tells you to follow the rule $(x - 2, y + 2)$.

738. $r_{y\text{-axis}}$ **and** $T_{0,2}$

Reflecting over the y-axis negates the x coordinate. $T_{0,2}$ tells you to follow the rule $(x, y + 2)$.

739. $r_{x\text{-axis}}$ **and** $T_{-3,0}$

Reflecting over the x-axis negates the y coordinate. $T_{-3,0}$ tells you to follow the rule $(x - 3, y)$.

740.

$r_{y=1}$ and $T_{2,-1}$

To reflect each point over the line $y = 1$, determine how far each point is from $y = 1$ and move that amount past $y = 1$. M (-4, -3) becomes $(-4, 5)$. N $(-2, -1)$ becomes $(-2, 3)$. O $(0, -3)$ becomes $(0, 5)$. $T_{2,-1}$ tells you to follow the rule $(x + 2, y - 1)$.

741.

$(1, 5)$

When doing the composition of transformations, you perform the transformation closest to the point first. Therefore, you perform the transformations from right to left.

In this question, you rotate Point P 90° counterclockwise first. Rotating 90° counterclockwise about the origin is the same as reflecting over the line $y = x$ and then reflecting over the y-axis. This means that the point (x, y) becomes the point $(-y, x)$:

$$R_{90°}(P) = (-1, 5)$$

After completing the rotation, you reflect the point over the y-axis. Reflecting over the y-axis negates the x coordinate:

$$r_{y\text{ axis}}(-1, 5) = (1, 5)$$

742.

$(1, 12)$

First dilate Point P by a scale factor of 2. A dilation by a scale factor of 2 multiplies the x and y coordinates by 2:

$$D_2(5, 1) = (10, 2)$$

Next, you perform the translation. $T_{2,-3}$ follows the rule $(x + 2, y - 3)$:

$$T_{2,-3}(10, 2) = (12, -1)$$

Last, rotate the point 90° counterclockwise. Rotating 90° counterclockwise about the origin is the same as reflecting over the line $y = x$ and then reflecting over the y-axis. This means that the point (x, y) becomes the point $(-y, x)$:

$$R_{90°}(12, -1) = (1, 12)$$

743.

$(0, 3)$

You perform the translation first. $T_{4,-1}$ follows the rule $(x + 4, y - 1)$:

$$T_{4,-1}(5, 1) = (9, 0)$$

Next, perform the dilation by a scale factor of $\frac{1}{3}$. A dilation by a scale factor of $\frac{1}{3}$ multiplies the x and y coordinates by $\frac{1}{3}$:

$$D_{\frac{1}{3}}(9, 0) = (3, 0)$$

Last, reflect the point over the line $y = x$. When reflecting over $y = x$, the x and y coordinates switch places:

$$r_{y=x}(3, 0) = (0, 3)$$

744. (1, 7)

Perform the reflection over $y = -3$ first. To do this, find out how far each point is away from $y = -3$ and then move that amount beyond $y = -3$:

$$r_{y=-3}(5, 1) = (5, -7)$$

Next, perform the reflection over $x = 2$. To do this, find out how far each point is away from $x = 2$ and move that amount beyond $x = 2$:

$$r_{x=2}(5, -7) = (-1, -7)$$

Last, rotate the point 180° counterclockwise. Rotating 180° is the same as reflecting over the origin. Both the x and y coordinates are negated:

$$R_{180°}(-1, -7) = (1, 7)$$

745. (–11, –3)

Perform the rotation of 90° clockwise first. One full revolution is 360°. This means that rotating 90° clockwise is the same as rotating 270° counterclockwise. Rotating 270° counterclockwise is the same as reflecting over the line $y = x$ and then reflecting over the x-axis. This means that the image becomes $(y, -x)$:

$$R_{270°}(5, 1) = (1, -5)$$

Next, perform the reflection over $y = x$. Reflecting over the line $y = x$ switches the x and y coordinates:

$$r_{y=x}(1, -5) = (-5, 1)$$

The next transformation is the translation. $T_{-6,2}$ follows the rule $(x - 6, y + 2)$:

$$T_{-6,2}(-5, 1) = (-11, 3)$$

Last, reflect the point over the x-axis. Reflecting over the x-axis negates the y coordinate:

$$r_{x\text{-axis}}(-11, 3) = (-11, -3)$$

746. (2, 6)

Perform the reflection over $y = x$ first. Reflecting over the line $y = x$ switches the x and y coordinates:

$$r_{y=x}(5, 1) = (1, 5)$$

The next transformation is the translation. $T_{3,7}$ follows the rule $(x+3,\ y+7)$:

$$T_{3,7}(1,5) = (4,12)$$

Last, you dilate by a scale factor of $\frac{1}{2}$. Dilating by a scale factor of $\frac{1}{2}$ multiplies both the x and y coordinates by $\frac{1}{2}$:

$$D_{\frac{1}{2}}(4,12) = (2,6)$$

747. True

A *glide reflection* is a composition of a line reflection and a translation parallel to the line of reflection.

748. False

A *glide reflection* is a composition of a line reflection and a translation parallel to the line of reflection. Although this composition of transformations is a reflection followed by a translation, the translation is not parallel to the line of reflection.

749. True

A *glide reflection* is a composition of a line reflection and a translation parallel to the line of reflection. Glide reflections observe the commutative property. This means that the order in which the transformations are performed doesn't matter.

750. False

A *direct isometry* is a transformation that preserves distance and orientation. A *glide reflection* is a composition of a line reflection and a translation parallel to the line of reflection. Line reflections do not preserve orientation; therefore, a glide reflection isn't a direct isometry.

751. True

A *direct isometry* is a transformation that preserves distance and orientation. A translation does not change the size between the points, nor does it change the order in which the points are arranged; therefore, a translation is a direct isometry.

752. False

An *opposite isometry* is a transformation that preserves distance but does not preserve orientation. Although a line reflection changes the orientation of the points, another line reflection will put the points back in their original order. This means that the orientation of the points is preserved. Therefore, the compositions in this question represent a direct isometry rather than an opposite isometry.

753. False

A *direct isometry* is a transformation that preserves distance and orientation. Although a rotation of 90° preserves both distance and orientation, a reflection over the line $y = x$ preserves only distance, not orientation. Because this transformation does not preserve orientation, it's an example of an *opposite isometry,* not a direct isometry.

754. True

An *isometry* is a transformation that preserves distance. Both rotations and line reflections preserve distance; therefore, this composition of transformations is an isometry.

755. True

Both rotations and line reflections preserve distance; therefore, these segments are congruent.

756. False

Rotate Point H 90° counterclockwise first. Rotating 90° counterclockwise about the origin is the same as reflecting over the line $y = x$ and then reflecting over the y-axis. This means that the point (x, y) becomes the point $(-y, x)$:

$$R_{90°}(2, 3) = (-3, 2)$$

After completing the rotation, you reflect the point over the line $y = x$. Reflecting over the line $y = x$ switches the x and y coordinates:

$$r_{y=x}(-3, 2) = (2, -3)$$

757. True

Rotate Point A 90° counterclockwise first. Rotating 90° counterclockwise about the origin is the same as reflecting over the line $y = x$ and then reflecting over the y-axis. This means that the point (x, y) becomes the point $(-y, x)$:

$$R_{90°}(6, 7) = (-7, 6)$$

After completing the rotation, you reflect the point over the line $y = x$. Reflecting over the line $y = x$ switches the x and y coordinates:

$$r_{y=x}(-7, 6) = (6, -7)$$

758.

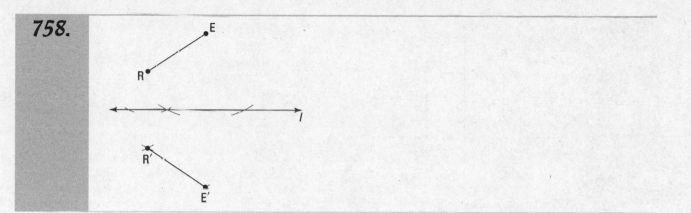

To construct the reflection of a segment over a line, you must construct a perpendicular line through each point. Place the compass point at R and draw arcs through line l. With the same width of the compass, place the point at both locations where the arcs intersect the lines and draw new arcs. The location where the arcs intersect is R':

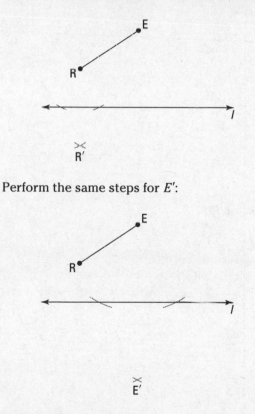

Perform the same steps for E':

To finish, connect R' and E'.

759.

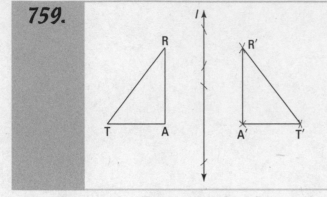

To construct the reflection of a triangle over a line, you must construct a perpendicular line through each vertex. Place the compass point at T and draw arcs through line l. With the same width of the compass, place the point at both locations where the arcs intersect the lines and draw new arcs. The location where the arcs intersect is T':

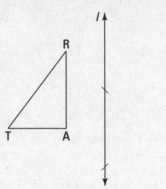

Perform the same steps for R':

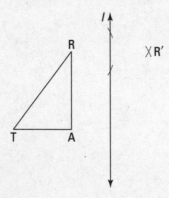

And for A':

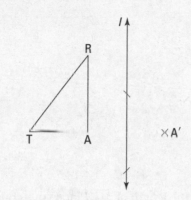

To finish, connect T', R', and A'.

760.

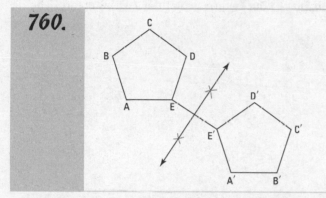

To construct a line of reflection, you need to draw a perpendicular bisector through the line segment that contains a point and its image. To create the perpendicular bisector for $\overline{EE'}$, Connect E and E'. Place the compass point at E and measure the length of $\overline{EE'}$. Using that compass width, keep the compass point at E and draw two arcs, one above and one below the line. Repeat the same steps, putting the point of the compass at E'. Connect the points where the arcs intersect. This line is the reflection line.

761.

To construct a line of reflection, you need to draw a perpendicular bisector through the segment containing a point and its image. To create the perpendicular bisector for $\overline{RR'}$, connect R and R'. Place the compass point at R and measure the length of $\overline{RR'}$. Using that compass width, keep the compass point at R and draw two arcs, one above and one below the line. Repeat the same steps, putting the point of the compass at R'. Connect the points where the arcs intersect. This line is the reflection line.

762.

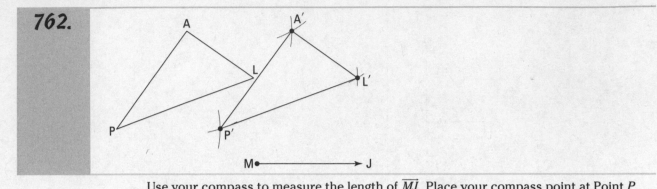

Use your compass to measure the length of $\overline{MJ}$. Place your compass point at Point P and draw an arc. Place your compass point at Point M and measure the distance from M to P. Keeping that compass width, place your compass point at J. Draw an arc. The place where the arcs intersect is P':

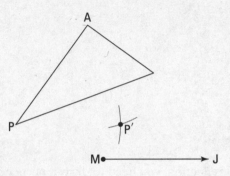

Repeat these steps at Point L:

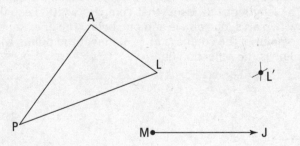

And at Point A:

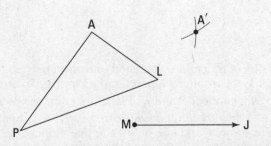

To finish, connect P', L', and A'.

763.

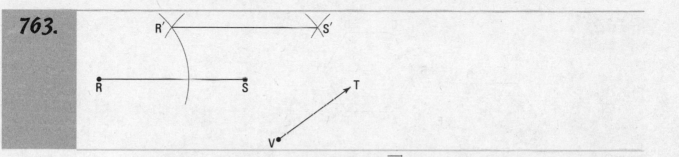

Use your compass to measure the length of $\overrightarrow{VT}$. Place your compass point at Point R and draw an arc. Place your compass point at Point V and measure the distance from V to R. Keeping that compass width, place your compass point at T. Draw an arc. The place where the arcs intersect is R':

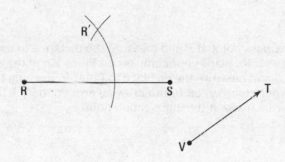

Repeat these steps using Point S as well:

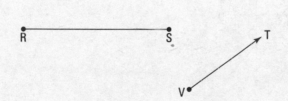

To finish, connect R' and S'.

764.

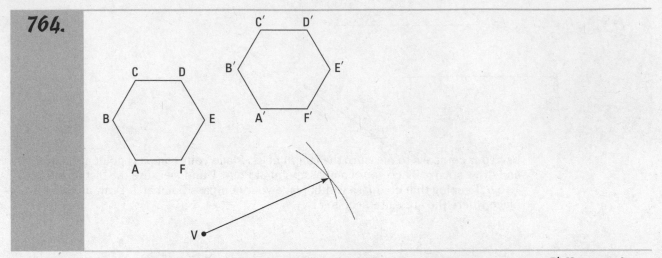

Place your compass point at *B* and measure the distance to its image, *B'*. Keeping the same compass width, place your compass on Point *V* and draw an arc. Place your compass point at *B* and measure the distance to Point *V*. Keeping that same compass width, place your compass on *B'* and draw an arc. Connect *V* to the point of intersection and place an arrow at the intersection point.

765.

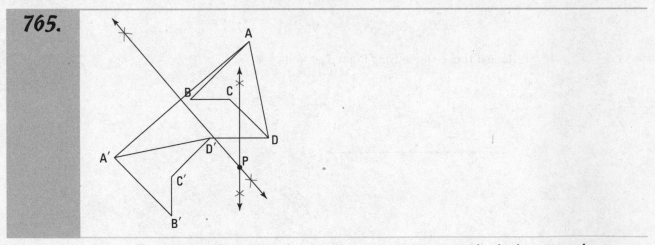

To construct the center of rotation, construct two perpendicular bisectors of segments between a point and its image point. You can start by connecting *A* and *A'*. Use your compass to measure the distance between *A* and *A'*. Using this distance, place your compass at Point *A* and draw two arcs. Keeping the same distance, place your compass at *A'* and draw two more arcs. Connect the points where the arcs intersect. Here's the construction of the first perpendicular bisector:

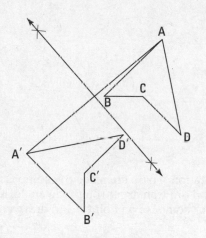

Follow the same steps for *D* and *D'*:

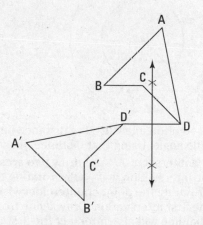

The point where the two bisectors intersect is the center of rotation.

766.

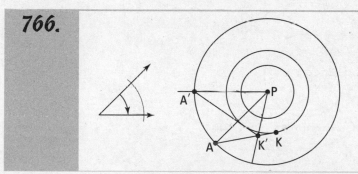

Connect Point *A* to Point *P* and measure the distance with your compass. Draw a circle with your compass point at *P* using that radius.

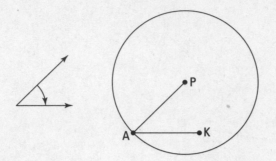

Now copy the angle of rotation. To do so, place the compass point on the vertex of the angle of rotation and draw an arc through the angle. Keeping the same compass width, place your compass on Point *P* and draw a circle:

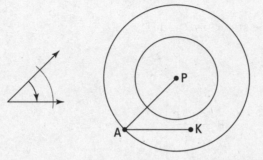

Going back to the angle of rotation, measure the distance between both places where the arc intersects the angle. Using that distance, place your compass on the point where your circle intersects $\overline{AP}$ and draw two arcs. They should intersect your circle twice. One intersection is if you're rotating the point clockwise, and the other is if you're rotating the point counterclockwise. You're rotating the point clockwise, so use the first intersection. Draw a line from Point *P* through that intersection point. That line will also intersect the first circle drawn in your construction. The point where the line intersects that circle is *A′*.

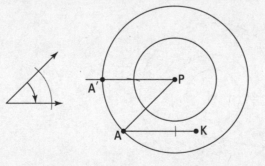

Follow the same steps for Point *K* and connect *A′* to *K′*.

767. **8π in.**

The formula for the circumference of a circle is $C = \pi d$, where d is the diameter of the circle:

$$C = \pi(8) = 8\pi$$

768. **20π cm**

The formula for the circumference of a circle is $C = \pi d$, where d is the diameter of the circle. Because the diameter of a circle is twice the radius, you can also write this formula as $C = 2\pi r$, where r is the circle's radius. In this problem, the radius is 10, which means that the circumference of the circle is

$$C = 2\pi(10) = 20\pi$$

769. **7 units**

The formula for the circumference of a circle is $C = \pi d$, where d is the diameter of the circle. Because the diameter of a circle is twice the radius, you can also write this formula as $C = 2\pi r$, where r is the circle's radius. In this problem, you're given that the circumference is 14π:

$$C = 2\pi r$$
$$14\pi = 2\pi r$$
$$\frac{14\pi}{2\pi} = \frac{2\pi r}{2\pi}$$
$$7 = r$$

770. **16π units²**

The formula for the area of a circle is $A = \pi r^2$, where r is the radius of the circle. In this problem, you're given that the area of the circle is 64π:

$$A = \pi r^2$$
$$64\pi = \pi r^2$$
$$\frac{64\pi}{\pi} = \frac{\pi r^2}{\pi}$$
$$64 = r^2$$
$$\sqrt{64} = \sqrt{r^2}$$
$$8 = r$$

The formula for the circumference of a circle is $C = \pi d$, where d is the diameter of the circle. Because the diameter of a circle is twice the radius, you can also write this formula as $C = 2\pi r$, where r is the circle's radius:

$$C = 2\pi r = 2\pi(8) = 16\pi$$

771. 12π **in.**

You calculate the area of a square with the formula $A = s^2$. Start by solving for side s:

$$A = s^2$$
$$144 = s^2$$
$$\sqrt{144} = \sqrt{s^2}$$
$$12 = s$$

Because the circle is inscribed in the square, the diameter of the circle is equal to the side of the square, which you just found to be 12 inches.

The formula for the circumference of a circle is $C = \pi d$, where d is the diameter of the circle:

$$C = \pi(12) = 12\pi$$

772. $3.5\pi + 5$ **units**

To find the perimeter of the figure, you need to find the sum of the borders of both semicircles and the hypotenuse of the triangle. Use the Pythagorean theorem to find the hypotenuse of the triangle. It states that $a^2 + b^2 = c^2$, where a and b are the legs of the right triangle and c is the hypotenuse:

$$a^2 + b^2 = c^2$$
$$3^2 + 4^2 = c^2$$
$$25 = c^2$$
$$\sqrt{25} = \sqrt{c^2}$$
$$5 = c$$

The formula for the circumference of a circle is $C = \pi d$, where d is the diameter of the circle. Find the circumference of a circle whose diameter is 3. Because you're dealing with a semicircle, you'll have to cut the circumference in half:

$$\frac{C}{2} = \frac{\pi d}{2} = \frac{\pi(3)}{2} = 1.5\pi$$

For the other semicircle, find the circumference of a circle whose diameter is 4 and cut the circumference in half:

$$\frac{C}{2} = \frac{\pi d}{2} = \frac{\pi(4)}{2} = 2\pi$$

The perimeter of the figure is

$$1.5\pi + 2\pi + 5 = 3.5\pi + 5$$

773. **$16\sqrt{2}$ units**

The formula for the circumference of a circle is $C = \pi d$, where d is the diameter of the circle. In this problem, you're given that the circumference is 8π:

$$C = \pi d$$
$$8\pi = \pi d$$
$$\frac{8\pi}{\pi} = \frac{\pi d}{\pi}$$
$$8 = d$$

The diameter of a circle is twice the radius:

$$d = 2r$$
$$8 = 2r$$
$$4 = r$$

The radius of a circle extends from the center to any point on the circle. The radius of this circle is 4, which means that

$$EB = EC = EA = ED = 4$$

Diameters $\overline{AC}$ and $\overline{BD}$ divide the square into four congruent right triangles. Find the length of a side of the square by using the Pythagorean theorem to find the hypotenuse:

$$4^2 + 4^2 = s^2$$
$$16 + 16 = s^2$$
$$s = \sqrt{32}$$

Because all four sides of a square are congruent, you can find the perimeter of the square by multiplying the side length by 4:

$$P = 4s = 4\sqrt{32} = 4\sqrt{16}\sqrt{2} = 4\left(4\sqrt{2}\right) = 16\sqrt{2}$$

774. **100π units^2**

The formula for the area of a circle is $A = \pi r^2$, where r represents the radius of the circle. In this problem, the radius of the circle is 10, and you're looking for the area:

$$A = \pi r^2 = \pi\left(10^2\right) = 100\pi$$

775. **196π ft^2**

The formula for the area of a circle is $A = \pi r^2$, where r represents the radius of the circle. The diameter of a circle is equal to twice the radius:

$$d = 2r$$
$$28 = 2r$$
$$14 = r$$

Now that you know the radius is 14, you can use the area formula to calculate the area:

$$A = \pi r^2 = \pi\left(14^2\right) = 196\pi$$

776.   **625π units^2**

The formula for the circumference of a circle is $C = \pi d$, where d is the diameter of the circle. In this problem, you're given that the circumference is 50π :

$$C = \pi d$$
$$50\pi = \pi d$$
$$\frac{50\pi}{\pi} = \frac{\pi d}{\pi}$$
$$50 = d$$

The diameter of a circle is twice the radius:

$$d = 2r$$
$$50 = 2r$$
$$25 = r$$

Now that you know the radius is 25, you can use the area formula to calculate the area:

$$A = \pi r^2 = \pi\left(25^2\right) = 625\pi$$

777.   **34 units**

The formula for the area of a circle is $A = \pi r^2$, where r represents the radius of the circle. In this problem, you're given that the area of the circle is 289π units^2. Use the area formula to find the radius of the circle:

$$A = \pi r^2$$
$$289\pi = \pi r^2$$
$$\frac{289\pi}{\pi} = \frac{\pi r^2}{\pi}$$
$$289 = r^2$$
$$\sqrt{289} = \sqrt{r^2}$$
$$17 = r$$

The diameter of a circle is twice the radius:

$$d = 2r = 2\left(17\right) = 34$$

778. **100π units^2**

A square has four equal sides. If the perimeter of the square is 80, each side of the square measures

$$\frac{80}{4} = 20$$

A side of the square is the same length as the diameter of the circle, and the diameter of a circle is twice the radius. This means that if the diameter of the circle is 20 units, the radius is 10 units.

The formula for the area of a circle is $A = \pi r^2$, where r represents the radius of the circle. In this problem, the radius of the circle is 10, and you're looking for the area:

$$A = \pi r^2 = \pi\left(10^2\right) = 100\pi$$

779. **50π units^2**

The formula for the area of a square is $A = s^2$, so

$$A = s^2$$
$$100 = s^2$$
$$\sqrt{100} = \sqrt{s^2}$$
$$10 = s$$

When a square is inscribed in a circle, the diagonal of the square is equal in length to the diameter of the circle. The diagonal splits the square into two right triangles, with the diagonal as the hypotenuse, so you can use the Pythagorean theorem to solve for the diagonal of the square:

$$a^2 + b^2 = c^2$$
$$10^2 + 10^2 = c^2$$
$$200 = c^2$$
$$\sqrt{200} = \sqrt{c^2}$$
$$\sqrt{100}\sqrt{2} = c$$
$$10\sqrt{2} = c$$

The diameter of a circle is twice the radius:

$$d = 2r$$
$$10\sqrt{2} = 2r$$
$$5\sqrt{2} = r$$

The formula for the area of a circle is $A = \pi r^2$, where r represents the radius of the circle. In this problem, the radius of the circle is $5\sqrt{2}$, and you're looking for the area:

$$A = \pi r^2 = \pi\left(5\sqrt{2}\right)^2 = \pi(25)(2) = 50\pi$$

780. 132π units^2

To find the area of the shaded region, you first have to find the area of the larger circle, whose radius is 14:

$$A = \pi r^2 = \pi\left(14^2\right) = 196\pi$$

Now you need to find the area of the smaller circle, whose radius is 8:

$$A = \pi r^2 = \pi\left(8^2\right) = 64\pi$$

Subtract the area of the smaller circle from the area of the larger circle to get the area of the shaded region:

$$196\pi - 64\pi = 132\pi$$

781. 80 ft^2

The formula for the area of the sector of a circle is $A = \frac{1}{2}r^2\theta$, where r is the radius of the circle and θ is measured in radians:

$$A = \frac{1}{2}r^2\theta = \frac{1}{2}\left(8^2\right)(2.5) = \frac{1}{2}(64)(2.5) = 80$$

782. 36 ft^2

The formula for the area of the sector of a circle is $A = \frac{1}{2}r^2\theta$, where r is the radius of the circle and θ is measured in radians. The diameter of a circle is twice the radius. If the diameter of a circle is 12, the radius of the circle must be 6. Therefore,

$$A = \frac{1}{2}r^2\theta = \frac{1}{2}\left(6^2\right)(2) = \frac{1}{2}(36)(2) = 36$$

783. 1,728 units^2

The formula for the circumference of a circle is $C = \pi d$, so

$$
\begin{aligned}
C &= \pi d \\
48\pi &= \pi d \\
\frac{48\pi}{\pi} &= \frac{\pi d}{\pi} \\
48 &= d
\end{aligned}
$$

The diameter of a circle is twice the radius. If the diameter of a circle is 48, the radius of the circle must be 24.

The formula for the area of the sector of a circle is $A = \frac{1}{2}r^2\theta$, where r is the radius of the circle and θ is measured in radians. Therefore,

$$A = \frac{1}{2}r^2\theta = \frac{1}{2}\left(24^2\right)(6) = \frac{1}{2}(576)(6) = 1{,}728$$

784. **14 units**

The formula for the area of the sector of a circle is $A = \frac{1}{2}r^2\theta$, where r is the radius of the circle and θ is measured in radians. Therefore,

$$A = \frac{1}{2}r^2\theta$$

$$147\pi = \frac{1}{2}\left(r^2\right)\left(\frac{3\pi}{2}\right)$$

$$147\pi = \frac{3\pi}{4}r^2$$

$$\left(\frac{4}{3\pi}\right)147\pi = \frac{3\pi}{4}r^2\left(\frac{4}{3\pi}\right)$$

$$196 = r^2$$

$$\sqrt{196} = \sqrt{r^2}$$

$$14 = r$$

785. $\frac{5\pi}{3}$ **radians**

The formula for the area of the sector of a circle is $A = \frac{1}{2}r^2\theta$, where r is the radius and θ is measured in radians. Therefore,

$$A = \frac{1}{2}r^2\theta$$

$$120\pi = \frac{1}{2}\left(12^2\right)\theta$$

$$120\pi = 72\theta$$

$$\frac{120\pi}{72} = \frac{72\theta}{72}$$

$$\frac{5\pi}{3} = \theta$$

786. **80 cm²**

Use the formula $s = r\theta$, where s is the length of the intercepted arc, r is the radius of the circle, and θ is the measure of the central angle in radians. Start by solving for θ:

$$s = r\theta$$

$$20 = (8)\theta$$

$$\frac{20}{8} = \theta$$

$$2.5 = \theta$$

The formula for the area of the sector of a circle is $A = \frac{1}{2}r^2\theta$, where r is the radius and θ is measured in radians. Therefore, the area is

$$A = \frac{1}{2}\left(8^2\right)(2.5) = 80$$

787. **12π units**

To determine the length of an arc of a circle, use the formula $s = r\theta$, where s represents the length of the intercepted arc, r represents the radius of the circle, and θ represents the central angle of the circle in radians:

$$s = r\theta = 16\left(\frac{3\pi}{4}\right) = 12\pi$$

788. **20.94 km**

To determine the length of an arc of a circle, use the formula $s = r\theta$, where s represents the length of the intercepted arc, r represents the radius of the circle, and θ represents the central angle of the circle in radians. Round your answer to the nearest hundredth.

$$s = r\theta = 10\left(\frac{2\pi}{3}\right) \approx 20.94$$

789. **10.19 cm**

To determine the length of the radius of a circle, use the formula $s = r\theta$, where s represents the length of the intercepted arc, r represents the radius of the circle, and θ represents the central angle of the circle in radians. You first have to convert the angle measure from degrees to radians. To do so, multiply the degree measure by $\frac{\pi}{180°}$:

$$45°\left(\frac{\pi}{180°}\right) = \frac{\pi}{4}$$

Now plug all your information into the formula and round your answer to the nearest hundredth:

$$s = r\theta$$
$$8 = r\left(\frac{\pi}{4}\right)$$
$$\left(\frac{4}{\pi}\right)8 = r\left(\frac{\pi}{4}\right)\left(\frac{4}{\pi}\right)$$
$$\frac{32}{\pi} = r$$
$$10.19 \approx r$$

790. $\frac{4\pi}{3}$ **m**

To determine the length of the intercepted arc of the circle, use the formula $s = r\theta$, where s represents the length of the intercepted arc, r represents the radius of the circle, and θ represents the central angle of the circle in radians. You first have to convert the angle measure from degrees to radians. To do so, multiply the degree measure by $\frac{\pi}{180°}$:

$$120°\left(\frac{\pi}{180°}\right) = \frac{2\pi}{3}$$

Now you can plug all your information into the formula:

$$s = r\theta = 2\left(\frac{2\pi}{3}\right) = \frac{4\pi}{3}$$

791.   **1.6 radians**

To determine the radian measure of the central angle of the circle, use the formula $s = r\theta$, where s represents the length of the intercepted arc, r represents the radius of the circle, and θ represents the central angle of the circle in radians. Round your answer to the nearest tenth.

$$s = r\theta$$
$$36 = (22)\theta$$
$$\frac{36}{22} = \frac{(22)\theta}{22}$$
$$1.6 \approx \theta$$

792.   **1.5 radians**

To determine the radian measure of the central angle of the circle, use the formula $s = r\theta$, where s represents the length of the intercepted arc, r represents the radius of the circle, and θ represents the central angle of the circle in radians.

You need to make sure that the length of the radius and the length of the arc are measured in the same unit. Because there are 12 inches in a foot, a radius of 3 feet is the same as 36 inches. Now you can plug your information into the formula to determine the radian measure of the central angle:

$$s = r\theta$$
$$54 = (36)\theta$$
$$\frac{54}{36} = \frac{(36)\theta}{36}$$
$$1.5 = \theta$$

793.   **Center: (–4, 0); radius: 5**

The standard form of a circle is $(x-h)^2 + (y-k)^2 = r^2$, where (h,k) is the center of the circle and r is the radius of the circle. In this equation, h must equal –4 and y must equal 0, making the center $(-4,0)$. Now solve for the radius if $r^2 = 25$:

$$r^2 = 25$$
$$\sqrt{r^2} = \sqrt{25}$$
$$r = 5$$

794. Center: $\left(3, -\frac{2}{3}\right)$; radius: $5\sqrt{2}$

The standard form of a circle is $(x-h)^2 + (y-k)^2 = r^2$, where (h,k) is the center of the circle and r is the radius of the circle. In this equation, h must equal 3 and y must equal $\frac{-2}{3}$, making the center $\left(3, -\frac{2}{3}\right)$. Now solve for the radius if $r^2 = 50$:

$$r^2 = 50$$
$$\sqrt{r^2} = \sqrt{50}$$
$$r = \sqrt{25}\sqrt{2}$$
$$r = 5\sqrt{2}$$

795. Center: (0, 6); radius: 12

The standard form of a circle is $(x-h)^2 + (y-k)^2 = r^2$, where (h,k) is the center of the circle and r is the radius of the circle. In this equation, h must equal 0 and y must equal 6, making the center $(0,6)$. Now solve for the radius if $r^2 = 144$:

$$r^2 = 144$$
$$\sqrt{r^2} = \sqrt{144}$$
$$r = 12$$

796. Center: (1, –10); radius: $3\sqrt{5}$

The standard form of a circle is $(x-h)^2 + (y-k)^2 = r^2$, where (h,k) is the center of the circle and r is the radius of the circle. In this equation, h must equal 1 and y must equal –10, making the center $(1,-10)$. Now solve for the radius if $r^2 = 45$:

$$r^2 = 45$$
$$\sqrt{r^2} = \sqrt{45}$$
$$r = \sqrt{9}\sqrt{5}$$
$$r = 3\sqrt{5}$$

797. Center: (*h*, *k*); radius: *r*

The standard form of a circle is $(x-h)^2 + (y-k)^2 = r^2$, where (h,k) is the center of the circle and r is the radius of the circle.

798. $x^2 + y^2 = 49$

The standard form of a circle is $(x-h)^2 + (y-k)^2 = r^2$, where (h,k) is the center of the circle and r is the radius of the circle. If the center is the origin, then $h = 0$ and $k = 0$. The radius is 7, so

$$(x-h)^2 + (y-k)^2 = r^2$$
$$(x-0)^2 + (y-0)^2 = 7^2$$
$$x^2 + y^2 = 49$$

799. $(x-9)^2+(y+8)^2=\dfrac{4}{9}$

The standard form of a circle is $(x-h)^2+(y-k)^2=r^2$, where (h,k) is the center of the circle and r is the radius of the circle. If the center is $(9,-8)$, then $h=9$ and $k=-8$. The radius is $\dfrac{2}{3}$, so

$$(x-h)^2+(y-k)^2=r^2$$
$$(x-9)^2+\left(y-(-8)\right)^2=\left(\dfrac{2}{3}\right)^2$$
$$(x-9)^2+(y+8)^2=\dfrac{4}{9}$$

800. $(x+7)^2+(y-13)^2=25$

The standard form of a circle is $(x-h)^2+(y-k)^2=r^2$, where (h,k) is the center of the circle and r is the radius of the circle. If the center is $(-7, 13)$, then $h=-7$ and $k=13$. The diameter of a circle is twice the radius, so if the diameter is 10, the radius must be 5.

$$(x-h)^2+(y-k)^2=r^2$$
$$\left(x-(-7)\right)^2+(y-13)^2=5^2$$
$$(x+7)^2+(y-13)^2=25$$

801. $(x-3)^2+(y+4)^2=104$

The standard form of a circle is $(x-h)^2+(y-k)^2=r^2$, where (h, k) is the center of the circle and r is the radius of the circle. If the center is $(3, -4)$, then $h=3$ and $k=-4$:

$$(x-h)^2+(y-k)^2=r^2$$
$$(x-3)^2+\left(y-(-4)\right)^2=r^2$$
$$(x-3)^2+(y+4)^2=r^2$$

You're given that the point $(5,6)$ lies on the circle, so you can plug 5 in for x and 6 in for y to determine what r^2 is:

$$(x-3)^2+(y+4)^2=r^2$$
$$(5-3)^2+(6+4)^2=r^2$$
$$2^2+10^2=r^2$$
$$4+100=r^2$$
$$104=r^2$$

The equation of the circle is therefore

$$(x-3)^2+(y+4)^2=104$$

802. $(x-4)^2+(y-8)^2=4$

Find the center of the circle by calculating the midpoint of the endpoints of the diameter:

$$M=\left(\frac{x_1+x_2}{2},\frac{y_1+y_2}{2}\right)=\left(\frac{2+6}{2},\frac{8+8}{2}\right)=\left(\frac{8}{2},\frac{16}{2}\right)=(4,8)$$

The standard form of a circle is $(x-h)^2+(y-k)^2=r^2$, where (h,k) is the center of the circle and r is the radius of the circle. If the center is $(4, 8)$, then $h=4$ and $k=8$:

$$(x-4)^2+(y-8)^2=r^2$$

You're given that the point $(2,8)$ lies on the circle, so you can plug 2 in for x and 8 in for y to determine what r^2 is:

$$(x-4)^2+(y-8)^2=r^2$$
$$(2-4)^2+(8-8)^2=r^2$$
$$(-2)^2+0^2=r^2$$
$$4+0=r^2$$
$$4=r^2$$

The equation of the circle is therefore

$$(x-4)^2+(y-8)^2=4$$

803. $(x+7)^2+(y-6)^2=17$

Find the center of the circle by calculating the midpoint of the endpoints of the diameter:

$$M=\left(\frac{x_1+x_2}{2},\frac{y_1+y_2}{2}\right)=\left(\frac{-11+(-3)}{2},\frac{5+7}{2}\right)=\left(\frac{-14}{2},\frac{12}{2}\right)=(-7,6)$$

The standard form of a circle is $(x-h)^2+(y-k)^2=r^2$, where (h,k) is the center of the circle and r is the radius of the circle. If the center is $(-7,6)$, then $h=-7$ and $k=6$:

$$(x-(-7))^2+(y-6)^2=r^2$$
$$(x+7)^2+(y-6)^2=r^2$$

You're given that the point (–11, 5) lies on the circle, so you can plug –11 in for x and 5 in for y to determine what r^2 is:

$$\left(x+7\right)^2+\left(y-6\right)^2=r^2$$
$$\left(-11+7\right)^2+\left(5-6\right)^2=r^2$$
$$\left(-4\right)^2+\left(-1\right)^2=r^2$$
$$16+1-r^2$$
$$17=r^2$$

The equation of the circle is therefore

$$\left(x+7\right)^2+\left(y-6\right)^2=17$$

804. $\left(x+1\right)^2+\left(y-10\right)^2=29$

Find the center of the circle by calculating the midpoint of the endpoints of the diameter:

$$M=\left(\frac{x_1+x_2}{2},\frac{y_1+y_2}{2}\right)=\left(\frac{4+(-6)}{2},\frac{12+8}{2}\right)=\left(\frac{-2}{2},\frac{20}{2}\right)=(-1,\,10)$$

The standard form of a circle is $\left(x-h\right)^2+\left(y-k\right)^2=r^2$, where (h,k) is the center of the circle and r is the radius of the circle. If the center is $(-1,10)$, then $h=-1$ and $k=10$:

$$\left(x-(-1)\right)^2+\left(y-10\right)^2=r^2$$
$$\left(x+1\right)^2+\left(y-10\right)^2=r^2$$

You're given that the point (4, 12) lies on the circle, so you can plug 4 in for x and 12 in for y to determine what r^2 is:

$$\left(x+1\right)^2+\left(y-10\right)^2=r^2$$
$$\left(4+1\right)^2+\left(12-10\right)^2=r^2$$
$$5^2+2^2=r^2$$
$$25+4=r^2$$
$$29=r^2$$

The equation of the circle is therefore

$$\left(x+1\right)^2+\left(y-10\right)^2=29$$

805. Center: $(-1, 0)$; radius: 4

To complete the square, first bring the constant to the other side and rearrange the equation so that the x's are together and the y's are together:

$$x^2 + y^2 + 2x - 15 = 0$$
$$x^2 + 2x + y^2 = 15$$

Next, add $\left(\dfrac{b}{2}\right)^2$ to the x's. In a quadratic expression, b represents the coefficient of the variable to the first power. For the x's in this equation, $b = 2$. You also need to add this value to the other side of the equation to keep both sides equal:

$$x^2 + 2x + \left(\dfrac{2}{2}\right)^2 + y^2 = 15 + \left(\dfrac{2}{2}\right)^2$$
$$x^2 + 2x + \left(1\right)^2 + y^2 = 15 + \left(1\right)^2$$
$$x^2 + 2x + 1 + y^2 = 16$$

Now factor the x's separately from the y's to put the equation in standard form:

$$(x+1)(x+1) + y^2 = 16$$
$$(x+1)^2 + y^2 = 16$$

The standard form of a circle is $(x - h)^2 + (y - k)^2 = r^2$, where (h, k) is the center of the circle and r is the radius of the circle. In this equation, h must equal -1 and y must equal 0, making the center $(-1, 0)$. Finally, solve for the radius:

$$r^2 = 16$$
$$\sqrt{r^2} = \sqrt{16}$$
$$r = 4$$

806. Center: $(-2, -3)$; radius: 3

To complete the square, first bring the constant to the other side and rearrange the equation so that the x's are together and the y's are together:

$$x^2 + y^2 + 4x + 6y + 4 = 0$$
$$x^2 + 4x + y^2 + 6y = -4$$

Next, add $\left(\dfrac{b}{2}\right)^2$ to the x's and also to the y's. In a quadratic expression, b represents the coefficient of the variable to the first power. For the x's in this equation, $b = 4$, and for the y's, $b = 6$. You also need to add these values to the other side of the equation to keep both sides equal:

$$x^2 + 4x + \left(\dfrac{4}{2}\right)^2 + y^2 + 6y + \left(\dfrac{6}{2}\right)^2 = -4 + \left(\dfrac{4}{2}\right)^2 + \left(\dfrac{6}{2}\right)^2$$
$$x^2 + 4x + \left(2\right)^2 + y^2 + 6y + \left(3\right)^2 = -4 + \left(2\right)^2 + \left(3\right)^2$$
$$x^2 + 4x + 4 + y^2 + 6y + 9 = 9$$

Now factor the x's separately from the y's to put the equation in standard form:

$$(x+2)(x+2)+(y+3)(y+3)=9$$
$$(x+2)^2+(y+3)^2=9$$

The standard form of a circle is $(x-h)^2+(y-k)^2=r^2$, where (h,k) is the center of the circle and r is the radius of the circle. In this equation, h must equal -2 and y must equal -3, making the center $(-2,-3)$. Finally, solve for the radius:

$$r^2=9$$
$$\sqrt{r^2}=\sqrt{9}$$
$$r=3$$

---

807.   **Center: (5, –1); radius: 10**

To complete the square, first bring the constant to the other side and rearrange the equation so that the x's are together and the y's are together:

$$x^2+y^2-10x+2y-74=0$$
$$x^2-10x+y^2+2y=74$$

Next, add $\left(\dfrac{b}{2}\right)^2$ to the x's and also to the y's. In a quadratic expression, b represents the coefficient of the variable to the first power. For the x's in this equation, $b=-10$, and for the y's, $b=2$. You also need to add these values to the other side of the equation to keep both sides equal:

$$x^2-10x+\left(\frac{-10}{2}\right)^2+y^2+2y+\left(\frac{2}{2}\right)^2=74+\left(\frac{-10}{2}\right)^2+\left(\frac{2}{2}\right)^2$$
$$x^2-10x+(-5)^2+y^2+2y+(1)^2=74+(-5)^2+(1)^2$$
$$x^2-10x+25+y^2+2y+1=100$$

Now factor the x's separately from the y's to put the equation in standard form:

$$(x-5)(x-5)+(y+1)(y+1)=100$$
$$(x-5)^2+(y+1)^2=100$$

The standard form of a circle is $(x-h)^2+(y-k)^2=r^2$, where (h,k) is the center of the circle and r is the radius of the circle. In this equation, h must equal 5 and y must equal -1, making the center $(5,-1)$. Finally, solve for the radius:

$$r^2=100$$
$$\sqrt{r^2}=\sqrt{100}$$
$$r=10$$

808. Center: (–3, 7); radius: $\sqrt{6}$

Divide each term by 2 so that the x^2 and y^2 have a coefficient of 1:

$$2x^2 + 2y^2 + 12x - 28y + 104 = 0$$

$$\frac{2x^2}{2} + \frac{2y^2}{2} + \frac{12x}{2} - \frac{28y}{2} + \frac{104}{2} = \frac{0}{2}$$

$$x^2 + y^2 + 6x - 14y + 52 = 0$$

To complete the square, first bring the constant to the other side and rearrange the equation so that the x's are together and the y's are together:

$$x^2 + 6x + y^2 - 14y = -52$$

Next, add $\left(\frac{b}{2}\right)^2$ to the x's and also to the y's. In a quadratic expression, b represents the coefficient of the variable to the first power. For the x's in this equation, $b = 6$, and for the y's, $b = -14$. You also need to add these values to the other side of the equation to keep both sides equal:

$$x^2 + 6x + \left(\frac{6}{2}\right)^2 + y^2 - 14y + \left(\frac{-14}{2}\right)^2 = -52 + \left(\frac{6}{2}\right)^2 + \left(\frac{-14}{2}\right)^2$$

$$x^2 + 6x + \left(3\right)^2 + y^2 - 14y + \left(-7\right)^2 = -52 + \left(3\right)^2 + \left(-7\right)^2$$

$$x^2 + 6x + 9 + y^2 - 14y + 49 = 6$$

Now factor the x's separately from the y's to put the equation in standard form:

$$(x+3)(x+3) + (y-7)(y-7) = 6$$

$$(x+3)^2 + (y-7)^2 = 6$$

The standard form of a circle is $(x-h)^2 + (y-k)^2 = r^2$, where (h,k) is the center of the circle and r is the radius of the circle. In this equation, h must equal –3 and y must equal 7, making the center (–3, 7). Finally, solve for the radius:

$$r^2 = 6$$

$$\sqrt{r^2} = \sqrt{6}$$

$$r = \sqrt{6}$$

809. $x^2 + (y+4)^2 = 100$

The center of a circle is located where the perpendicular bisectors of any two chords of the circle intersect. First look at the chord containing the points $(0,6)$ and $(6,4)$. To determine the equation of the perpendicular bisector, you must first find the midpoint and then find the slope of the chord. The midpoint is

$$M = \left(\frac{x_1 + x_2}{2}, \frac{y_1 + y_2}{2}\right) = \left(\frac{0+6}{2}, \frac{6+4}{2}\right) = \left(\frac{6}{2}, \frac{10}{2}\right) = (3, 5)$$

The slope is

$$m = \frac{y_2 - y_1}{x_2 - x_1} = \frac{4-6}{6-0} = \frac{-2}{6} = \frac{-1}{3}$$

The line perpendicular to the chord must have a slope that's the negative reciprocal, which is 3.

Plug your information into the equation of a line to determine the equation of the perpendicular bisector of the chord that contains these points:

$$y = mx + b$$
$$5 = 3(3) + b$$
$$5 = 9 + b$$
$$-4 = b$$

$$y = 3x - 4$$

Now look at the chord containing the points $(0,6)$ and $(-8,2)$. To determine the equation of the perpendicular bisector, first find the midpoint and then find the slope. The midpoint is

$$M = \left(\frac{x_1 + x_2}{2}, \frac{y_1 + y_2}{2} \right) = \left(\frac{0 + (-8)}{2}, \frac{6 + 2}{2} \right) = \left(\frac{-8}{2}, \frac{8}{2} \right) = (-4, 4)$$

The slope is

$$m = \frac{y_2 - y_1}{x_2 - x_1} = \frac{2 - 6}{-8 - 0} = \frac{-4}{-8} = \frac{1}{2}$$

The line perpendicular to this chord must have a slope that's the negative reciprocal, which is –2.

Plug your information into the equation of a line to determine the equation of the perpendicular bisector of the chord containing these points:

$$y = mx + b$$
$$4 = -2(-4) + b$$
$$4 = 8 + b$$
$$-4 = b$$

$$y = -2x - 4$$

Now find where these two perpendicular bisectors intersect by setting the line equations equal to each other:

$$3x - 4 = -2x - 4$$
$$5x - 4 = -4$$
$$5x = 0$$
$$x = 0$$

$$y = 3x - 4 = 3(0) - 4 = -4$$

This means that $(0,-4)$ is the center of the circle. The standard form of a circle is $(x-h)^2+(y-k)^2=r^2$, where (h,k) is the center of the circle and r is the radius of the circle. If the center is $(0,-4)$, then $h=0$ and $k=-4$:

$$(x-0)^2+\left(y-(-4)\right)^2=r^2$$
$$x^2+(y+4)^2=r^2$$

You're given that the point $(0,6)$ lies on the circle. This means that you can plug 0 in for x and 6 in for y to determine what r^2 is:

$$x^2+(y+4)^2=r^2$$
$$0^2+(6+4)^2=r^2$$
$$100=r^2$$

The equation of the circle is therefore

$$x^2+(y+4)^2=100$$

810. $\quad x^2+y^2+8y-84=0$

The general form of a circle is $Ax^2+By^2+Cx+Dy+E=0$. To determine the general form of the circle, expand the standard form (from Question 809) and set the equation equal to zero:

$$x^2+(y+4)(y+4)=100$$
$$x^2+y^2+8y+16=100$$
$$x^2+y^2+8y-84=0$$

811. $\quad$ **Center: (0, –4); radius: 10**

The standard form of a circle is $(x-h)^2+(y-k)^2=r^2$, where (h,k) is the center of the circle and r is the radius of the circle. If the center is $(0,-4)$, as you found in Question 809, then $h=0$ and $k=-4$:

$$(x-0)^2+\left(y-(-4)\right)^2=r^2$$
$$x^2+(y+4)^2=r^2$$

You're given that the point $(0,6)$ lies on the circle. You can plug 0 in for x and 6 in for y to determine what r is:

$$x^2+(y+4)^2=r^2$$
$$0^2+(6+4)^2=r^2$$
$$100=r^2$$
$$\sqrt{100}=\sqrt{r^2}$$
$$10=r$$

812. 50°

The central angle of a circle is congruent to the arc that it intercepts.

813. 64°

The central angle of a circle is congruent to the arc that it intercepts.

814. 285°

The central angle of a circle is congruent to the arc that it intercepts. $\angle DOE = 75°$, which makes $m\widehat{DE} = 75°$.

$$m\widehat{DE} + m\widehat{DFE} = 360°$$
$$75° + m\widehat{DFE} = 360°$$
$$m\widehat{DFE} = 285°$$

815. 60°

$$m\widehat{DE} + m\widehat{DFE} = 360°$$
$$m\widehat{DE} + 300° = 360°$$
$$m\widehat{DE} = 60°$$

The central angle of a circle is congruent to the arc that it intercepts. $m\widehat{DE} = 60°$, which makes $m\angle DOE = 60°$.

816. 70°

$$m\widehat{DF} + m\widehat{EF} + m\widehat{DE} = 360°$$
$$150° + 140° + m\widehat{DE} = 360°$$
$$290° + m\widehat{DE} = 360°$$
$$m\widehat{DE} = 70°$$

The central angle of a circle is congruent to the arc that it intercepts. $m\widehat{DE} = 70°$, which makes $m\angle DOE = 70°$.

817. 160°

The central angle of a circle is congruent to the arc that it intercepts. $m\angle DOE = 40°$, which makes $m\overset{\frown}{DE} = 40°$. $m\overset{\frown}{DF} = m\overset{\frown}{EF} = x$, so

$$m\overset{\frown}{DF} + m\overset{\frown}{EF} + m\overset{\frown}{DE} = 360°$$
$$x + x + 40° = 360°$$
$$2x = 320°$$
$$x = 160°$$

Therefore, $m\overset{\frown}{DF} = 160°$.

818. 120°

$\triangle EQU$ is equilateral, which means all sides of the triangle are equal. This also means that $\overset{\frown}{EQ} \cong \overset{\frown}{QU} \cong \overset{\frown}{EU}$. If $m\overset{\frown}{EQ} = x$, then

$$m\overset{\frown}{EQ} + m\overset{\frown}{QU} + m\overset{\frown}{EU} = 360°$$
$$x + x + x = 360°$$
$$3x = 360°$$
$$x = 120°$$

Therefore, $m\overset{\frown}{EQ} = 120°$.

819. 240°

$\triangle EQU$ is equilateral, which means all sides of the triangle are equal. This also means that $\overset{\frown}{EQ} \cong \overset{\frown}{QU} \cong \overset{\frown}{EU}$. If $m\overset{\frown}{EQ} = x$, then

$$m\overset{\frown}{EQ} + m\overset{\frown}{QU} + m\overset{\frown}{EU} = 360°$$
$$x + x + x = 360°$$
$$3x = 360°$$
$$x = 120°$$

Therefore, $m\overset{\frown}{EQ} = m\overset{\frown}{QU} = m\overset{\frown}{EU} = 120°$. Now find $m\overset{\frown}{QUE}$:

$$m\overset{\frown}{QUE} = m\overset{\frown}{QU} + m\overset{\frown}{EU}$$
$$= 120° + 120°$$
$$= 240°$$

---

820. **60°**

The inscribed angle of a circle is half the measure of the intercepted arc:

$$m\angle IRT = \frac{1}{2}m\widehat{IT}$$
$$= \frac{1}{2}(120°)$$
$$= 60°$$

---

821. **30°**

$\overline{RI}$ is the diameter of the circle, so it divides the circle into two semicircles whose arcs total 180°:

$$m\widehat{RT} + m\widehat{IT} = 180°$$
$$m\widehat{RT} + 120° = 180°$$
$$m\widehat{RT} = 60°$$

The inscribed angle of a circle is half the measure of the intercepted arc:

$$m\angle RIT = \frac{1}{2}m\widehat{RT}$$
$$= \frac{1}{2}(60°)$$
$$= 30°$$

---

822. **20**

The inscribed angle of a circle is half the measure of the intercepted arc:

$$m\angle RTI = \frac{1}{2}m\widehat{RI} = \frac{1}{2}(180°) = 90°$$
$$m\angle IRT = \frac{1}{2}m\widehat{IT} = \frac{1}{2}(120°) = 60°$$

$\overline{OR}$ and $\overline{OI}$ are radii in Circle O. $OR = OI = 20$, so diameter $RI = 40$.

$\triangle RTI$ is a right triangle, so you can use trigonometric ratios to determine the length of $\overline{RT}$. You know that the measure of the hypotenuse is 40 and that $m\angle IRT = 60°$. $\overline{RT}$ is the side adjacent to $\angle IRT$; therefore, you can use the cosine function to solve for the length of $\overline{RT}$:

$$\cos\theta = \frac{\text{adjacent}}{\text{hypotenuse}}$$
$$\cos 60° = \frac{x}{40}$$
$$40\cos 60° = x$$
$$20 = x$$

823. $20\sqrt{3}$

The inscribed angle of a circle is half the measure of the intercepted arc:

$$m\angle RTI = \tfrac{1}{2}\,m\widehat{RI} = \tfrac{1}{2}\left(180°\right) = 90°$$

$$m\angle IRT = \tfrac{1}{2}\,m\widehat{IT} = \tfrac{1}{2}\left(120°\right) = 60°$$

$\overline{OR}$ and $\overline{OI}$ are radii in Circle O. $OR = OI = 20$, so diameter $RI = 40$.

$\triangle RTI$ is a right triangle, so you can use trigonometric ratios to determine the length of $\overline{IT}$. In this right triangle, the hypotenuse is 40 and $m\angle IRT = 60°$. $\overline{IT}$ is the side opposite $\angle IRT$; therefore, you can use the sine function to solve for IT:

$$\sin\theta = \frac{\text{opposite}}{\text{hypotenuse}}$$

$$\sin 60° = \frac{x}{40}$$

$$40\sin 60° = x$$

$$40\left(\frac{\sqrt{3}}{2}\right) = x$$

$$20\sqrt{3} = x$$

824. 54°

Let

$$m\widehat{IM} = x$$

$$m\widehat{IX} = 3x - 9$$

Chords $\overline{IX}$ and $\overline{MX}$ are congruent, which means that $m\widehat{IX} = m\widehat{MX} = 3x - 9$. The arcs of a circle total to 360°, so

$$m\widehat{IM} + m\widehat{IX} + m\widehat{MX} = 360°$$

$$x + 3x - 9 + 3x - 9 = 360$$

$$7x - 18 = 360$$

$$7x = 378$$

$$x = 54$$

Therefore, $m\widehat{IM} = 54°$.

825. 54°

Let

$$m\widehat{IM} = x$$

$$m\widehat{IX} = 3x - 9$$

Chords $\overline{IX}$ and $\overline{MX}$ are congruent, which means that $m\widehat{IX} = m\widehat{MX} = 3x - 9$. The arcs of a circle total to 360°, so

$$m\widehat{IM} + m\widehat{IX} + m\widehat{MX} = 360°$$
$$x + 3x - 9 + 3x - 9 = 360$$
$$7x - 18 = 360$$
$$7x = 378$$
$$x = 54$$

Therefore, $m\widehat{IM} = 54°$.

$\angle MOI$ is a central angle in Circle O. The central angle of a circle is equal to the arc that it intersects, so $m\angle MOI = 54°$.

---

826.   **27°**

Let

$$m\widehat{IM} = x$$
$$m\widehat{IX} = 3x - 9$$

Chords $\overline{IX}$ and $\overline{MX}$ are congruent, which means that $m\widehat{IX} = m\widehat{MX} = 3x - 9$. The arcs of a circle total to 360°, so

$$m\widehat{IM} + m\widehat{IX} + m\widehat{MX} = 360°$$
$$x + 3x - 9 + 3x - 9 = 360$$
$$7x - 18 = 360$$
$$7x = 378$$
$$x = 54$$

Therefore, $m\widehat{IM} = 54°$.

$\angle X$ is an inscribed angle in Circle O. The inscribed angle of a circle is equal to half of the arc that it intersects, so

$$m\angle X = \frac{1}{2}m\widehat{IM} = \frac{1}{2}(54°) = 27°$$

---

827.   **50°**

The measure of a pair of vertical angles in the circle is equal to the average of the arcs they intersect:

$$m\angle LEU = \frac{m\widehat{LU} + m\widehat{PS}}{2} = \frac{43° + 57°}{2} = 50°$$

828. 95°

The measure of a pair of vertical angles in the circle is equal to the average of the arcs they intersect:

$$m\angle LEP = \frac{m\widehat{PL} + m\widehat{US}}{2} = \frac{114° + 76°}{2} = 95°$$

829. 70°

The measure of a pair of vertical angles in the circle is equal to the average of the arcs they intersect:

$$m\angle SEU = \frac{m\widehat{PL} + m\widehat{US}}{2}$$

$$100° = \frac{130° + m\widehat{US}}{2}$$

$$200° = 130° + m\widehat{US}$$

$$70° = m\widehat{US}$$

830. 50°

$\angle LEP$ and $\angle LEU$ form a linear pair, so they add up to 180°:

$$m\angle LEP + m\angle LEU = 180°$$

$$132° + m\angle LEU = 180°$$

$$m\angle LEU = 48°$$

The measure of a pair of vertical angles in the circle is equal to the average of the arcs they intersect:

$$m\angle LEU = \frac{m\widehat{LU} + m\widehat{PS}}{2}$$

$$48° = \frac{46° + m\widehat{PS}}{2}$$

$$96° = 46° + m\widehat{PS}$$

$$50° = m\widehat{PS}$$

831. 60°

The total of the arcs forming a circle is 360°:

$$m\overset{\frown}{AB} + m\overset{\frown}{BC} + m\overset{\frown}{CD} + m\overset{\frown}{AD} = 360°$$
$$65° + 85° + 150° + m\overset{\frown}{AD} = 360°$$
$$300° + m\overset{\frown}{AD} = 360°$$
$$m\overset{\frown}{AD} = 60°$$

832. 72.5°

The total of the arcs forming a circle is 360°:

$$m\overset{\frown}{AB} + m\overset{\frown}{BC} + m\overset{\frown}{CD} + m\overset{\frown}{AD} = 360°$$
$$65° + 85° + 150° + m\overset{\frown}{AD} = 360°$$
$$300° + m\overset{\frown}{AD} = 360°$$
$$m\overset{\frown}{AD} = 60°$$

$\angle AED$ is a vertical angle in Circle O. The measure of a pair of vertical angles in the circle is equal to the average of the arcs they intersect:

$$m\angle AED = \frac{m\overset{\frown}{AD} + m\overset{\frown}{BC}}{2} = \frac{60° + 85°}{2} = \frac{145°}{2} = 72.5°$$

833. 107.5°

The total of the arcs forming a circle is 360°:

$$m\overset{\frown}{AB} + m\overset{\frown}{BC} + m\overset{\frown}{CD} + m\overset{\frown}{AD} = 360°$$
$$65° + 85° + 150° + m\overset{\frown}{AD} = 360°$$
$$300° + m\overset{\frown}{AD} = 360°$$
$$m\overset{\frown}{AD} = 60°$$

$\angle AED$ is a vertical angle in Circle O. The measure of a pair of vertical angles in the circle is equal to the average of the arcs they intersect:

$$m\angle AED = \frac{m\overset{\frown}{AD} + m\overset{\frown}{BC}}{2} = \frac{60° + 85°}{2} = 72.5°$$

$\angle AED$ and $\angle CED$ form a linear pair, so they add up to 180°:

$$m\angle AED + m\angle CED = 180°$$
$$72.5° + m\angle CED = 180°$$
$$m\angle CED = 107.5°$$

834. **79°**

Let

$$m\widehat{TH} = x$$
$$m\widehat{MH} = 2x$$

The sum of the arcs of a circle is 360°, so

$$m\widehat{AM} + m\widehat{AT} + m\widehat{TH} + m\widehat{MH} = 360°$$
$$64° + 59° + x + 2x = 360°$$
$$123° + 3x = 360°$$
$$3x = 237°$$
$$x = 79°$$

Therefore, $m\widehat{TH} = 79°$.

835. **158°**

Let

$$m\widehat{TH} = x$$
$$m\widehat{MH} = 2x$$

The sum of the arcs of a circle is 360°, so

$$m\widehat{AM} + m\widehat{AT} + m\widehat{TH} + m\widehat{MH} = 360°$$
$$64° + 59° + x + 2x = 360°$$
$$123° + 3x = 360°$$
$$3x = 237°$$
$$x = 79°$$

Therefore, $m\widehat{MH} = 2x = 2(79°) = 158°$.

836. **71.5°**

Let

$$m\widehat{TH} = x$$
$$m\widehat{MH} = 2x$$

The sum of the arcs of a circle is 360°, so

$$m\widehat{AM} + m\widehat{AT} + m\widehat{TH} + m\widehat{MH} = 360°$$
$$64° + 59° + x + 2x = 360°$$
$$123° + 3x = 360°$$
$$3x = 237°$$
$$x = 79°$$

Therefore, $m\widehat{TH} = 79°$.

$\angle AEM$ is a vertical angle inside the circle. You find its measure by calculating the average of the two arcs it intersects:

$$m\angle AEM = \frac{m\widehat{AM} + m\widehat{TH}}{2} = \frac{64° + 79°}{2} = \frac{143°}{2} = 71.5°$$

837.   **108.5°**

Let

$$m\widehat{TH} = x$$
$$m\widehat{MH} = 2x$$

The sum of the arcs of a circle is 360°, so

$$m\widehat{AM} + m\widehat{AT} + m\widehat{TH} + m\widehat{MH} = 360°$$
$$64° + 59° + x + 2x = 360°$$
$$123° + 3x = 360°$$
$$3x = 237°$$
$$x = 79°$$

Therefore, $m\widehat{MH} = 2x = 2(79°) = 158°$.

$\angle AET$ is a vertical angle inside the circle. You find its measure by calculating the average of the two arcs it intersects:

$$m\angle AET = \frac{m\widehat{AT} + m\widehat{MH}}{2} = \frac{59° + 158°}{2} = \frac{217°}{2} = 108.5°$$

838.   **40°**

$\angle Y$ is an exterior angle. Its measure is half the difference between the major and minor arcs that it intersects:

$$m\angle Y = \frac{m\widehat{GM} - m\widehat{ET}}{2} = \frac{100° - 20°}{2} = \frac{80°}{2} = 40°$$

839.   **56°**

$\angle Y$ is an exterior angle. Its measure is half the difference between the major and minor arcs that it intersects:

$$m\angle Y = \frac{m\widehat{GM} - m\widehat{ET}}{2} = \frac{155° - 43°}{2} = \frac{112°}{2} = 56°$$

840. 21°

$\angle Y$ is an exterior angle. Its measure is half the difference between the major and minor arcs that it intersects:

$$m\angle Y = \frac{m\overset{\frown}{GM} - m\overset{\frown}{ET}}{2}$$

$$39° = \frac{99° - m\overset{\frown}{ET}}{2}$$

$$78° = 99° - m\overset{\frown}{ET}$$

$$-21° = -m\overset{\frown}{ET}$$

$$21° = m\overset{\frown}{ET}$$

841. 25°

$\angle Y$ is an exterior angle. Its measure is half the difference between the major and minor arcs that it intersects:

$$m\angle Y = \frac{m\overset{\frown}{GM} - m\overset{\frown}{ET}}{2}$$

$$58.5° = \frac{142° - m\overset{\frown}{ET}}{2}$$

$$117° = 142° - m\overset{\frown}{ET}$$

$$-25° = -m\overset{\frown}{ET}$$

$$25° = m\overset{\frown}{ET}$$

842. 174°

$\angle Y$ is an exterior angle. Its measure is half the difference between the major and minor arcs that it intersects:

$$m\angle Y = \frac{m\overset{\frown}{GM} - m\overset{\frown}{ET}}{2}$$

$$52° = \frac{m\overset{\frown}{GM} - 70°}{2}$$

$$104° = m\overset{\frown}{GM} - 70°$$

$$174° = m\overset{\frown}{GM}$$

843. 183.5°

$\angle Y$ is an exterior angle. Its measure is half the difference between the major and minor arcs that it intersects:

$$m\angle Y = \frac{m\overset{\frown}{GM} - m\overset{\frown}{ET}}{2}$$

$$73.25° = \frac{m\overset{\frown}{GM} - 37°}{2}$$

$$146.5° = m\overset{\frown}{GM} - 37°$$

$$183.5° = m\overset{\frown}{GM}$$

844. 120°

The sum of the arcs of a circle is 360°:

$$m\overset{\frown}{AGN} + m\overset{\frown}{AN} = 360°$$

$$300° + m\overset{\frown}{AN} = 360°$$

$$m\overset{\frown}{AN} = 60°$$

$\angle T$ is an exterior angle. Its measure is half the difference between the major and minor arcs that it intersects:

$$m\angle T = \frac{m\overset{\frown}{AGN} - m\overset{\frown}{AN}}{2} = \frac{300° - 60°}{2} = \frac{240°}{2} = 120°$$

845. 70°

The sum of the arcs of a circle is 360°:

$$m\overset{\frown}{AGN} + m\overset{\frown}{AN} = 360°$$

$$250° + m\overset{\frown}{AN} = 360°$$

$$m\overset{\frown}{AN} = 110°$$

$\angle T$ is an exterior angle. Its measure is half the difference between the major and minor arcs that it intersects:

$$m\angle T = \frac{m\overset{\frown}{AGN} - m\overset{\frown}{AN}}{2} = \frac{250° - 110°}{2} = \frac{140°}{2} = 70°$$

846. 75°

The sum of the arcs of a circle is 360°:

$$m\overset{\frown}{AGN} + m\overset{\frown}{AN} = 360°$$
$$m\overset{\frown}{AGN} + 105° = 360°$$
$$m\overset{\frown}{AGN} = 255°$$

$\angle T$ is an exterior angle. Its measure is half the difference between the major and minor arcs that it intersects:

$$m\angle T = \frac{m\overset{\frown}{AGN} - m\overset{\frown}{AN}}{2} = \frac{255° - 105°}{2} = \frac{150°}{2} = 75°$$

847 240°

The sum of the arcs of a circle is 360°. If you let $m\overset{\frown}{AGN} = x$, then

$$m\overset{\frown}{AGN} + m\overset{\frown}{AN} = 360°$$
$$x + m\overset{\frown}{AN} = 360°$$
$$m\overset{\frown}{AN} = 360° - x$$

$\angle T$ is an exterior angle. Its measure is half the difference between the major and minor arcs that it intersects:

$$m\angle T = \frac{m\overset{\frown}{AGN} - m\overset{\frown}{AN}}{2}$$
$$60° = \frac{x - (360° - x)}{2}$$
$$120° = x - 360° + x$$
$$120° = 2x - 360°$$
$$480° = 2x$$
$$240° = x$$

Therefore, $m\overset{\frown}{AGN} = 240°$.

848. 99°

The sum of the arcs of a circle is 360°. If you let $m\overset{\frown}{AN} = x$, then

$$m\overset{\frown}{AGN} + m\overset{\frown}{AN} = 360°$$
$$m\overset{\frown}{AGN} + x = 360°$$
$$m\overset{\frown}{AGN} = 360° - x$$

$\angle T$ is an exterior angle. Its measure is half the difference between the major and minor arcs that it intersects:

$$m\angle T = \frac{m\overset{\frown}{AGN} - m\overset{\frown}{AN}}{2}$$

$$81° = \frac{(360° - x) - x}{2}$$

$$162° = 360° - 2x$$

$$-198° = -2x$$

$$\frac{-198°}{-2} = \frac{-2x}{-2}$$

$$99° = x$$

Therefore, $m\overset{\frown}{AN} = 99°$.

---

849.   **119°**

Let

$$m\overset{\frown}{AD} = x$$

$$m\overset{\frown}{DC} = 2x - 3$$

The sum of the arcs of a circle is 360°. Diameter $\overline{AOC}$ divides the circle into the two congruent arcs $\overset{\frown}{ADC}$ and $\overset{\frown}{AC}$, each measuring 180°, so

$$m\overset{\frown}{AD} + m\overset{\frown}{DC} = 180°$$

$$x + 2x - 3 = 180°$$

$$3x - 3 = 180°$$

$$3x = 183°$$

$$x = 61°$$

Therefore, $m\overset{\frown}{DC} = 2x - 3 = 2(61°) - 3 = 119°$.

---

850.   **59.5°**

Let

$$m\overset{\frown}{AD} = x$$

$$m\overset{\frown}{DC} = 2x - 3$$

The sum of the arcs of a circle is 360°. Diameter $\overline{AOC}$ divides the circle into the two congruent arcs $\overset{\frown}{ADC}$ and $\overset{\frown}{AC}$, each measuring 180°, so

$$m\overset{\frown}{AD} + m\overset{\frown}{DC} = 180°$$

$$x + 2x - 3 = 180°$$

$$3x - 3 = 180°$$

$$3x = 183°$$

$$x = 61°$$

Therefore, $m\widehat{DC} = 2(61°) - 3 = 119°$. $\angle A$ is an inscribed angle, so it's half the measure of the arc it intercepts:

$$m\angle A = \tfrac{1}{2}m\widehat{DC} = \tfrac{1}{2}(119°) = 59.5°$$

---

851. 30.5°

Let

$$m\widehat{AD} = x$$
$$m\widehat{DC} = 2x - 3$$

The sum of the arcs of a circle is 360°. Diameter $\overline{AOC}$ divides the circle into the two congruent arcs $\widehat{ADC}$ and $\widehat{AC}$, each measuring 180°:

$$m\widehat{AD} + m\widehat{DC} = 180°$$
$$x + 2x - 3 = 180°$$
$$3x - 3 = 180°$$
$$3x = 183°$$
$$x = 61°$$

Therefore, $m\widehat{DC} = 2(61°) - 3 = 119°$.

$\angle B$ is an exterior angle, so its measure is half the difference between the major and minor arcs that it intercepts:

$$m\angle B = \tfrac{1}{2}\left(m\widehat{AC} - m\widehat{DC}\right) = \tfrac{1}{2}(180° - 119°) = \tfrac{1}{2}(61°) = 30.5°$$

---

852 90°

A line tangent to a circle is perpendicular to the radius drawn to the point of tangency. In this problem, $\overline{BC}$ is a tangent and C is the point of tangency. Radius $\overline{OC}$ meets $\overline{BC}$ at Point C. This means that $\overline{OC} \perp \overline{BC}$. Perpendicular lines form right angles, so $m\angle C = 90°$.

---

853. $\overline{IE}$

When two chords intersect in a circle, the products of their segments are equal:

$$RE \times LE = CE \times IE$$

---

854. $\overline{RE}$

When two chords intersect in a circle, the products of their segments are equal:

$$CE \times IE = RE \times LE$$

855. 4

When two chords intersect in a circle, the products of their segments are equal:

$$CE \times IE = RE \times LE$$
$$(x)(6) = (3)(8)$$
$$6x = 24$$
$$x = 4$$

856. 4

When two chords intersect in a circle, the products of their segments are equal:

$$CE \times IE = RE \times LE$$
$$(6)(8) = (x)(12)$$
$$48 = 12x$$
$$4 = x$$

857. 20

When two chords intersect in a circle, the products of their segments are equal:

$$CE \times IE = RE \times LE$$
$$(8)(12.5) = (5)(x)$$
$$100 = 5x$$
$$20 = x$$

858. 5

When two chords intersect in a circle, the products of their segments are equal:

$$CE \times IE = RE \times LE$$
$$(x)(x+1) = (3)(10)$$
$$x^2 + x = 30$$
$$x^2 + x - 30 = 0$$
$$(x+6)(x-5) = 0$$
$$\cancel{x = -6} \text{ or } x = 5$$

859. 8

Let $IE = CE = x$. When two chords intersect in a circle, the products of their segments are equal:

$$CE \times IE = RE \times LE$$
$$(x)(x) = (4)(16)$$
$$x^2 = 64$$
$$x^2 - 64 = 0$$
$$(x+8)(x-8) = 0$$
$$\cancel{x = -8} \text{ or } x = 8$$

860. 4

When two chords intersect in a circle, the products of their segments are equal:

$$CE \times IE = RE \times LE$$
$$(7)(8) = (x)(x+10)$$
$$56 = x^2 + 10x$$
$$0 = x^2 + 10x - 56$$
$$0 = (x-4)(x+14)$$
$$x = 4 \text{ or } \cancel{x = -14}$$

861. 10

When two chords intersect in a circle, the products of their segments are equal:

$$CE \times IE = RE \times LE$$
$$(x)(x+2) = (x+5)(8)$$
$$x^2 + 2x = 8x + 40$$
$$x^2 - 6x - 40 = 0$$
$$(x+4)(x-10) = 0$$
$$\cancel{x = -4} \text{ or } x = 10$$

862. 15

When the diameter of a circle is perpendicular to a chord, the chord is bisected. This divides the chord into two equal segments.

863. 12

When the diameter of a circle is perpendicular to a chord, the chord is bisected, cutting the chord into two equal segments. And when two chords intersect in a circle, the products of their segments are equal, so

$$(9)(16)=(x)(x)$$
$$144=x^2$$
$$0=x^2-144$$
$$0=(x+12)(x-12)$$
$$x=-12 \text{ or } x=12$$

864. 3

The radius of this circle is $4+1=5$. This means that radius $OB=5$ as well. Diameter $\overline{AB}$ is divided into two segments measuring 1 and 9.

When the diameter of a circle is perpendicular to a chord, the chord is bisected, cutting the chord into two equal segments. And when two chords intersect in a circle, the products of their segments are equal, so

$$(1)(9)=(x)(x)$$
$$9=x^2$$
$$0=x^2 \ 9$$
$$0=(x+3)(x-3)$$
$$x=-3 \text{ or } x=3$$

865. 4

The radius of this circle is $3+2=5$. This means that radius $OB=5$ as well. Diameter $\overline{AB}$ is divided into two segments measuring 2 and 8.

When the diameter of a circle is perpendicular to a chord, the chord is bisected, cutting the chord into two equal segments. And when two chords intersect in a circle, the products of their segments are equal, so

$$(2)(8)=(x)(x)$$
$$16=x^2$$
$$0=x^2-16$$
$$0=(x+4)(x-4)$$
$$x=-4 \text{ or } x=4$$

866. 6

The radius of this circle is $4 + 2.5 = 6.5$. This means that the other radius drawn is also equal to 6.5. The diameter is divided into two segments measuring 4 and $2.5 + 6.5 = 9$.

When the diameter of a circle is perpendicular to a chord, the chord is bisected. This means the chord is cut into two equal segments. And when two chords intersect in a circle, the products of their segments are equal, so

$$(4)(9) = (x)(x)$$
$$36 = x^2$$
$$0 = x^2 - 36$$
$$0 = (x+6)(x-6)$$
$$x = -6 \text{ or } x = 6$$

867. 20

You first need to determine the length of the radius of this circle. To do so, connect Point O to Point C. This is a radius in Circle O as well as a hypotenuse in right $\triangle OEC$. Use the Pythagorean theorem to determine the length of radius $\overline{OC}$:

$$a^2 + b^2 = c^2$$
$$6^2 + 8^2 = c^2$$
$$100 = c^2$$
$$\sqrt{100} = \sqrt{c^2}$$
$$10 = c$$

The diameter of a circle is twice the measure of the radius, so the diameter of this circle is $2(10) = 20$.

868. TA

When a tangent and a secant are drawn from the same exterior point, you can determine the lengths with the following rule:

$$(\text{exterior segment of secant})(\text{whole length of secant}) = (\text{tangent})(\text{tangent})$$

Therefore, $x = TA$:

$$(TE)(TS) = (TA)(TA)$$

869. 6

When a tangent and a secant are drawn from the same exterior point, you can determine the lengths with the following rule:

$$(\text{exterior segment of secant})(\text{whole length of secant}) = (\text{tangent})(\text{tangent})$$

Therefore,

$$(TE)(TS) = (TA)(TA)$$
$$(3)(12) = (x)(x)$$
$$36 = x^2$$
$$\sqrt{36} = \sqrt{x^2}$$
$$6 = x$$

870.   4

When a tangent and a secant are drawn from the same exterior point, you can determine the lengths with the following rule:

$$(\text{exterior segment of secant})(\text{whole length of secant}) = (\text{tangent})(\text{tangent})$$

Therefore,

$$(TE)(TS) = (TA)(TA)$$
$$(x)(16) = (8)(8)$$
$$16x = 64$$
$$x = 4$$

871.   7

When a tangent and a secant are drawn from the same exterior point, you can determine the lengths with the following rule:

$$(\text{exterior segment of secant})(\text{whole length of secant}) = (\text{tangent})(\text{tangent})$$

Therefore,

$$(TE)(TE + SE) = (TA)(TA)$$
$$(9)(9 + x) = (12)(12)$$
$$81 + 9x = 144$$
$$9x = 63$$
$$x = 7$$

872.   5

When a tangent and a secant are drawn from the same exterior point, you can determine the lengths with the following rule:

$$(\text{exterior segment of secant})(\text{whole length of secant}) = (\text{tangent})(\text{tangent})$$

Therefore,

$$(TE)(TE + SE) = (TA)(TA)$$
$$(x)(x + x + 10) = (10)(10)$$
$$(x)(2x + 10) = 100$$
$$2x^2 + 10x = 100$$
$$2x^2 + 10x - 100 = 0$$
$$2(x^2 + 5x - 50) = 0$$
$$2(x + 10)(x - 5) = 0$$
$$x = -10 \text{ or } x = 5$$

873. *EB*

When two secants are drawn from the same exterior point, you can determine their lengths with the following rule:

$$\left(\begin{matrix}\text{exterior segment} \\ \text{of secant}\end{matrix}\right)\left(\begin{matrix}\text{whole length} \\ \text{of secant}\end{matrix}\right) = \left(\begin{matrix}\text{exterior segment} \\ \text{of secant}\end{matrix}\right)\left(\begin{matrix}\text{whole length} \\ \text{of secant}\end{matrix}\right)$$

Therefore, $x = EB$:

$$(EB)(EA) = (ED)(EC)$$

874. *DE*

When two secants are drawn from the same exterior point, you can determine their lengths with the following rule:

$$\left(\begin{matrix}\text{exterior segment} \\ \text{of secant}\end{matrix}\right)\left(\begin{matrix}\text{whole length} \\ \text{of secant}\end{matrix}\right) = \left(\begin{matrix}\text{exterior segment} \\ \text{of secant}\end{matrix}\right)\left(\begin{matrix}\text{whole length} \\ \text{of secant}\end{matrix}\right)$$

Therefore, $x = DE$:

$$(DE)(CE) = (BE)(AE)$$

875. 4

When two secants are drawn from the same exterior point, you can determine their lengths with the following rule:

$$\left(\begin{matrix}\text{exterior segment} \\ \text{of secant}\end{matrix}\right)\left(\begin{matrix}\text{whole length} \\ \text{of secant}\end{matrix}\right) = \left(\begin{matrix}\text{exterior segment} \\ \text{of secant}\end{matrix}\right)\left(\begin{matrix}\text{whole length} \\ \text{of secant}\end{matrix}\right)$$

Therefore,

$$(DE)(CE) = (BE)(AE)$$
$$(x)(24) = (6)(16)$$
$$24x = 96$$
$$x = 4$$

876. 15

When two secants are drawn from the same exterior point, you can determine their lengths with the following rule:

$$\left(\begin{array}{c}\text{exterior segment}\\\text{of secant}\end{array}\right)\left(\begin{array}{c}\text{whole length}\\\text{of secant}\end{array}\right)=\left(\begin{array}{c}\text{exterior segment}\\\text{of secant}\end{array}\right)\left(\begin{array}{c}\text{whole length}\\\text{of secant}\end{array}\right)$$

Therefore,

$$(DE)(CE)=(BE)(AE)$$
$$(x)(20)=(10)(30)$$
$$20x=300$$
$$x=15$$

877. 3

When two secants are drawn from the same exterior point, you can determine their lengths with the following rule:

$$\left(\begin{array}{c}\text{exterior segment}\\\text{of secant}\end{array}\right)\left(\begin{array}{c}\text{whole length}\\\text{of secant}\end{array}\right)=\left(\begin{array}{c}\text{exterior segment}\\\text{of secant}\end{array}\right)\left(\begin{array}{c}\text{whole length}\\\text{of secant}\end{array}\right)$$

Therefore,

$$(ED)(ED+DC)=(EB)(EB+BA)$$
$$(x)(x+5)=(2)(2+10)$$
$$x^2+5x=24$$
$$x^2+5x-24=0$$
$$(x+8)(x-3)=0$$
$$x=-8 \text{ or } x=3$$

878. 13

When two secants are drawn from the same exterior point, you can determine their lengths with the following rule:

$$\left(\begin{array}{c}\text{exterior segment}\\\text{of secant}\end{array}\right)\left(\begin{array}{c}\text{whole length}\\\text{of secant}\end{array}\right)=\left(\begin{array}{c}\text{exterior segment}\\\text{of secant}\end{array}\right)\left(\begin{array}{c}\text{whole length}\\\text{of secant}\end{array}\right)$$

Therefore,

$$(ED)(ED+DC)=(EB)(EB+BA)$$
$$(3)(3+x)=(4)(4+8)$$
$$3x+9=48$$
$$3x=39$$
$$x=13$$

879. 5

When two secants are drawn from the same exterior point, you can determine their lengths with the following rule:

$$\left(\begin{matrix} \text{exterior segment} \\ \text{of secant} \end{matrix}\right)\left(\begin{matrix} \text{whole length} \\ \text{of secant} \end{matrix}\right) = \left(\begin{matrix} \text{exterior segment} \\ \text{of secant} \end{matrix}\right)\left(\begin{matrix} \text{whole length} \\ \text{of secant} \end{matrix}\right)$$

Therefore,

$$(ED)(ED + DC) = (EB)(EB + BA)$$
$$(x-2)(x-2+22) = (x)(x+10)$$
$$(x-2)(x+20) = x^2 + 10x$$
$$x^2 + 18x - 40 = x^2 + 10x$$
$$8x - 40 = 0$$
$$8x = 40$$
$$x = 5$$

880. 28

Two tangents drawn to a circle from the same exterior point are congruent. $HA = 28$, so $HT = 28$.

881. 24

The radius of a circle is half the length of the diameter, so the radius of this circle is $\frac{14}{2} = 7$. The radius of a circle is perpendicular to a tangent line at the point of tangency. This means $\triangle AOB$ is a right triangle. You can use the Pythagorean theorem to solve for the length of $\overline{AB}$:

$$a^2 + b^2 = c^2$$
$$7^2 + b^2 = 25^2$$
$$49 + b^2 = 625$$
$$b^2 = 576$$
$$\sqrt{b^2} = \sqrt{576}$$
$$b = 24$$

882.   62

You find the perimeter of a triangle by adding the three sides together. Two tangents drawn to a circle from the same exterior point are congruent, so

$$AC = AB = 12$$
$$TB = TD = 8$$
$$NC = ND = 11$$

Adding the three sides gives you $P = (12+12)+(8+8)+(11+11) = 62$.

883.   13

You find the perimeter of a triangle by adding the three sides together. Two tangents drawn to a circle from the same exterior point are congruent, so

$$SA = SB = 4$$

$ST = 11$ and $SB = 4$, so $TB = TC = 7$. Let $RA = RC = x$ and use the perimeter formula to find the value of x:

$$(x+x)+(7+7)+(4+4) = 34$$
$$2x+22 = 34$$
$$2x = 12$$
$$x = 6$$

Therefore, $RT = x + 7 = 6 + 7 = 13$.

884.   Yes

The radius of a circle is perpendicular to a tangent line at the point of tangency. If $\overline{TA}$ is tangent to Circle O at Point A, then $\angle OAT$ would be a right angle and $\triangle OAT$ would satisfy the Pythagorean theorem:

$$a^2 + b^2 = c^2$$
$$10^2 + 24^2 \stackrel{?}{=} 26^2$$
$$100 + 576 \stackrel{?}{=} 676$$
$$676 = 676$$

Because the Pythagorean theorem has proven true, $\triangle OAT$ is a right triangle. That means $\overline{TA} \perp \overline{AO}$ and $\overline{TA}$ is therefore tangent to Circle O.

885. Yes

The radius of a circle is perpendicular to a tangent line at the point of tangency. If $\overline{TA}$ is tangent to Circle O at Point A, then $\angle OAT$ would be a right angle and $\triangle OAT$ would satisfy the Pythagorean theorem:

$$a^2 + b^2 = c^2$$
$$\left(3\sqrt{41}\right)^2 + 16^2 \overset{?}{=} 25^2$$
$$369 + 256 \overset{?}{=} 625$$
$$625 = 625$$

Because the Pythagorean theorem has proven true, $\triangle OAT$ is a right triangle. That means $\overline{TA} \perp \overline{AO}$ and $\overline{TA}$ is therefore tangent to Circle O.

886. Yes

The radius of a circle is perpendicular to a tangent line at the point of tangency. If $\overline{TA}$ is tangent to Circle O at Point A, $\angle OAT$ would be a right angle and $\triangle OAT$ would satisfy the Pythagorean theorem:

$$a^2 + b^2 = c^2$$
$$12^2 + 15^2 \overset{?}{=} \left(3\sqrt{41}\right)^2$$
$$144 + 225 \overset{?}{=} 369$$
$$369 = 369$$

Because the Pythagorean theorem has proven true, $\triangle OAT$ is a right triangle. That means $\overline{TA} \perp \overline{AO}$ and $\overline{TA}$ is therefore tangent to Circle O.

887. No

The radius of a circle is perpendicular to a tangent line at the point of tangency. If $\overline{TA}$ is tangent to Circle O at Point A, $\angle OAT$ would be a right angle and $\triangle OAT$ would satisfy the Pythagorean theorem:

$$a^2 + b^2 = c^2$$
$$5^2 + 12^2 \overset{?}{=} 15^2$$
$$25 + 144 \overset{?}{=} 225$$
$$169 \neq 225$$

Because the Pythagorean theorem hasn't proven true, $\triangle OAT$ is not a right triangle. $\overline{TA}$ is not perpendicular to $\overline{AO}$, and $\overline{TA}$ is therefore not tangent to Circle O.

888.   18°

The diameter of a circle divides a circle into two arcs, each measuring 180°. You're given that $m\overset{\frown}{WY}$ and $m\overset{\frown}{YR}$ are in a ratio of 1:9, so the sum of x and $9x$ has to be 180°:

$$m\overset{\frown}{WY} + m\overset{\frown}{YR} = 180°$$
$$x + 9x = 180°$$
$$10x = 180°$$
$$x = 18°$$

889.   82°

The diameter of a circle divides a circle into two arcs, each measuring 180°. You're given that $m\overset{\frown}{WY}$ and $m\overset{\frown}{YR}$ are in a ratio of 1:9, so the sum of x and $9x$ has to be 180°:

$$m\overset{\frown}{WY} + m\overset{\frown}{YR} = 180°$$
$$x + 9x = 180°$$
$$10x = 180°$$
$$x = 18°$$

Therefore, $m\overset{\frown}{YR} = 9x = 9(18°) = 162°$.

$\angle N$ is an exterior angle. Its measure is half the difference between the major and minor arcs that it intersects:

$$m\angle N = \frac{m\overset{\frown}{YR} - m\overset{\frown}{ER}}{2}$$
$$40° = \frac{162° - m\overset{\frown}{ER}}{2}$$
$$80° = 162° - m\overset{\frown}{ER}$$
$$-82° = -m\overset{\frown}{ER}$$
$$82° = m\overset{\frown}{ER}$$

890.   50°

The diameter of a circle divides a circle into two arcs, each measuring 180°. You're given that $m\overset{\frown}{WY}$ and $m\overset{\frown}{YR}$ are in a ratio of 1:9, so the sum of x and $9x$ has to be 180°:

$$\overset{\frown}{WY} + \overset{\frown}{YR} = 180°$$
$$x + 9x = 180°$$
$$10x = 180°$$
$$x = 18°$$

Therefore, $m\overset{\frown}{YR} = 9x = 9(18°) = 162°$.

$\angle N$ is an exterior angle. Its measure is half the difference between the major and minor arcs that it intersects:

$$m\angle N = \frac{m\widehat{YR} - m\widehat{ER}}{2}$$

$$40° = \frac{162° - m\widehat{ER}}{2}$$

$$80° = 162° - m\widehat{ER}$$

$$-82° = -m\widehat{ER}$$

$$82° = m\widehat{ER}$$

$\angle WKY$ is a vertical angle inside the circle. Find its measure by calculating the average of the two arcs it intersects:

$$m\angle WKY = \frac{m\widehat{WY} + m\widehat{ER}}{2} = \frac{18° + 82°}{2} = \frac{100°}{2} = 50°$$

891.   49°

The diameter of a circle divides a circle into two arcs, each measuring 180°. You're given that $m\widehat{WY}$ and $m\widehat{YR}$ are in a ratio of 1:9, so the sum of x and $9x$ has to be 180°:

$$\widehat{WY} + \widehat{YR} = 180°$$

$$x + 9x = 180°$$

$$10x = 180°$$

$$x = 18°$$

Therefore, $m\widehat{YR} = 9x = 9(18°) = 162°$.

$\angle N$ is an exterior angle. Its measure is half the difference between the major and minor arcs that it intersects:

$$m\angle N = \frac{m\widehat{YR} - m\widehat{ER}}{2}$$

$$40° = \frac{162° - m\widehat{ER}}{2}$$

$$80° = 162° - m\widehat{ER}$$

$$-82° = -m\widehat{ER}$$

$$82° = m\widehat{ER}$$

Now consider $\widehat{WER}$:

$$m\widehat{WE} + m\widehat{ER} = 180°$$

$$m\widehat{WE} + 82° = 180°$$

$$m\widehat{WE} = 98°$$

$\angle ERW$ is an inscribed angle. The inscribed angle of a circle is half the measure of the intercepted arc:

$$m\angle ERW = \tfrac{1}{2}m\widehat{EW} = \tfrac{1}{2}(98°) = 49°$$

892.   **99°**

The diameter of a circle divides a circle into two arcs, each measuring 180°. You're given that $m\widehat{WY}$ and $m\widehat{YR}$ are in a ratio of 1:9, so the sum of x and $9x$ has to be 180°:

$$m\widehat{WY} + m\widehat{YR} = 180°$$
$$x + 9x = 180°$$
$$10x = 180°$$
$$x = 18°$$

Therefore, $m\widehat{YR} = 9x = 9(18°) = 162°$.

$\angle YER$ is an inscribed angle. The inscribed angle of a circle is half the measure of the intercepted arc:

$$m\angle YER = \tfrac{1}{2}m\widehat{YR} = \tfrac{1}{2}(162°) = 81°$$

$\angle YER$ and $\angle NER$ form a linear pair, so

$$m\angle YER + m\angle NER = 180°$$
$$81° + m\angle NER = 180°$$
$$m\angle NER = 99°$$

893.   **60°**

$\overline{IC}$ is parallel to diameter $\overline{BE}$, which means that $\widehat{BI} \cong \widehat{CE}$. You're given that $\widehat{IC} \cong \widehat{CE}$, so $\widehat{BI} \cong \widehat{CE} \cong \widehat{IC}$:

$$m\widehat{BI} + m\widehat{CE} + m\widehat{IC} = 180°$$
$$x + x + x = 180°$$
$$3x = 180°$$
$$x = 60°$$

894.   **83°**

The diameter of a circle divides a circle into two arcs, each measuring 180°, so

$$m\widehat{BL} + m\widehat{EL} = 180°$$
$$m\widehat{BL} + 74° = 180°$$
$$m\widehat{BL} = 106°$$

$\angle BRL$ is a vertical angle inside the circle. You find its measure by calculating the average of the two arcs it intersects:

$$m\angle BRL = \frac{m\widehat{BL} + m\widehat{EC}}{2} = \frac{106° + 60°}{2} = \frac{166°}{2} = 83°$$

---

895. **83°**

The diameter of a circle divides a circle into two arcs, each measuring 180°, so

$$m\widehat{BL} + m\widehat{EL} = 180°$$
$$m\widehat{BL} + 74° = 180°$$
$$m\widehat{BL} = 106°$$

$\angle BRL$ is a vertical angle inside the circle. You find its measure by calculating the average of the two arcs it intersects:

$$m\angle BRL = \frac{m\widehat{BL} + m\widehat{EC}}{2} = \frac{106° + 60°}{2} = \frac{166°}{2} = 83°$$

$\overline{IC}$ is parallel to diameter $\overline{BE}$ and is cut by transversal $\overline{LG}$. When two parallel lines are cut by a transversal, corresponding angles are formed. Corresponding angles are congruent. $\angle BRL$ and $\angle ICL$ are corresponding angles, so

$$m\angle BRL = m\angle ICL$$
$$83° = m\angle ICL$$

---

896. **53°**

The diameter of a circle divides a circle into two arcs, each measuring 180°, so

$$m\widehat{BL} + m\widehat{EL} = 180°$$
$$m\widehat{BL} + 74° = 180°$$
$$m\widehat{BL} = 106°$$

$\overline{IC}$ is parallel to diameter $\overline{BE}$, which means that $\widehat{BI} \cong \widehat{CE}$. You're given that $\widehat{IC} \cong \widehat{CE}$. This means that $\widehat{BI} \cong \widehat{CE} \cong \widehat{IC}$, so

$$m\widehat{BI} + m\widehat{CE} + m\widehat{IC} = 180°$$
$$x + x + x = 180°$$
$$3x = 180°$$
$$x = 60°$$

$\angle G$ is an exterior angle. Its measure is half the difference between the major and minor arcs that it intersects:

$$m\angle G = \frac{\widehat{LI} - \widehat{IC}}{2} = \frac{(106° + 60°) - 60°}{2} = \frac{106°}{2} = 53°$$

897. 30°

∠*ICL* and ∠*ICG* form a linear pair, so they add up to 180°:

$$m\angle ICL + m\angle ICG = 180°$$
$$83° + m\angle ICG = 180°$$
$$m\angle ICG = 97°$$

The sum of the angles of a triangle is 180°. In △*ICG*, you know $m\angle ICG = 97°$ and $m\angle G = 53°$, which means $m\angle GIC = 30°$.

898. **Radii in a circle are congruent.**

A radius is the distance from the center of a circle to any point on the circle. All radii in a circle are congruent.

899. $\overline{PR} \cong \overline{PF}$

$\overline{PR}$ and $\overline{PF}$ are two tangents drawn from the same exterior point. Because tangents drawn from the same exterior point are congruent, $\overline{PR} \cong \overline{PF}$.

900. $\overline{PO} \cong \overline{PO}$

The reflexive property states that a segment is congruent to itself.

901. SSS

In △*POF* and △*POR*, you've determined that

$$\overline{OF} \cong \overline{OR}$$
$$\overline{PF} \cong \overline{PR}$$
$$\overline{PO} \cong \overline{PO}$$

Because all three sides of the triangles have proven congruent, you'd use SSS (side-side-side) to prove the two triangles congruent.

902. **Right**

The radius of a circle is perpendicular to a tangent line at the point of tangency. Perpendicular lines form right angles. $\overline{PR}$ and $\overline{PF}$ are tangents, and $\overline{OF}$ and $\overline{OR}$ are radii. This means that ∠*PRO* and ∠*PFO* are right angles.

903. **Diameters in a circle are congruent.**

$\overline{AC}$ and $\overline{DB}$ are diameters of Circle *O*. Diameters of a circle are congruent to each other, which means $\overline{AC} \cong \overline{DB}$.

904. **Reflexive property**

The reflexive property states that a segment is congruent to itself.

905. **Inscribed angles that intercept half the circle form right angles.**

Diameters $\overline{AC}$ and $\overline{BD}$ both divide the circle into two arcs measuring 180°. $\angle ABC$ and $\angle DCB$ are inscribed angles. Inscribed angles are half the measure of the arcs they intersect, so both $\angle ABC$ and $\angle DCB$ measure $\frac{1}{2}(180°) = 90°$.

906. **Inscribed angles that intercept the same arc are congruent.**

Inscribed angles are half the measure of the arcs they intersect. If both inscribed angles intercept the same arc, they're both equal to half the measure of that arc, making them congruent to each other.

907. **True**

In a right rectangular prism, the sides of the prism are perpendicular to the base. $\overline{QR}$ is parallel to $\overline{ST}$, and $\overline{ST}$ is parallel to $\overline{ZY}$; therefore, $\overline{QR}$ must be parallel to $\overline{ZY}$.

908. **False**

In order for two edges to be perpendicular, they have to form right angles at their point of intersection. $\overline{WX}$ and $\overline{ZT}$ will never intersect; therefore, they cannot be perpendicular. Lines that don't intersect and aren't parallel are called *skew lines*.

909. **False**

In a right rectangular prism, the sides of the prism are perpendicular to the base. $\overline{RS}$ and $\overline{YS}$ are perpendicular to each other, not parallel.

910. **False**

In a right rectangular prism, the sides of the prism are perpendicular to the base. Plane *XRSY* is perpendicular, not parallel, to plane *STZY*.

911. **True**

In a right rectangular prism, the sides of the prism are perpendicular to the base. This means that plane *QRXW* is perpendicular to plane *QRST*.

912. **True**

In a right rectangular prism, the sides of the prism are perpendicular to the base. This means that edge $\overline{XR}$ is perpendicular to plane *QRST*.

913. **False**

In a right rectangular prism, the sides of the prism are perpendicular to the base. There are four planes perpendicular to plane *QRXW*. They are planes *WXYZ*, *QRST*, *QWZT*, and *RXYS*.

914. **True**

Skew lines are nonparallel lines that do not intersect in three-dimensional space. $\overline{QW}$ and *ZY* are not parallel and will never intersect; therefore, they're skew lines.

915. **True**

Segments that are coplanar lie on the same plane. Although the figure doesn't show a plane that contains both $\overline{QW}$ and $\overline{SY}$, such a plane exists. Imagine the diagonal plane *QWYS*. Now you can see that $\overline{QW}$ and $\overline{SY}$ lie along the same plane and are therefore coplanar.

916. **True**

If a point is not on a given plane, only one line can be drawn through the point that will be perpendicular to the given plane:

917. **False**

If two planes are perpendicular to a line, the two planes must be parallel to each other:

918. **True**

If a line is perpendicular to a plane, then every plane containing the line is perpendicular to the given plane.

919. **False**

The line could be either parallel or perpendicular to the other plane.

920. True

Two points that lie along the same plane are *coplanar*. This also means that the line joining the two points also lies in the same plane.

921. True

Skew lines are lines that are not parallel but do not intersect in three-dimensional space.

922. False

There's only one plane that is perpendicular to line n and passes through Point P.

923. 726 in.²

A cube is a three-dimensional shape with six equal square faces. You find the surface area of a cube by finding the sum of the areas of each square face. In this problem, the edge of the cube is 11. Because the area of a square is $A = s^2$, the area of one face of the cube is

$$A = 11^2 = 121$$

There are six square faces, and they're all equal; therefore, the surface area of the cube is $6(121) = 726$ in.²

924. 128 units²

The figure shows a rectangular solid that has four faces that are 4 units by 6 units. It also has two faces that are 4 units by 4 units. To calculate the surface area, find the sum of the areas of each face. Each face that's 4 units by 6 units has an area of $4(6) = 24$. There are four of those faces, which gives you a total area of $4(24) = 96$.

Each face that's 4 units by 4 units has an area of $4(4) = 16$. There are two of those faces, for a total area of $2(16) = 32$.

The total surface area of the rectangular prism is $96 + 32 = 128$ units².

925.   **40 ft²**

The figure shows a rectangular prism that has four faces that are 4 feet by 2 feet. It also has two faces that are 2 feet by 2 feet. To calculate the surface area, find the sum of the areas of each face. Each face that is 4 feet by 2 feet has an area of $4(2)=8$. There are four of those faces, which gives you a total area of $4(8)=32$.

Each face that is 2 feet by 2 feet has an area of $2(2)=4$. There are two of those faces, for a total area of $2(4)=8$.

The total surface area of the rectangular prism is $32+8=40$ feet^2.

926.   **126 cm²**

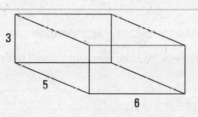

There are two faces that are 6 centimeters by 5 centimeters, there are two faces that are 3 centimeters by 5 centimeters, and there are two faces that are 6 centimeters by 3 centimeters. To calculate the surface area, find the sum of the areas of each face. Each face that is 6 centimeters by 5 centimeters has an area of $6(5)=30$. There are two of those faces, for a total area of $2(30)=60$.

Each face that is 3 centimeters by 5 centimeters has an area of $3(5)=15$. There are two of those faces, for a total area of $2(15)=30$.

Each face that is 6 centimeters by 3 centimeters has an area of $6(3)=18$. There are two of those faces, for a total area of $2(18)=36$.

The total surface area of the rectangular prism is $60+30+36=126$ centimeters^2.

927.   **24 ft**

A rectangular prism is a three-dimensional object containing six faces, each of which is a rectangle. Let x equal the length of the prism. There are two faces that are 20 feet by 15 feet, two faces that are 20 feet by x feet, and two faces that are 15 feet by x feet. Each face that is 20 feet by 15 feet has an area of $20(15)=300$. There are two of those faces, for a total area of $2(300)=600$.

Each face that is 20 feet by x feet has an area of $20(x)=20x$. There are two of those faces, for a total area of $2(20x)=40x$.

Each face that is 15 feet by x feet has an area of $15(x)=15x$. There are two of those faces, for a total area of $2(15x)=30x$.

The total surface area of the rectangular solid was given to be 2,280. Therefore,

$$600+30x+40x=2,280$$
$$600+70x=2,280$$
$$70x=1,680$$
$$x=24 \text{ ft}$$

928. 1,256.64 in.²

The lateral surface area of a right circular cylinder is $L = 2\pi rh$, where r is the radius of the circular base and h is the height of the cylinder. In this problem, the radius is 20 inches and the height is 10 inches. Plugging that information into the formula gives you

$$L = 2\pi(20)(10) \approx 1,256.64 \text{ in.}^2$$

929. 1,578.7 in.²

The formula $2\pi r^2 + 2\pi rh$, where h is the height of the cylinder and r is the radius of the circular base, allows you to quickly calculate the surface area of any cylinder. The height of the cylinder is 26, and the radius of the cylinder is half of the diameter, or $\frac{15}{2} = 7.5$. The surface area of the cylinder is therefore

$$SA = 2(\text{area of base}) + (\text{circumference of base})(\text{height})$$
$$= 2\pi r^2 + 2\pi rh$$
$$= 2\pi(7.5^2) + 2\pi(7.5)(26)$$
$$\approx 1,578.7 \text{ in.}^2$$

930. 448π mm²

The formula $2\pi r^2 + 2\pi rh$, where h is the height of the cylinder and r is the radius of the circular base, allows you to quickly calculate the surface area of any cylinder. The height of this cylinder is 20, and the radius of the cylinder is 8. Therefore, the surface area of the cylinder is

$$SA = 2(\text{area of base}) + (\text{circumference of base})(\text{height})$$
$$= 2\pi r^2 + 2\pi rh$$
$$= 2\pi(8^2) + 2\pi(8)(20)$$
$$= 320\pi + 128\pi$$
$$= 448\pi \text{ mm}^2$$

931. 201.06 units²

The lateral surface area of a right circular cylinder is $L = 2\pi rh$, where r is the radius of the circular base and h is the height of the cylinder. In this problem, the radius is 4 units and the height is 8 units. Plugging that information into the formula and rounding to the nearest hundredth gives you

$$L = 2\pi rh = 2\pi(4)(8) = 64\pi \approx 201.06 \text{ units}^2$$

932. **1,017.9 cm²**

You find the surface area of a cone by using the formula $SA = \pi r^2 + \pi rl$, where r is the radius of the circular base and l is the slant height of the cone. In this problem, the radius of the circular base is 20 and the slant height is 22.4. Plug this information into the formula and round your answer to the nearest tenth:

$$SA = \pi r^2 + \pi rl$$
$$= \pi(10^2) + \pi(10)(22.4)$$
$$= 100\pi + 224\pi$$
$$= 324\pi$$
$$\approx 1{,}017.9 \text{ cm}^2$$

933. **35π m²**

You find the lateral area of a right circular cone by using the formula $L = \pi rl$, where r is the radius of the circular base and l is the slant height. In this problem, the radius of the base is 5 and the slant height is 7. Plugging into the formula, you get

$$L = \pi rl = \pi(5)(7) = 35\pi \text{ m}^2$$

934. **60π in.²**

You find the lateral area of a right circular cone by using the formula $L = \pi rl$, where r is the radius of the circular base and l is the slant height. In this problem, the diameter of the base is 12. Because a radius is half the length of the diameter, the radius of the base is $\frac{12}{2} = 6$.

The altitude is 8. The following figure shows how the altitude and the radius form a right triangle, allowing you to use the Pythagorean theorem to solve for the slant height.

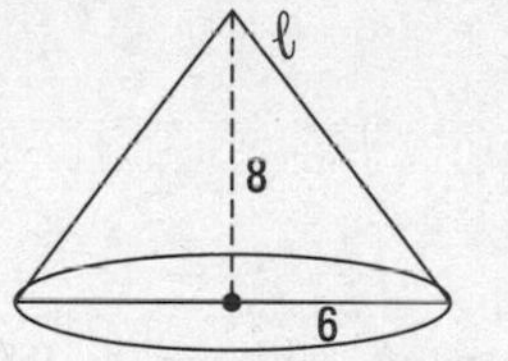

$$a^2 + b^2 = c^2$$
$$6^2 + 8^2 = c^2$$
$$100 = c^2$$
$$\sqrt{100} = \sqrt{c^2}$$
$$10 = c$$

You now know that the slant height is 10 and can plug the radius and slant height into the formula to get the lateral area of the cone:

$$L = \pi r l = \pi(6)(10) = 60\pi \text{ in.}^2$$

935. **90π ft^2**

You find the lateral area of a right circular cone by using the formula $L = \pi r l$, where r is the radius of the circular base and l is the slant height. In this problem, the radius of the base is 20. The following figure shows how the altitude and the radius form a right triangle, allowing you to use the Pythagorean theorem to solve for the slant height.

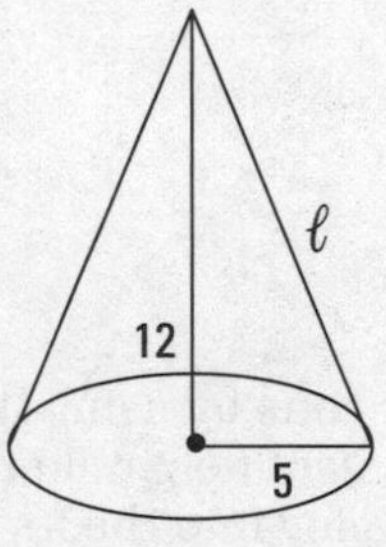

$$a^2 + b^2 = c^2$$
$$5^2 + 12^2 = c^2$$
$$169 = c^2$$
$$\sqrt{169} = \sqrt{c^2}$$
$$13 = c$$

You now know that the slant height is 13 and can plug the radius and slant height into the formula to get the surface area of the cone:

$$SA = \pi r^2 + \pi r l$$
$$= \pi(5^2) + \pi(5)(13)$$
$$= 25\pi + 65\pi$$
$$= 90\pi \text{ ft}^2$$

936. **50.27 ft^2**

You find the surface area of a sphere using the formula $SA = 4\pi r^2$, where r is the radius of the sphere. In this problem, you're given that the radius is 2. When you plug that into the formula and round your answer to the nearest hundredth, you get

$$SA = 4\pi r^2 = 4\pi(2^2) = 16\pi \approx 50.27 \text{ ft}^2$$

937. **8 units**

You find the surface area of a sphere using the formula $SA = 4\pi r^2$, where r is the radius of the sphere. In this problem, you're given that the surface area is 256π. When you plug that into the formula, you get

$$SA = 4\pi r^2$$
$$256\pi = 4\pi r^2$$
$$\frac{256\pi}{4\pi} = \frac{4\pi r^2}{4\pi}$$
$$64 = r^2$$
$$\sqrt{64} = \sqrt{r^2}$$
$$8 = r$$

938. **$4,000,000\pi \text{ km}^2$**

You find the surface area of a sphere using the formula $SA = 4\pi r^2$, where r is the radius of the sphere. In this problem, you're given that the radius is 1,000. When you plug that into the formula, you get

$$SA = 4\pi r^2 = 4\pi\left(1,000^2\right) = 4,000,000\pi \text{ km}^2$$

939. **91.125 ft³**

You find the volume of a cube using the formula $V = s^3$, where s is the length of the edge of the cube. In this problem, an edge of the cube is 4.5:

$$V = s^3 = 4.5^3 = 91.125 \text{ ft}^3$$

940. **71.4 m³**

You find the volume of a rectangular prism using the formula $V = lwh$, where l, w, and h represent the length, width, and height of the prism. In this problem, the length, width, and height are 8.5, 3.5, and 2.4:

$$V = lwh = \left(8.5\right)\left(3.5\right)\left(2.4\right) = 71.4 \text{ m}^3$$

941. **240 units³**

You find the volume of a rectangular prism using the formula $V = lwh$, where l, w, and h represent the length, width, and height of the prism. In this figure, the length, width, and height of the prism are 5, 12, and 4:

$$V = lwh = \left(5\right)\left(12\right)\left(4\right) = 240 \text{ units}^3$$

942. **32 units³**

You find the volume of a triangular prism by multiplying the area of the triangular base by the height of the prism. To find the area of the triangular base, use the formula $A = \frac{1}{2}bh$, where b represents the base of the triangle and h represents the height of the triangle. In this figure, the base of the triangle is 8 and the height of the triangle is 2:

$$A = \frac{1}{2}(8)(2) = 8 \text{ units}^2$$

To find the volume of the prism, you now have to multiply the area of the triangle by the height of the prism, which equals 4:

$$V = 8(4) = 32 \text{ units}^3$$

943. **192π in.³**

You find the volume of a cylinder using the formula $V = \pi r^2 h$, where r represents the radius of the circular base and h represents the height of the cylinder. You're given that the radius is 4 and the height is 12, so

$$V = \pi r^2 h = \pi\left(4^2\right)(12) = 192\pi \text{ in.}^3$$

944. **1,800π in.³**

You find the volume of a cylinder using the formula $V = \pi r^2 h$, where r represents the radius of the circular base and h represents the height of the cylinder. You're given that the diameter of the circular base is 20. Because a radius is half the length of a diameter, the radius must be $\frac{20}{2} = 10$.

You're also given that the height is 1.5 feet. The answer has to be in terms of cubic inches, so convert the feet to inches. There are 12 inches in a foot, so the height of the cylinder is $1.5(12) = 18$ inches.

When you plug that information into the volume formula, you get

$$V = \pi r^2 h = \pi\left(10^2\right)(18) = 1,800\pi \text{ in.}^3$$

945. **20,499 ft³**

You find the volume of a cylinder using the formula $V = \pi r^2 h$, where r represents the radius of the circular base and h represents the height of the cylinder. In this figure, you're given that the radius is 15 and the height is 29, so

$$V = \pi r^2 h = \pi\left(15^2\right)(29) = 6,525\pi \approx 20,499 \text{ ft}^3$$

946. **8 ft**

You find the volume of a cylinder using the formula $V = \pi r^2 h$, where r represents the radius of the circular base and h represents the height of the cylinder. In this question, you're given that the volume is 392π and that the radius is 7. When you plug that information into the volume formula, you get the following:

$$V = \pi r^2 h$$
$$392\pi = \pi \left(7^2\right)(h)$$
$$392\pi = 49\pi h$$
$$\frac{392\pi}{49\pi} = \frac{49\pi h}{49\pi}$$
$$8 \text{ ft} = h$$

947. **320π km³**

The formula for the volume of a right circular cone is $V = \frac{1}{3}\pi r^2 h$, where r represents the radius of the circular base and h represents the height of the cone. In this problem, you're given that the radius is 8 and that the height is 15. When you plug that information into the formula, you get

$$V = \frac{1}{3}\pi r^2 h = \frac{1}{3}\pi \left(8^2\right)(15) = 320\pi \text{ km}^3$$

948. **15 in.**

The formula for the volume of a right circular cone is $V = \frac{1}{3}\pi r^2 h$, where r represents the radius of the circular base and h represents the height of the cone. In this problem, you're given that the volume is $2,700\pi$ and that the height is 36. When you plug that information into the formula, you get the following:

$$V = \frac{1}{3}\pi r^2 h$$
$$2,700\pi = \frac{1}{3}\pi \left(r^2\right)(36)$$
$$2,700\pi = 12\pi r^2$$
$$\frac{2,700\pi}{12\pi} = \frac{12\pi r^2}{12\pi}$$
$$225 = r^2$$
$$\sqrt{225} = \sqrt{r^2}$$
$$15 \text{ in.} = r$$

949.   **400 ft³**

The formula for the volume of a square pyramid is $V = \frac{1}{3}s^2h$, where s represents a side of the square base and h represents the height of the pyramid. In this figure, you're given that the side of the square base is 10. You're also given that the slant height is 13. You can use the slant height to find the actual height of the pyramid. If you draw the altitude of the pyramid, it touches the base of the square directly in the middle and forms a right triangle with the slant height as the hypotenuse. You already know the hypotenuse is 13, and because the side of the square is 10, you know that a leg of the right triangle is 5. Use the Pythagorean theorem to solve for the height of the prism:

$$a^2 + b^2 = c^2$$
$$5^2 + x^2 = 13^2$$
$$25 + x^2 = 169$$
$$x^2 = 144$$
$$\sqrt{x^2} = \sqrt{144}$$
$$x = 12$$

You can now plug your information into the formula for the volume of a pyramid:

$$V = \frac{1}{3}\left(10^2\right)(12) = 400 \text{ ft}^3$$

950.   **18,432π in.³**

You calculate the volume of a sphere using the formula $V = \frac{4}{3}\pi r^3$, where r represents the radius of the sphere. In this problem, you're given that the radius of the sphere is 24. When you plug that into the formula, you get

$$V = \frac{4}{3}\pi r^3 = \frac{4}{3}\pi\left(24^3\right) = 18,432\pi \text{ in.}^3$$

951.   **381.7 ft³**

You calculate the volume of a sphere using the formula $V = \frac{4}{3}\pi r^3$, where r represents the radius of the sphere. In this problem, you're given that the diameter of the sphere is 9. Because a radius is half the length of the diameter, the radius of the sphere is $\frac{9}{2} = 4.5$. When you plug that into the formula and round your answer to the nearest tenth, you get

$$V = \frac{4}{3}\pi r^3 = \frac{4}{3}\pi\left(4.5^3\right) = 121.5\pi = 381.7 \text{ ft}^3$$

952. Cone

When a triangle is rotated about an axis of rotation that bisects the triangle, a cone is formed.

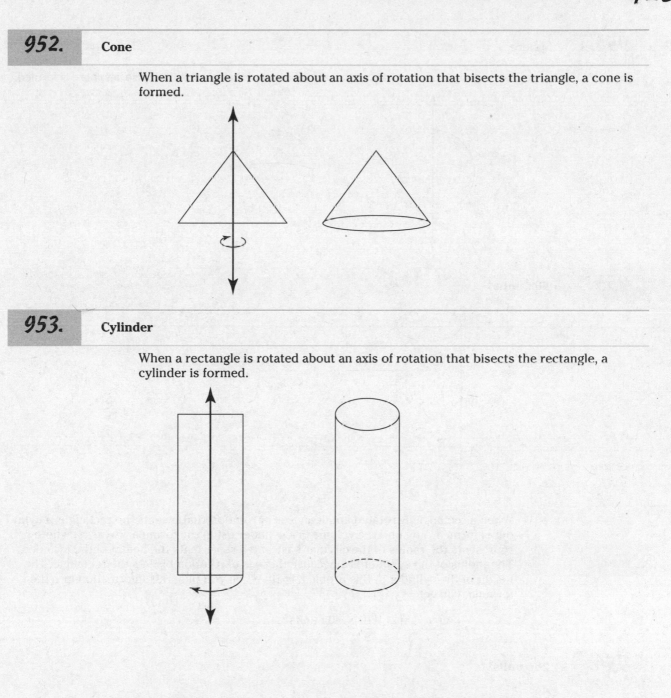

953. Cylinder

When a rectangle is rotated about an axis of rotation that bisects the rectangle, a cylinder is formed.

954. **Sphere**

When a circle is rotated about an axis of rotation that bisects the circle, a sphere is formed.

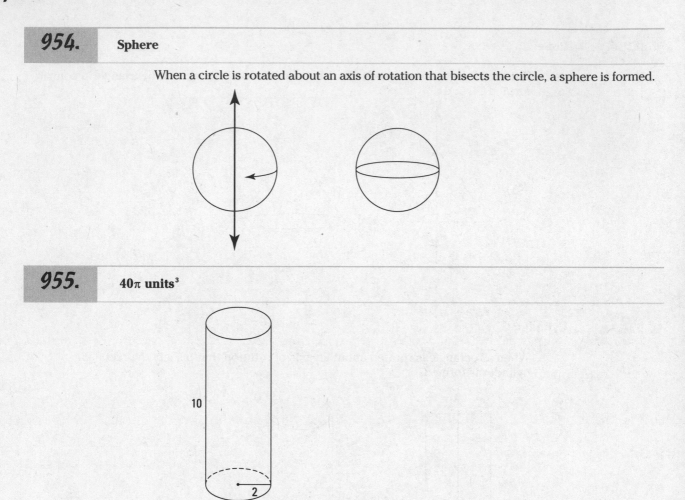

955. **40π units^3**

When a rectangle is rotated about an axis of rotation that bisects the rectangle, a cylinder is formed. You find the volume of a cylinder using the formula $V = \pi r^2 h$, where r represents the radius of the circular base and h represents the height of the cylinder. The radius of the cylinder is 2 because the axis of rotation bisects the rectangle. The height of the cylinder in this problem is 10. When you plug this information into the formula, you get

$$V = \pi r^2 h = \pi \left(2^2\right)(10) = 40\pi \text{ units}^3$$

956. **36π units^3**

When a circle is rotated about an axis of rotation that bisects the circle, a sphere is formed. You calculate the volume of a sphere using the formula $V = \frac{4}{3}\pi r^3$, where r represents the radius of the sphere. In this problem, you're given that the radius of the circle is 3. This means that the radius of the sphere is also 3. When you plug that into the volume formula, you get

$$V = \frac{4}{3}\pi r^3 = \frac{4}{3}\pi(3)^3 = 36\pi \text{ units}^3$$

957.   Circle with center at *P*

The locus of points equidistant from a given point is a circle whose center is the given point.

958.   The perpendicular bisector of $\overline{AB}$

The locus of points equidistant from two given points is the perpendicular bisector of the segment whose endpoints are the given points.

959.   Two parallel lines

The locus of points equidistant from a given line is two lines parallel to the given line.

960.   One line parallel to the given lines

The locus of points equidistant from two given parallel lines is one line parallel to the given lines.

961.   Parabola

The locus of points equidistant from a line and a point not on the line is a parabola. The given line is called the *directrix,* and the given point is called the *focus.*

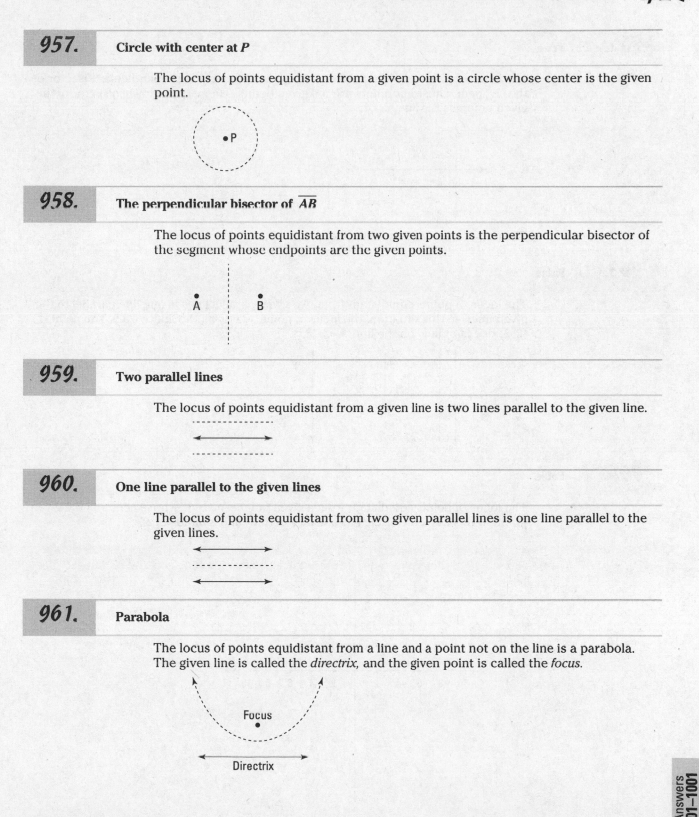

962. **True**

The locus of points equidistant from two given points is the perpendicular bisector of the segment whose endpoints are the given points. The perpendicular bisector of the given points is the line $x = 4$:

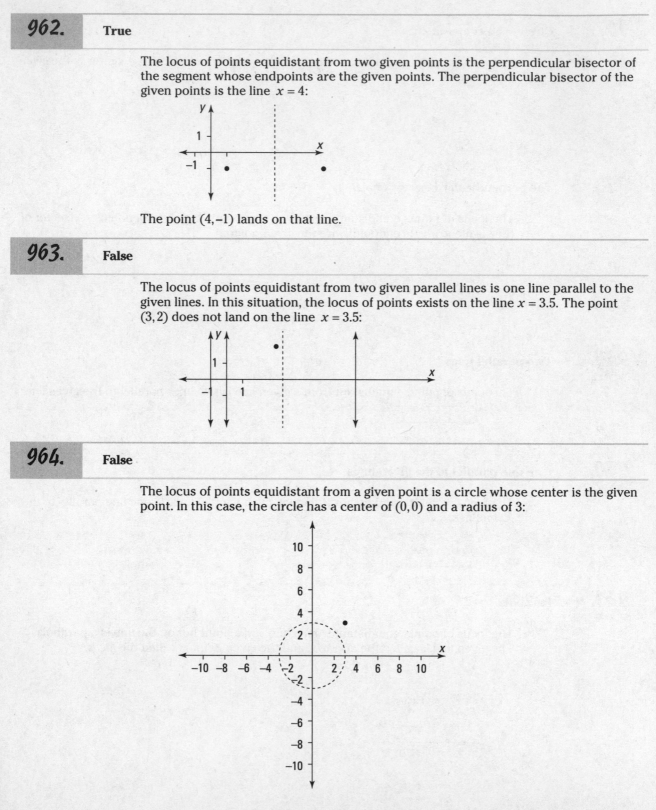

The point $(4, -1)$ lands on that line.

963. **False**

The locus of points equidistant from two given parallel lines is one line parallel to the given lines. In this situation, the locus of points exists on the line $x = 3.5$. The point $(3, 2)$ does not land on the line $x = 3.5$:

964. **False**

The locus of points equidistant from a given point is a circle whose center is the given point. In this case, the circle has a center of $(0, 0)$ and a radius of 3:

The equation of this circle is $x^2 + y^2 = 9$. To determine whether $(3,3)$ lands on the locus, see whether the equation of the circle proves true when you plug 3 in for x and plug 3 in for y:

$$x^2 + y^2 \overset{?}{=} 9$$
$$3^2 + 3^2 \overset{?}{=} 9$$
$$18 \neq 9$$

Because $(3,3)$ doesn't satisfy the equation of the circle, it doesn't land on the locus of points.

---

965.   **True**

The locus of points equidistant from a given point is a circle. In this case, the circle's center is $(2,1)$, and its radius is 5:

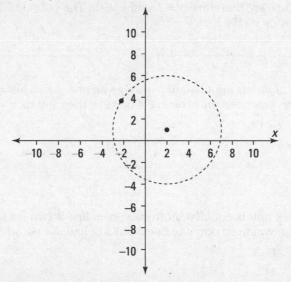

The equation of this circle is $(x-2)^2 + (y-1)^2 = 25$. To determine whether $(-2,4)$ lands on the locus, see whether the equation of a circle proves true when you plug -2 in for x and plug 4 in for y:

$$(x-2)^2 + (y-1)^2 \overset{?}{=} 25$$
$$(-2-2)^2 + (4-1)^2 \overset{?}{=} 25$$
$$25 = 25$$

Because $(-2,4)$ satisfies the equation of the circle, it does land on the locus of points.

966. **True**

The locus of points equidistant from a given line is two lines parallel to the given line. In this case, the two lines are 4 units above and 4 units below the given line:

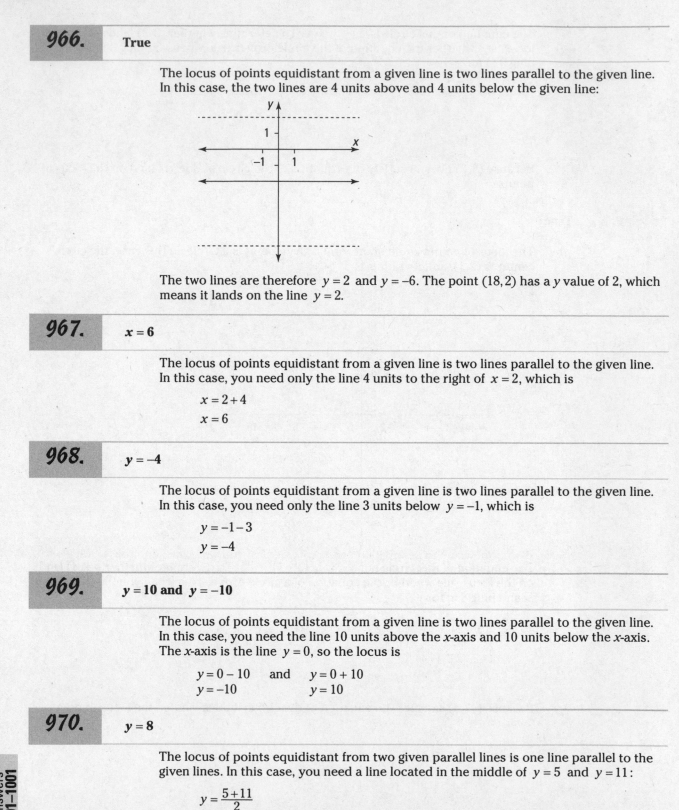

The two lines are therefore $y = 2$ and $y = -6$. The point $(18, 2)$ has a y value of 2, which means it lands on the line $y = 2$.

967. $x = 6$

The locus of points equidistant from a given line is two lines parallel to the given line. In this case, you need only the line 4 units to the right of $x = 2$, which is

$$x = 2 + 4$$
$$x = 6$$

968. $y = -4$

The locus of points equidistant from a given line is two lines parallel to the given line. In this case, you need only the line 3 units below $y = -1$, which is

$$y = -1 - 3$$
$$y = -4$$

969. $y = 10$ and $y = -10$

The locus of points equidistant from a given line is two lines parallel to the given line. In this case, you need the line 10 units above the x-axis and 10 units below the x-axis. The x-axis is the line $y = 0$, so the locus is

$$y = 0 - 10 \quad \text{and} \quad y = 0 + 10$$
$$y = -10 \quad\quad\quad\quad y = 10$$

970. $y = 8$

The locus of points equidistant from two given parallel lines is one line parallel to the given lines. In this case, you need a line located in the middle of $y = 5$ and $y = 11$:

$$y = \frac{5 + 11}{2}$$
$$y = 8$$

971. $x = 4$

The locus of points equidistant from two given parallel lines is one line parallel to the given lines. In this case, you need a line located in the middle of $x = -6$ and $x = 14$:

$$x = \frac{-6 + 14}{2}$$
$$x = 4$$

972. $y = x$ and $y = -x$

The locus of points equidistant from two intersecting lines is a pair of perpendicular lines that bisect both pairs of vertical angles formed by the intersecting lines:

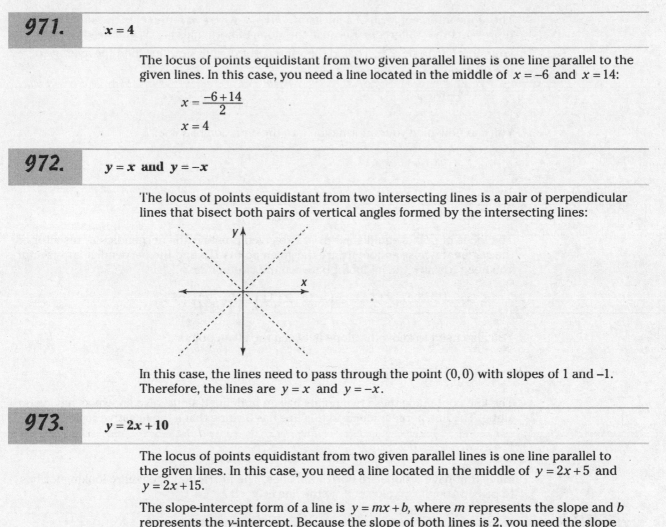

In this case, the lines need to pass through the point $(0, 0)$ with slopes of 1 and -1. Therefore, the lines are $y = x$ and $y = -x$.

973. $y = 2x + 10$

The locus of points equidistant from two given parallel lines is one line parallel to the given lines. In this case, you need a line located in the middle of $y = 2x + 5$ and $y = 2x + 15$.

The slope-intercept form of a line is $y = mx + b$, where m represents the slope and b represents the y-intercept. Because the slope of both lines is 2, you need the slope of your new line to be 2. To find the y-intercept of your new line, find the average of the y-intercepts of the given lines:

$$\frac{5 + 15}{2} = 10$$

You can now plug your information into the equation of a line:

$$y = mx + b = 2x + 10$$

974. $y = -\frac{1}{2}x + 1$

The locus of points equidistant from two given parallel lines is one line parallel to the given lines. In this case, you need a line located in the middle of $y = -\frac{1}{2}x - 4$ and $y = -\frac{1}{2}x + 6$.

The slope-intercept form of a line is $y = mx + b$, where m represents the slope and b represents the y-intercept. Because the slope of both lines is $-\frac{1}{2}$, you need the slope of your new line to be $-\frac{1}{2}$. To find the y-intercept of your new line, find the average of the y-intercepts of the given lines:

$$\frac{-4+6}{2} = 1$$

You can now plug your information into the equation of a line:

$$y = mx + b = -\frac{1}{2}x + 1$$

975. $y = 8$

The locus of points equidistant from two given points is the perpendicular bisector of the segment whose endpoints are the given points. To find the perpendicular bisector, you need to know the midpoint between the two points:

$$M = \left(\frac{x_1 + x_2}{2}, \frac{y_1 + y_2}{2} \right) = \left(\frac{3+3}{2}, \frac{5+11}{2} \right) = \left(\frac{6}{2}, \frac{16}{2} \right) = (3, 8)$$

You also need to know the slope between the given points:

$$m = \frac{y_2 - y_1}{x_2 - x_1} = \frac{11-5}{3-3} = \frac{6}{0}$$

The line containing these two points has an undefined slope, also known as having no slope. The line perpendicular to this line has a slope that's the negative reciprocal:

$$m = -\frac{0}{6} = 0$$

Lines that have 0 slope are horizontal lines. The horizontal line you're looking for has to pass through a y value of 8, so the line is $y = 8$.

976. $y = 14$

The locus of points equidistant from two given points is the perpendicular bisector of the segment whose endpoints are the given points. To find the perpendicular bisector, you need to know the midpoint between the two points:

$$M = \left(\frac{x_1 + x_2}{2}, \frac{y_1 + y_2}{2} \right) = \left(\frac{10+10}{2}, \frac{20+8}{2} \right) = \left(\frac{20}{2}, \frac{28}{2} \right) = (10, 14)$$

You also need to know the slope between the given points:

$$m = \frac{y_2 - y_1}{x_2 - x_1} = \frac{8-20}{10-10} = \frac{-12}{0}$$

The line containing these two points has an undefined slope, also known as having no slope. The line perpendicular to this has a slope that's the negative reciprocal:

$$m = \frac{0}{12} = 0$$

Lines that have 0 slope are horizontal lines. The horizontal line you're looking for has to pass through a y value of 14, so the line is $y = 14$.

977. $x = -4$

The locus of points equidistant from two given points is the perpendicular bisector of the segment whose endpoints are the given points. To find the perpendicular bisector, you need to know the midpoint between the two points:

$$M = \left(\frac{x_1 + x_2}{2}, \frac{y_1 + y_2}{2} \right) = \left(\frac{6 + (-14)}{2}, \frac{-2 + (-2)}{2} \right) = \left(\frac{-8}{2}, \frac{-4}{2} \right) = (-4, -2)$$

You also need to know the slope between the given points:

$$m = \frac{y_2 - y_1}{x_2 - x_1} = \frac{-2 - (-2)}{-14 - 6} = \frac{0}{-20}$$

The line containing these two points has 0 slope. The line perpendicular to this has a slope that's the negative reciprocal:

$$m = \frac{20}{0}$$

This means that your line needs to have undefined slope. Lines that have undefined slope are vertical lines. The vertical line you're looking for has to pass through an x value of –4, so the line is $x = -4$.

978. $x = -3.5$

The locus of points equidistant from two given points is the perpendicular bisector of the segment whose endpoints are the given points. To find the perpendicular bisector, you need to know the midpoint between the two points:

$$M = \left(\frac{x_1 + x_2}{2}, \frac{y_1 + y_2}{2} \right) = \left(\frac{-8 + 1}{2}, \frac{1 + 1}{2} \right) = \left(\frac{-7}{2}, \frac{2}{2} \right) = (-3.5, 1)$$

You also need to know the slope between the given points:

$$m = \frac{y_2 - y_1}{x_2 - x_1} = \frac{1 - 1}{1 - (-8)} = \frac{0}{9}$$

The line containing these two points has 0 slope. The line perpendicular to this has a slope that's the negative reciprocal:

$$m = -\frac{9}{0}$$

This means that your line needs to have undefined slope. Lines that have undefined slope are vertical lines. The vertical line you're looking for has to pass through an x value of –3.5, so the line is $x = -3.5$.

979. $y = -\frac{1}{2}x + 10$

The locus of points equidistant from two given points is the perpendicular bisector of the segment whose endpoints are the given points. To find the perpendicular bisector, you need to know the midpoint between the two points:

$$M = \left(\frac{x_1 + x_2}{2}, \frac{y_1 + y_2}{2} \right) = \left(\frac{2+6}{2}, \frac{4+12}{2} \right) = \left(\frac{8}{2}, \frac{16}{2} \right) = (4, 8)$$

You also need to know the slope between the given points:

$$m = \frac{y_2 - y_1}{x_2 - x_1} = \frac{12-4}{6-2} = \frac{8}{4} = 2$$

The line perpendicular to this has a slope that's the negative reciprocal:

$$m = -\frac{1}{2}$$

You can now plug your information into the slope-intercept form of a line to find the y-intercept:

$$y = mx + b$$
$$8 = \frac{-1}{2}(4) + b$$
$$8 = -2 + b$$
$$10 = b$$

Therefore, $y = -\frac{1}{2}x + 10$.

980. $y = 3x - 12$

The locus of points equidistant from two given points is the perpendicular bisector of the segment whose endpoints are the given points. To find the perpendicular bisector, you need to know the midpoint between the two given points:

$$M = \left(\frac{x_1 + x_2}{2}, \frac{y_1 + y_2}{2} \right) = \left(\frac{0+12}{2}, \frac{8+4}{2} \right) = \left(\frac{12}{2}, \frac{12}{2} \right) = (6, 6)$$

You also need to know the slope between the given points:

$$m = \frac{y_2 - y_1}{x_2 - x_1} = \frac{4-8}{12-0} = \frac{-4}{12} = -\frac{1}{3}$$

The line perpendicular to this has a slope that's the negative reciprocal:

$$m = 3$$

You can now plug your information into the slope-intercept form of a line to find the y-intercept:

$$y = mx + b$$
$$6 = 3(6) + b$$
$$6 = 18 + b$$
$$-12 = b$$

Therefore, $y = 3x - 12$.

---

981. $\quad y = \dfrac{3}{2}x + 18$

The locus of points equidistant from two given points is the perpendicular bisector of the segment whose endpoints are the given points. To find the perpendicular bisector, you need to know the midpoint between the two points:

$$M = \left(\frac{x_1 + x_2}{2}, \frac{y_1 + y_2}{2}\right) = \left(\frac{-2 + (-14)}{2}, \frac{2 + 10}{2}\right) = \left(\frac{-16}{2}, \frac{12}{2}\right) = (-8, 6)$$

You also need to know the slope between the given points:

$$m = \frac{y_2 - y_1}{x_2 - x_1} = \frac{10 - 2}{-14 - (-2)} = \frac{8}{-12} = -\frac{2}{3}$$

The line perpendicular to this has a slope that's the negative reciprocal:

$$m = \frac{3}{2}$$

You can now plug your information into the slope-intercept form of a line to find the y-intercept:

$$y = mx + b$$
$$6 = \frac{3}{2}(-8) + b$$
$$6 = -12 + b$$
$$18 = b$$

Therefore, $y = \dfrac{3}{2}x + 18$.

---

982. $\quad y = x + 4$

The origin is the point $(0, 0)$. The locus of points equidistant from two given points is the perpendicular bisector of the segment whose endpoints are the given points. To find the perpendicular bisector, you need to know the midpoint between the two points:

$$M = \left(\frac{x_1 + x_2}{2}, \frac{y_1 + y_2}{2}\right) = \left(\frac{0 + (-4)}{2}, \frac{0 + 4}{2}\right) = \left(\frac{-4}{2}, \frac{4}{2}\right) = (-2, 2)$$

You also need to know the slope between the given points:

$$m = \frac{y_2 - y_1}{x_2 - x_1} = \frac{4 - 0}{-4 - 0} = \frac{4}{-4} = -1$$

The line perpendicular to this has a slope that's the negative reciprocal:

$$m = 1$$

You can now plug your information into the slope-intercept form of a line to find the y-intercept:

$$y = mx + b$$
$$2 = 1(-2) + b$$
$$2 = -2 + b$$
$$4 = b$$

Therefore, $y = x + 4$.

983. $\quad x^2 + y^2 = 100$

The locus of points equidistant from a given point is a circle whose center is the given point and whose radius is the given distance. In this problem, the center is the origin, $(0,0)$, and the radius is 10. The equation of a circle is $(x - h)^2 + (y - k)^2 = r^2$, where (h, k) represents the center of the circle and r represents the radius of the circle. You can now plug in your information:

$$(x - 0)^2 + (y - 0)^2 = (10)^2$$
$$x^2 + y^2 = 100$$

984. $\quad (x + 2)^2 + (y - 8)^2 = 25$

The locus of points equidistant from a given point is a circle whose center is the given point and whose radius is the given distance. In this problem, the center is $(-2, 8)$ and the radius is 5. The equation of a circle is $(x - h)^2 + (y - k)^2 = r^2$, where (h, k) represents the center of the circle and r represents the radius of the circle. You can now plug in your information:

$$(x - (-2))^2 + (y - 8)^2 = 5^2$$
$$(x + 2)^2 + (y - 8)^2 = 25$$

985. $\quad (x + 3)^2 + (y - 15)^2 = 36$

The locus of points equidistant from a given point is a circle whose center is the given point and whose radius is the given distance. In this problem, the center is $(-3, 15)$ and the radius is 6. The equation of a circle is $(x - h)^2 + (y - k)^2 = r^2$, where (h, k) represents the center of the circle and r represents the radius of the circle. You can now plug in your information:

$$(x - (-3))^2 + (y - 15)^2 = 6^2$$
$$(x + 3)^2 + (y - 15)^2 = 36$$

986. $\quad (x+1)^2 + (y+2)^2 = 12.25$

The locus of points equidistant from a given point is a circle whose center is the given point and whose radius is the given distance. In this problem, the center is $(-1,-2)$ and the radius is 3.5. The equation of a circle is $(x-h)^2 + (y-k)^2 = r^2$, where (h,k) represents the center of the circle and r represents the radius of the circle. You can now plug in your information:

$$\left(x-(-1)\right)^2 + \left(y-(-2)\right)^2 = (3.5)^2$$
$$(x+1)^2 + (y+2)^2 = 12.25$$

987. $\quad x^2 + y^2 = 81$

The locus of points equidistant from two circles with the same center is a circle with the same center whose radius is the average of the original two radii. The equation of a circle is $(x-h)^2 + (y-k)^2 = r^2$, where (h,k) represents the center of the circle and r represents the radius of the circle. In this problem, $x^2 + y^2 = 49$ has the center $(0,0)$ and a radius of

$$r^2 = 49$$
$$\sqrt{r^2} = \sqrt{49}$$
$$r = 7$$

The circle represented by $x^2 + y^2 = 121$ has the center $(0,0)$ and a radius of

$$r^2 = 121$$
$$\sqrt{r^2} = \sqrt{121}$$
$$r = 11$$

The average of the radii is

$$\frac{7+11}{2} = 9$$

The equation of the circle is therefore

$$(x-0)^2 + (y-0)^2 = 9^2$$
$$x^2 + y^2 = 81$$

988. $\quad x^2 + y^2 = 132.25$

The locus of points equidistant from two circles with the same center is a circle with the same center whose radius is the average of the original two radii. The equation of a circle is $(x-h)^2 + (y-k)^2 = r^2$, where (h,k) represents the center of the circle and r represents the radius of the circle. In this problem, $x^2 + y^2 = 81$ has the center $(0,0)$ and a radius of

$$r^2 = 81$$
$$\sqrt{r^2} = \sqrt{81}$$
$$r = 9$$

The circle represented by $x^2 + y^2 = 196$ has the center $(0,0)$ and a radius of

$$r^2 = 196$$
$$\sqrt{r^2} = \sqrt{196}$$
$$r = 14$$

The average of the radii is

$$\frac{9+14}{2} = 11.5$$

The equation of the circle is therefore

$$(x-0)^2 + (y-0)^2 = 11.5^2$$
$$x^2 + y^2 = 132.25$$

989. **Four**

The locus of points equidistant from two intersecting lines is two perpendicular lines bisecting both pairs of vertical angles formed by the intersecting lines.

The locus of points equidistant from a given point is a circle whose center is the given point and whose radius is the given distance. In this problem, the center is $(0,0)$ and the radius is 5.

The points that satisfy both conditions are marked with an X in the following figure:

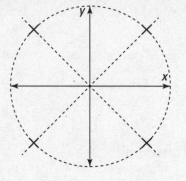

990. **Two**

The locus of points equidistant from a given line is two parallel lines. Because the given line is the y-axis and the given distance is 2 units, the two parallel lines are $x = 2$ and $x = -2$.

The locus of points equidistant from a given point is a circle whose center is the given point and whose radius is the given distance. In this problem, the center is $(0,0)$ and the radius is 2.

The points that satisfy both conditions are marked with an X in the following figure:

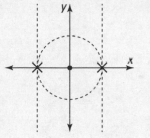

991. **One**

The locus of points equidistant from two given points is the perpendicular bisector of the segment whose endpoints are the given points. The average of the y values is $\frac{1+7}{2} = 4$, so $y = 4$ is the equation of the line.

The locus of points equidistant from a given point is a circle whose center is the given point and whose radius is the given distance. In this problem, the center is $(3, 1)$ and the radius is 3.

The point that satisfies both conditions is marked with an X in the following figure:

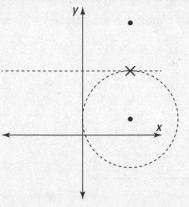

992. **Zero**

The locus of points equidistant from two parallel lines is one line parallel to the given lines. In this problem, your parallel line has to be in the middle of $x = -4$ and $x = 6$. This means your parallel line must be

$$x = \frac{-4+6}{2}$$
$$x = 1$$

The locus of points equidistant from a given point is a circle whose center is the given point and whose radius is the given value. In this problem, your center can be any point with an x value of 6, and your radius is 4.

The following figure shows that there are no points that satisfy these two conditions:

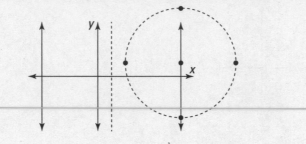

993.   Four

The locus of points equidistant from a given line is two lines parallel to the given line. In this problem, you need two lines parallel to $y = 3$ that are both 5 units above and below $y = 3$. Your equations would be

$$y = 3 + 5 \qquad \text{and} \qquad y = 3 - 5$$
$$y = 8 \qquad\qquad\qquad y = -2$$

The locus of points equidistant from two intersecting lines is two perpendicular lines that bisect the vertical angles created by the intersecting lines.

The points that satisfy both conditions are marked with an X in the following figure:

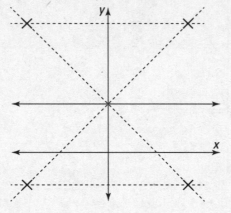

994.   One

The locus of points equidistant from two given points is the perpendicular bisector of the segment whose endpoints are the given points.

The locus of points equidistant from a given point is a circle whose center is the given point and whose radius is the given distance.

The point that satisfies both conditions is marked with an X in the following figure (*Note:* 1 box = 2 feet):

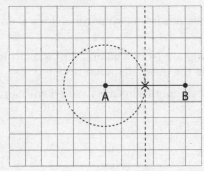

995. **One**

The locus of points equidistant from two parallel lines is one line parallel to the given lines.

The locus of points equidistant from two given points is the perpendicular bisector of the segment whose endpoints are the given points.

The points that satisfy both conditions are marked with an X in the following figure:

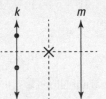

996. **Three**

The locus of points equidistant from a given line is two lines parallel to the given line.

The locus of points equidistant from a given point is a circle whose center is the given point and whose radius is the given distance.

The points that satisfy both conditions are marked with an X in the following figure (*Note:* 1 box = 2 feet):

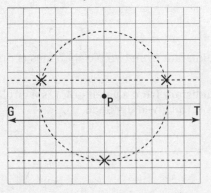

997. **Two**

The locus of points equidistant from two parallel lines is one line parallel to the given lines.

The locus of points equidistant from a given point is a circle whose center is the given point and whose radius is the given distance.

The points that satisfy both conditions are marked with an X in the following figure (*Note:* 1 box = 2 feet):

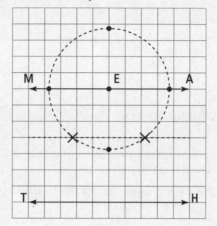

998. **One**

The locus of points equidistant from two parallel lines is one line parallel to the given lines.

The locus of points equidistant from two given points is the perpendicular bisector of the segment whose endpoints are the given points.

The point that satisfies both conditions is marked with an X in the following figure:

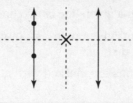

999. **Zero**

The locus of points equidistant from two parallel lines is one line parallel to the given lines.

The locus of points equidistant from a given point is a circle whose center is the given point and whose radius is the given distance.

The following figure shows that there are no points that satisfy these two conditions (*Note*: 1 box = 2 meters):

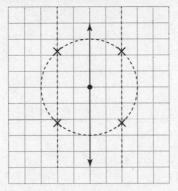

1,000.   Four

The locus of points equidistant from a given line is two lines parallel to the given line.

The locus of points equidistant from a given point is a circle whose center is the given point and whose radius is the given distance.

The points that satisfy both conditions are marked with an X in the following figure (*Note*: 1 box = 2 units):

1,001. $\dfrac{x^2}{25} + \dfrac{(y-1)^2}{9} = 1$

You first need to know the center of the ellipse. To locate the center, find the midpoint of the two fixed points *(foci)*:

$$M = \left(\frac{x_1 + x_2}{2}, \frac{y_1 + y_2}{2} \right) = \left(\frac{-4+4}{2}, \frac{1+1}{2} \right) = \left(\frac{0}{2}, \frac{2}{2} \right) = (0, 1)$$

Vertex *A* is equidistant from both fixed points. You know that the sum of the distance from both fixed points must be 10. If you call the distance from one fixed point *x*, then

$$x + x = 10$$
$$2x = 10$$
$$x = 5$$

You now know that Point *A* must be 5 units from both fixed points.

You also need to know how far Vertex B is from one of the fixed points (F_1). Call that distance x. You know that the sum of the distances from B to F_1 and from B to F_2 equals 10. This means that

$$x + x + 8 = 10$$
$$2x + 8 = 10$$
$$2x = 2$$
$$x = 1$$

This tells you that Vertex B must be 1 unit away from F_1. Because F_1 has an x coordinate of -4, Vertex B must have an x coordinate of -5, which means it's 5 units from the center of the ellipse.

You can create a right triangle to determine the distance from the center to Vertex A as well. Use the Pythagorean theorem to solve for the distance from the center to Vertex A:

$$a^2 + b^2 = c^2$$
$$4^2 + b^2 = 5^2$$
$$16 + b^2 = 25$$
$$b^2 = 9$$
$$\sqrt{b^2} = \sqrt{9}$$
$$b = 3$$

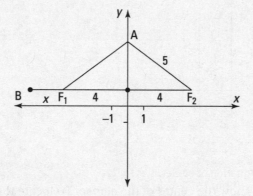

The general equation of an ellipse is

$$\frac{(x-h)^2}{a^2} + \frac{(y-k)^2}{b^2} = 1$$

where (h, k) represents the center of the ellipse, a represents the distance from Vertex B to the center of the ellipse, and b is the distance from Vertex A to the center of the ellipse.

You can now plug in your information to get the equation of the ellipse:

$$\frac{(x-0)^2}{5^2} + \frac{(y-1)^2}{3^2} = 1$$

$$\frac{x^2}{25} + \frac{(y-1)^2}{9} = 1$$

Index

• A •

acute angles, defined, 159
addition postulate, defined, 180
adjacent angles
  answers and explanations,
    161–163
  defined, 163
  parallel lines in, 84, 275–277
  practice questions, 10
alternate exterior angles
  answers and explanations, 270–272
  defined, 81
  practice questions, 82
alternate interior angles
  answers and explanations,
    270–272, 278–280
  avoiding mistakes with, 81, 89
  defined, 81
  practice questions, 82, 85–86
altitudes, of triangles
  avoiding mistakes with, 41
  defined, 158, 160
angle bisectors, 41
angles. *See also names of specific angles*
  arc measures and, 128–129,
    285–388
  avoiding mistakes with, 5, 17, 31,
    49, 67, 75
  in basic triangles, 11, 165–166
  bisectors in, 158–160, 208
  classifying triangles by, 34–35,
    83, 190–194, 273–274
  constructions with, 14, 169–170
  exterior angle theorem, 70–71,
    260–262
  forming linear pairs, 10, 161–163
  involving parallel lines, 82–86,
    270–282
  in polygons, 76–78, 265–269
  in quadrilaterals, 89
  relationships between sides and,
    68, 253–255
apothem, defined, 75
arcs
  central angles and, 128, 385–386
  of circles, 121, 124, 373–375
  constructions with, 13
  inscribed angles and, 128–129,
    386–388
area
  of circles, 121–123, 369–372
  of polygons, 75, 79, 269–270
  of sectors of circles, 121,
    123–124, 372–373
  of solid figures, 143–145,
    416–421

• B •

bisectors
  angle, 41, 208
  centers of a triangle, 41
  defined, 158–160, 180, 198–200
  perpendicular, 41, 158, 207–208

• C •

centers
  avoiding mistakes with, 41, 121
  centroids, 41, 43–45, 204–207,
    209–210
  of circles, 121
  circumcenters, 41, 45, 207–209
  incenters, 42, 198–200, 208–210
  orthocenters, 42–43, 201–204,
    209–210
  of triangles, 45–47, 208–210
central angles, in circles, 127–128,
  385–386
centroids, 41, 43–45, 204–207,
  209–210
circles
  arcs, 121, 124, 127, 373–375
  area of, 121–123, 369–372
  area of sectors of, 121, 123–124,
    372–373
  avoiding mistakes with, 121, 127
  centers of, 121
  central angles and arcs, 127–128,
    385–386
  circumference of, 121–122, 367–369
  diameter of, 121
  equation of circles in standard
    form, 124–126, 375–384
  inscribed angles, 128–129,
    386–388
  intersecting chords, 129–130,
    132–134, 388–393, 398–402
  proofs with, 138–140, 413–414
  radius of, 121, 136, 407–408
  secants, 127, 130–132, 134–136,
    393–398, 402–407
  tangents, 127, 130–132, 134–136,
    393–398, 402–408
  using loci to write equations of,
    152, 436–438
circumcenters, 41, 45, 207–209
circumference, 121–122, 367–369
collinear, defined, 315
completing the square, 380–382
cones, 144–146, 419–420, 423, 425
congruence theorems, 18–24,
  176–181
congruent angles, 17, 81

(continued)

congruent segments, 5
congruent triangles, 17, 31
constructions
  with angles and segments, 14,
    169–170
  avoiding mistakes with, 13
  congruent, 13, 166–168
  creative, 16, 173–176
  involving lines, 15, 170–173
coordinate geometry
  avoiding mistakes with, 97
  defined, 97
  distance formula in, 98, 309–312
  loci in, 150, 152–153, 428–430,
    438–440
  midpoint formula in, 98–99,
    312–314
  parallel lines in, 100–101, 317–320
  perpendicular lines in, 100–101,
    317–320
  proofs in, 102–103, 327–334
  slope formula in, 99–100, 315–317
  slope-intercept form in, 101–102,
    320–327
coplanar, defined, 416
corresponding angles
  avoiding mistakes with, 81
  defined, 81
  parallel lines in, 84–85, 275–279
cosine, 59, 247–252
cylinders, 144–146, 418, 422–423,
  425

• D •

degrees, converting to radians, 789
degrees, of rotations, 105
diagonals
  of quadrilaterals, 89
  of rectangles, 92, 291–295
  of rhombuses, 93–94, 297–302
diameter, 121
dilations, 113–114, 349–352
direct isometry, 105, 117, 357–358
directrix, defined, 427
distance formula, 97–98, 309–312
dodecagons, defined, 264

• E •

equilateral triangles, 185–186,
  189–190, 195, 241
Euler line, 41, 48, 210–211
exterior angle theorem, 70–71,
  260–262
extremes, in similar triangles,
  55–58, 221–222

• F •

focus of a parabola, defined, 427
formulas. *See names of specific formulas*

• G •

geometric definitions, 5–8, 157–160
glide reflections, 117, 357–358

• H •

height, slant, 141
heptagons, defined, 264
hexagons, defined, 264, 270
hypotenuse-leg (HL) theorem, 31
hypotenuses, 59

• I •

incenters, 41–42, 198–200, 208–210
indirect isometries, 117, 357–358
indirect proofs, 28–30, 183–185
inequalities, triangle
    exterior angle theorem, 70–71, 260–262
    inequality theorems, 69, 72–73, 255–256, 263–264
    isosceles triangles, 70, 258–259
    missing side lengths, 69, 257–258
    relationships between sides and angles, 68, 253–255
inequality theorem, 69, 255–256
inscribed angles, in circles, 127–129, 386–388
intersecting chord theorem, 132–134, 398–402
intersecting chords, angles formed by, 129–130, 388–393
isometries, 105, 117, 357–358
isosceles trapezoids, 89
isosceles triangles
    avoiding mistakes with, 31
    defined, 159, 185–186, 189–190
    inequalities in, 70, 258–259
    proofs with, 36–40, 195–198
    properties of, 33–34, 189–190

• L •

lateral area, 418–420
linear pairs, 10, 159, 161–163
lines
    avoiding mistakes with, 97
    bisectors in, 160
    parallel, 15, 81–87, 100–101, 170–173, 270–283, 317–320
    perpendicular, 15, 100–101, 157, 170–173, 317–320
    of reflection, 110, 341–344

skew, 416
in three-dimensional geometry, 141–143, 414–416
loci
    avoiding mistakes with, 149
    basic theorems, 150, 427
    compound and challenging problems, 153, 440–444
    drawing pictures to help with, 149
    locating and connecting points, 149
    of points equidistant from one line, 150–151, 430–431
    of points equidistant from two lines, 150–151, 430–431
    of points equidistant from two points, 151–152, 432–436
    tools for accuracy with, 149
    using coordinate geometry, 150, 152–153, 428–430, 438–440
    writing equations of circles, 152, 436–438
locus theorems, 150, 427

• M •

means and extremes, 55–58, 221–222
medians, 41, 158, 160, 174, 176
midpoint formula, 97–99, 312–315
midpoints, defined, 158
midsegments, 50–51, 213–215
motion. *See rigid motion*

• N •

nonagons, defined, 264–265

• O •

obtuse angles, defined, 160
opposite isometries, defined, 357–358
orthocenters, 41–43, 201–204, 209–210

• P •

parallel lines
    constructions with, 15, 170–173
    in coordinate geometry, 97, 100–101, 317–320
    finding angle measures involving, 83, 274–275
    forming adjacent angles, 84, 275–277
    forming alternate interior angles, 82, 270–272
    forming corresponding angles, 84–85, 275–279

forming vertical angles, 84, 275–277
    proofs incorporating, 86–87, 282–283
parallelograms, 89–91, 283–290
pentagons, defined, 264
perimeters, of triangles, 59
perpendicular bisectors, defined, 158, 207–208
perpendicular lines
    constructions with, 15, 170–173
    in coordinate geometry, 97, 100–101, 317–320
    defined, 157, 159
pi (π), 121
planes, 141–143, 414–416
point symmetry, 110–111, 345
points
    locus of, 150–152, 430–436
    midpoints, 158
    point symmetry, 110–111, 345
    reflecting, 109–110, 339–340
    in three-dimensional geometry, 141–143, 414–416
    translating points, 112, 347
polygons
    angles of, 76–77, 265–267
    apothem of, 75
    area of, 75, 79, 269–270
    avoiding mistakes with, 75
    naming, 76, 264
    sum of exterior angles in, 77–78, 268–269
    sum of interior angles in, 75, 77–78, 268–269
postulates
    addition, 180
    answers and explanations, 160–161
    practice questions, 8–9
    substitution, 161
prisms, rectangular, 142–145, 414–417, 421
proofs
    avoiding mistakes with, 17, 31, 81
    with circles, 138–140, 413–414
    in coordinate geometry, 102–103, 327–334
    incorporating parallel lines, 86–87, 282–283
    indirect, 28–30, 183–185
    with isosceles triangles, 36–38, 195–198
    overlapping triangles, 24–27, 181–183
    triangle congruence theorems, 18–24, 176–181
    triangle inequality theorems, 72–73, 263–264
properties
    avoiding mistakes with, 97
    in basic geometry, 8–9, 160–161
    of diagonals, 93–94, 297–302
    of parallel lines, 81–87, 270–283
    reflexive, 160–161

of squares, 94–95, 302–306
substitution, 5
transitive, 5
of trapezoids, 95, 306–308
of triangles, 33–34, 97, 189–190
proportions
centroid and median segments, 204, 206
means and extremes, 55–58, 221–222
in right triangles, 60–63, 226–237
in similar triangles, 246, 211–213, 215–220
trigonometric ratios, 65–66, 247–252
props, 141
Pythagorean theorem, 59–60, 67, 223–225

• Q •

quadrilaterals, 89. *See also* names of specific quadrilaterals

• R •

radians, converting degrees to, 789
radius, 121, 136, 407–408
ratios, 49, 59, 65–66, 247–252
rectangles
defined, 89
finding diagonals of, 92, 291–295
properties of, 91–92, 290–291
rectangular prisms, 142–145, 414–417, 421
reflecting points, 109–110, 339–340
reflections
equations for lines of, 110, 341–344
glide, 117, 357–358
reflecting points, 109–110, 339–340
reflexive property, 17, 160–161
regular hexagons, defined, 270
regular octagons, defined, 265
regular polygons, defined, 75
rhombuses
defined, 89
diagonal properties of, 93–94, 297–302
perimeter of, 300
properties of, 93, 295–296
right triangles
avoiding mistakes with, 59
proportions in, 60–63, 226–237
Pythagorean theorem, 60, 223–225
special, 63–65, 237–246
trigonometric ratios, 65–66, 247–252
rigid motion
answers and explanations, 334–339
avoiding mistakes with, 105

constructions of, 118–119, 358–365
defined, 105, 345
practice questions, 106–109
of triangles, 115–116, 354–355
rotations
rules for, 115, 353–354
in three-dimensional geometry, 146–147, 425–426
in transformational geometry, 105, 114, 352–353

• S •

scalene triangles, defined, 185
secants
angles formed by, 130–132, 393–398
avoiding mistakes with, 127
defined, 127
lengths of, 134–136, 402–407
sectors, area of, 121, 123–124, 372–373
segments. *See also* midsegments
avoiding mistakes with, 5, 17
bisecting, 5
congruent, 5
constructions with, 14–15, 169–170
midpoints in, 157–158
transformations of, 117, 358
sides. *See also* right triangles
classifying triangles by, 32–33, 185–188
of polygons, 75
of quadrilaterals, 89
of similar triangles, 49, 54, 220–221
and triangle inequalities, 67–69, 253–255, 257–258
similar triangles
avoiding mistakes with, 49
creating, 51–52, 215–218
defined, 49
midsegments in, 50–51, 213–215
proving proportion in corresponding sides, 54, 220–221
proving similarity, 53, 220
proving with means and extremes, 55–58, 221–222
understanding, 50, 211–213
word problems, 52, 218–220
sine, 59, 247–252
skew lines, defined, 416
slant height, 141
slope formula, 97, 99–100, 315–317
slope-intercept form, 101–102, 320–327
slopes, 97
solid figures
surface area of, 143–145, 416–421
volume of, 145–146, 421–424
special right triangles, 63–65, 237–246

spheres, 420–421, 424, 426
square, completing the, 380–382
squares, 89, 94–95, 302–306
substitution property, 5, 161
supplementary angles
in basic geometry, 10–11, 163–165
defined, 163–165
involving parallel lines, 85–86, 278, 280–282
surface area, of solid figures, 143–145, 416–421

• T •

tangent lines and segments
angles formed by, 130–132, 393–398
avoiding mistakes with, 59
defined, 127
lengths of, 134–136, 402–407
radii and, 136, 407–408
tangent (trigonometry), 247–252
theorems
basic locus, 150, 427
exterior angle, 70–71, 260–262
hypotenuse-leg, 31
inequality, 69, 255–256
intersecting chord, 132–134, 398–402
Pythagorean, 59–60, 67, 223–225
triangle congruence, 17–24, 176–181
triangle inequality, 72–73, 263–264
three-dimensional geometry
avoiding mistakes with, 141
lines, 142–143, 414–416
planes, 142–143, 414–416
points, 142–143, 414–416
rotations of two-dimensional figures, 146–147, 425–426
surface area, 143–145, 416–421
understanding different shapes, 141
volume, 145–146, 421–424
transformational geometry
avoiding mistakes with, 105
compositions of transformations, 116, 355–357
dilations, 113–114, 349–352
glide reflections, 117, 357–358
isometries, 105, 117, 357–358
lines of reflection, 110, 341–344
point symmetry, 110–111, 345
reflecting points over x- and y-axes, 109–110, 339–340
rigid motion, 105–109, 115–116, 118–119, 334–339, 354–355, 358–365
rotations, 114–115, 352–354
segments, 117, 358
translating points, 112, 347
translation rules, 113, 347–349
triangle translations, 111–112, 346

transitive property, 5
translations, 111–113, 346–349
transversals, 81, 89
trapezoids, 89, 95, 306–308
triangle inequality theorems, 72–73, 263–264
triangles. *See also names of specific triangles*
  altitudes of, 158, 160
  centers of, 41, 45–46, 208–209
  centroids, 43–45, 204–207, 209–210
  circumcenters, 41, 45, 207–209
  classifying, 32–35, 83, 185–188, 190–195, 273–274
  congruence theorems, 18–24, 176–181
  constructing centers in, 47, 209–210
  constructing special, 13

Euler line, 41, 48, 209–211
incenters, 41–42, 198–200, 208–210
inequalities (possible side lengths), 67–73, 253–264
medians, 158
orthocenters, 42–43, 201–204, 209–210
overlapping, 24–27, 181–183
proofs with, 17, 31, 36–40, 195–198
Pythagorean theorem, 59
similar, 49–58, 211–222
in transformational geometry, 111–112, 115–116, 346, 354–355
trigonometric ratios, 59
trigonometry
  avoiding mistakes with, 59

ratios in right triangles, 65–66, 247–252
significance of right triangles, 59

• V •

vertical angles, 10, 84, 161–163, 275–277
vertices, of quadrilaterals, 89
volume, of solid figures, 145–146, 421–424

• X •

x- and *y*-axes
  in coordinate geometry, 97
  in transformational geometry, 109–110, 339–340

About the Authors

Allen Ma has been a math teacher at John F. Kennedy High School in Bellmore, New York, for the past 18 years. He is the math team coach at Kennedy and was also the honors math research coordinator for many years. He has taught geometry for over 20 years.

Amber Kuang has been a math teacher at John F. Kennedy High School in Bellmore for the past 14 years. She has taught all levels of mathematics, spanning from algebra to calculus.

Dedication

We would like to dedicate this book to both of our families. This would have been an impossible task if it weren't for the infinite love and support from our children, spouses, and parents.

Authors' Acknowledgments

Writing the 1,001 geometry questions for this book was an amazing experience that required a tremendous amount of effort from many people. This would not have been possible without the help of so many wonderful professionals. We would like to thank Bill and Mary Ma for their advice and support throughout this project. We would like to extend our gratitude to Lindsay Lefevere and to our agent Grace Freedson for believing in us. We would also like to take the opportunity to thank the amazing Wiley team of editors, Danielle Voirol, Amy Nicklin, and Christina Guthrie, for their support, suggestions, and attention to detail. To our Kennedy High School family: We are so thankful to be part of such a great team of teachers, administrators, and students. You have all provided us with invaluable resources that have played a part in writing this book.

Publisher's Acknowledgments

Executive Editor: Lindsay Sandman Lefevere

Editorial Project Manager: Christina Guthrie

Copy Editor: Danielle Voirol

Technical Editor: Amy Nicklin

Art Coordinator: Alicia B. South

Project Coordinator: Melissa Cossell

Illustrator: Thomson Digital

Cover Image: ©iStockphoto.com/DNY59